Biology and the Foundation of Ethics

This is the first of a number of specially commissioned collaborative volumes on topics of great current interest to philosophers, biologists, and social scientists that will be appearing in *Cambridge Studies in Philosophy and Biology* in the coming years.

Much attention has been devoted in recent years to the question of whether our moral principles can be related to our biological nature. This collection of new essays focuses on the connections between biology, in particular evolutionary biology, and foundational questions in ethics. The book asks, for example, whether humans are innately selfish and whether there are particular facets of human nature that bear directly on social practices.

The volume is organized historically, beginning with Aristotle and covering such major figures as Hume and Darwin down to the present and the work of Harvard sociobiologist E. O. Wilson. It is one of the first efforts to provide historical perspective on the relationships between biology and ethics, and it has been written by some of the leading figures in the history and philosophy of science, authors whose work is very much at the cutting edge of these disciplines.

Jane Maienschein is Professor of Philosophy and Biology at Arizona State University.

Michael Ruse is Professor of Philosophy and Zoology at the University of Guelph, Ontario.

Biology and the
Foundation of Ethics

Edited by

JANE MAIENSCHEIN
MICHAEL RUSE

CAMBRIDGE
UNIVERSITY PRESS

PUBLISHED BY THE PRESS SYNDICATE OF THE UNIVERSITY OF CAMBRIDGE
The Pitt Building, Trumpington Street, Cambridge, United Kingdom

CAMBRIDGE UNIVERSITY PRESS
The Edinburgh Building, Cambridge CB2 2RU, UK   http://www.cup.cam.ac.uk
40 West 20th Street, New York, NY 10011-4211, USA   http://www.cup.org
10 Stamford Road, Oakleigh, Melbourne 3166, Australia

© Cambridge University Press 1999

First published 1999

Printed in the United States of America

Typeface Times Roman 10/12 pt.   *System* QuarkXPress™   [AG]

A catalog record for this book is available from the British Library.

Library of Congress Cataloging in Publication Data

Biology and the foundation of ethics / edited by Jane
Maienschein, Michael Ruse.
p.   cm. – (Cambridge studies in philosophy and biology)
Includes bibliographical references.
ISBN 0-521-55100-5 (hb). – ISBN 0-521-55923-5 (pbk.)
1. Biology – Moral and ethical aspects.   2. Ethics, evolutionary.
I. Maienschein, Jane.   II. Ruse, Michael.   III. Series.
BJ1311.B5   1999
171'.7–dc21   98-30711
                                                       CIP

ISBN 0 521 55100 5 hardback
ISBN 0 521 55923 5 paperback

Contents

Introduction   1
Jane Maienschein and Michael Ruse

1   Aristotle on the Biological Roots of Virtue: The Natural
      History of Natural Virtue   10
      James G. Lennox

2   The Moral Status of Animals in Eighteenth-Century
      British Philosophy   32
      Michael Bradie

3   From Natural Law to Evolutionary Ethics in Enlightenment
      French Natural History   52
      Phillip R. Sloan

4   French Evolutionary Ethics during the Third Republic:
      Jean de Lanessan   84
      Paul Lawrence Farber

5   The State and Nature of Unity and Freedom: German
      Romantic Biology and Ethics   98
      Myles W. Jackson

6   Darwin's Romantic Biology: The Foundation
      of His Evolutionary Ethics   113
      Robert J. Richards

7   Nietzsche and Darwin   154
      Jean Gayon

8   Evolutionary Ethics in the Twentieth Century:
      Julian Sorell Huxley and George Gaylord Simpson   198
      Michael Ruse

Contents

9   The Laws of Inheritance and the Rules of Morality:
    Early Geneticists on Evolution and Ethics   225
    Marga Vicedo

10 Scientific Responsibility and Political Context:
    The Case of Genetics under the Swastika   257
    Diane B. Paul and Raphael Falk

11 The Case against Evolutionary Ethics Today   276
    Peter G. Woolcock

12 Biology and Value Theory   307
    Robert J. McShea and Daniel W. McShea

    Notes on Contributors   329
    Index   332

Introduction

Jane Maienschein and Michael Ruse

> It is at this point, I think, that we can best make the comparison between ethics
> and science, and the insurmountable barrier between them seems to me to lie
> in this fact, that in science we have such a source of conviction and in ethics
> we have not. Science rests ultimately on a basis of absolute certainty; ethics, so
> far at least, has not in general found any basis at all.

Thus asserted "Prof. H. Dingle" (Herbert Dingle) in an article in *Nature* in
1946, as quoted in "50 Years Ago" in 1996. He continued:

> Science can therefore advance with confidence that although it may make mis-
> takes they are not irreparable, and that even though its most trusted structures
> may come tumbling about its ears, it cannot finally collapse because underneath
> are the everlasting arms. They are two – reason and experience; on these twin
> supports science has an indestructible foundation. [Dingle 1996]

Although not everyone would endorse Professor Dingle's confidence about
the absoluteness of the certainty in science, few would disagree with the proposition that science and ethics rest on different bases. And many would agree
that attempts to provide a compelling epistemic warrant for ethical theory
have failed. Indeed, moral theorists have often been willing to give up the
search and engage in descriptive and normative ethical discussions, leaving
the metatheoretical search to others. Biologists and philosophers of biology
have eagerly taken up the challenge. Thus an unabashed program for natu-
ralizing ethics has gained enthusiastic supporters in recent decades. Sociobiology, evolutionary ethics, and genetic determinism have all played their
parts.

This interest in bringing together biology and ethics, traditionally thought
to inhabit C. P. Snow's "two cultures" and to remain in separate domains pursued by separate lines of thought, is not new but has expanded considerably
in recent decades. Until recently, philosophers had little interest in the life
sciences, and moral philosophers even less.

There are several reasons for the recent reevaluation. From the philosophic side, a major factor has been the general move to "naturalism," meaning by this a drive to make philosophical thought more empirical, more in line with the methods and techniques of the physical sciences. This is something that has occurred particularly in the area of the theory of knowledge, epistemology, especially under the influence of major thinkers like the Harvard philosopher Willard van Orman Quine. In parallel with this has been a significant move to naturalism by philosophers of science, a move that was in major respects sparked by the stimulating influence of Thomas Kuhn's *Structure of Scientific Revolutions*. There has been a shift from prescriptive idealizations about the way in which philosophers think that science should behave to more modest attempts at describing the actual workings of real science, past and present. The fact that so many of today's philosophers of science devote their energies to the life sciences magnifies the significance of these changes.

Ethics and moral theorists have also been major causal factors in raising the new interest in morality and biology. In recent decades, thinking about morality has been drastically revamped. There was a time when the applications of ethics were not considered central issues within the domain of moral philosophers. There the concerns remained theoretical, primarily about foundations. Then a series of external factors began to contribute to changing all that: The war in Vietnam, the struggle of African Americans for equality, the rise of feminism, major advances in medical technology, and changing job markets that rewarded those who explored new niches all changed the consciousness of the community of moral philosophers. Specialists in moral theory became aware of the significance of the world beyond their own theoretical domain, and with that came the counterpart to epistemological naturalism, as more and more people speculated on the physical bases of moral thought and action. Noteworthy in this respect was Quine's Harvard colleague John Rawls. In his magisterial *A Theory of Justice* he found opportunity (if only in a footnote) to speculate on the evolutionary origins of his basic claims about moral principles. Others have expanded from that footnote to place biological considerations at the center of moral theorizing.

Another major set of contributions promoting consideration of biology and ethics has come from historians of science. Although the history of science has had distinguished practitioners in the past, such as the nineteenth-century English polymath William Whewell, it is only in the past few decades that it has been fully professionalized, with training in historical methods. Now we have scholars who know how to read texts, who are aware of the importance of archival research, who can relate science to the more general and broader cultural contexts in which science is pursued. With this has come

2

an interest in and willingness to examine the connections of science to such issues as moral behavior and the things that have been said, formally and informally, publicly or privately, deliberately or spontaneously, by practicing scientists on such subjects. Most particularly has come an interest in the many things that students of the life sciences have had to say about ethics, because much hinges on biological considerations of human nature. Philosophers, and especially moral philosophers, have until recently been reluctant to study science. Fortunately, that has begun to change.

There have also been scientific developments motivating a closer look at the connections between biology and society. It seems fair to say that the period since World War II really has been an era of biology. Studies in genetics and evolution have provided new data and new theories and have raised new questions relevant to discussions of human behavior, including ethics. Sociobiology has suggested that human behavior is controlled by genetic underpinnings and evolutionary adaptations. Some advocates, most prominently and most persistently represented by Harvard entomologist Edward O. Wilson (1978), have argued that biology now speaks directly to human nature, explaining and illuminating our most intimate and essential aspects:

> The genes hold culture on a leash. The leash is very long, but inevitably values will be constrained in accordance with their effects on the human gene pool. The brain is a product of evolution. Human behavior – like the deepest capacities for emotional response which drive and guide it – is the circuitous technique by which human genetic material has been and will be kept intact. Morality has no other demonstrable ultimate function. [p. 167]

Genetic determinists also argue for the efficacy of inherited genes. This view has gained tremendous popular support, so that hardly a week goes by that does not bring news of some other "gene for" some other aspect of human behavior. Rarely do we hear of the limitations of genetic determinism or of the long leash to which Wilson referred. Undoubtedly, the enormous advances in biology and the enthusiastic popular interest have fueled attention from philosophers and moral theorists alike – whether or not any of these claims is valid. Scientists and philosophers have even begun to collaborate to explore the issues of common interest and the implications of the biological basis of human nature.

Going beyond the flush of enthusiasm for sociobiology and genetic determinism, scholars need to look carefully at the extent to which and the way in which biology really does inform ethics. And we can gain much by close scholarly attention to historical efforts to link biology and ethics, for, as the essays in this volume show, this effort is not new, even though we see the past efforts in new light.

We have put together a volume that can be read profitably by someone who knows little or nothing about the subject, but it also can inform those trying to delve more deeply into the relationships between the life sciences and morality. Generally, it has not been our aim to look at the applications of biology in moral situations, such as the implications of the Human Genome Project. Nor do we consider ethical conduct of science. This is certainly not because we think that such issues are unimportant. Fortunately, these areas are being reasonably well served already. It is also not our intention to be comprehensive or to guide the discussion along particular tracks. We invited the contributors to provide essays that they felt would best address what they saw as the important issues in biology and society. We expected a rich range of offerings, and we are not disappointed.

Theodosius Dobzhansky said that nothing in biology makes sense except in the light of evolution. None of us involved in the project would want to deny that claim, but there is more to biology than evolution, and that is reflected in the contributions here. Though it is certainly the case that much of the literature on biology and ethics (and perhaps most of what has been written in the past century) has some sort of evolutionary flavor or perspective, that has not been our organizing theme. Evolution dominates some of the essays, but is notably absent or of less importance in others.

The proof of the pudding is in the eating, but we first provide an overview of what is on the menu. As so often is the case, we can trace the tradition of seeking ethical foundations to Aristotle. James Lennox makes the case that Aristotle sought to explain how individual humans gain their sets of moral virtues. Rather than issues of epistemic warrant, Aristotle focused on issues of acquisition and how the development of the moral virtues is grounded in the interaction of practical intelligence and natural virtues. This is, as Lennox puts it, a "natural history for virtue ethics."

Michael Bradie shows that the eighteenth-century British debates about the moral status of animals raised fundamental questions about the moral nature of man as well. The question shifted from "Do animals have souls?" to "Do they have the requisite epistemic and cognitive capacities to have moral standing?" to "Do they suffer?" The shifting focus led to ambiguous answers and to a blurring of the distinction between man and animals, such that the fundamental grounding for claims about the moral standing of man came into question as well, though Bradie does not explore those latter questions at length in his essay.

Those changing assumptions about what is natural, in the context of increasing attention to the natural, provided for lively exchanges of ideas in the late eighteenth and nineteenth centuries. Phillip Sloan carries us through the French Enlightenment, from Buffon's natural historical thinking to Lamarck's transformism. He focuses on the natural-law tradition grounded in "the assumption of an intimate connection between a dynamic biological conception of human nature and an objective normative ground for ethical reasoning." The precise nature of that intimate connection could vary, and indeed it did. Increasingly, biologists saw human nature as collapsing into the natural, rather than rising above nature, such that humans became increasingly the subjects of scientific investigation as natural animals. The discussion centered, then, on the extent to which human ethics derives from their biological nature and the extent to which some larger external agency also plays a role. For Lamarck, at least, ethical principles gained objectivity from their grounding in nature. That raised new questions and new possibilities.

Paul Farber takes up the specific possibility of an evolutionary grounding for ethics in his discussion of the French thinker Lanessan, who wrote extensively on ethics. Lanessan believed that he had derived a set of ethical principles from nature and that proper raising of children would carry those values on in the society. He apparently did not acknowledge that he had provided no compelling account of why humans *ought* to embrace those behaviors, but assumed nonetheless that his set of values was clearly preferable to others. Farber suggests that Lanessan provides an important instructive example that bears on all discussions of evolutionary ethics. The objections to Lanessan's views hold for other evolutionary ethical arguments as well, he insists, and thus "contemporary authors who look to evolution for knowledge applicable to ethics need to be mindful of past attempts to use evolution as a foundation for ethical systems and to avoid duplicating the mistakes of the past." This bears particularly on the discussions by Robert Richards of Darwin's evolutionary ethics.

For the nineteenth century, we have a set of concerns of varying importance to different writers: To what extent and in what ways can ethics be informed by nature, derived from nature, or actually also justified by nature? And what is the conception of nature within which any of these discussions can be framed?

Myles Jackson introduces us to the German romantic vision of nature, emphasizing the unity of nature and its implications for the unity of ethics and politics with nature. Implicitly, he raises some of the same caveats as Farber, because the two thinkers he discusses in detail, Oken and Goethe, used their considerations of nature and ethics to reach quite different views. Both emphasized the unity of nature, including man, such that knowledge of nature was also both moral and political knowledge. Yet, for Oken, the freedom inherent in nature suggested a freedom of individual action and a political system that would lead to revolutionary action. For Goethe, in contrast, enlightened despotism was necessary to retain law and order in nature. Their

5

divergent attempts at grounding ethics in the unity of nature show the complexities and uncertainties of such projects.

Robert Richards provides a meticulous picture of Darwin as inspired and influenced by the German romantics. Richards objects to the usual picture of Darwin as inhabiting a mechanistic, cruel world of blind struggle and death. Instead, Darwin saw a natural world rich with moral values and intelligence. Darwin's evolutionary ethics, then, involves not a jerking of values out of nature, as Richards puts it; rather, nature is naturally imbued with values already, and Darwin simply discovers them. Nature, for Darwin, is a source of moral value. Thus Darwin does not commit the naturalistic fallacy of moving from a descriptive fact to moral value. Rather, the value is there all along. Richards obviously admires Darwin's "accomplishment" and defends it against various criticisms, even while he admits that it is a view that "must fail under stronger metaethical analysis typical of our time." He does not, however, provide that analysis, leaving open the question whether or not Richards's reconstructed Darwin can muster a defense.

Jean Gayon gives us a major analysis of the thinking of the German philosopher Friedrich Nietzsche on the subjects of evolution and ethical thought and action. Nietzsche's stock as a philosopher has risen dramatically in recent years, and Gayon confirms that this is no fluke advance. He does argue that Nietzsche himself had confused ideas about the nature of evolutionary theory, Darwinian theory in particular, and that when these confusions are revealed and rectified (though there are still major differences) the philosophy and the science mesh more closely than one might initially have expected.

Michael Ruse takes us further into the search for an evolutionary grounding for ethics with his study of Julian Huxley and G. G. Simpson. Again we have two divergent efforts starting on similar searches. Both sought evolutionary foundations for ethics, but Huxley asked for (though not in so many words) epistemological justification, and Simpson settled for "grounding" in a more descriptive sense. Huxley demanded objective ethical truths, whereas Ruse believes that Simpson would have been satisfied with a solid account of what happens in nature and why it works to keep society functioning morally. Simpson did not need moral objectivity, then, but could explain the existence and persistence of moral norms as resulting from evolution: Society evolves, and we who are part of the evolutionary system evolve to develop a shared set of norms, which we perpetuate through training – or something like that, but with the result that we need not commit the naturalistic fallacy nor invoke the existence of absolute, objective norms outside ourselves. Ruse favors Simpson's approach, but remains sympathetic to the attempt to ground ethics (including normative ethics) in evolution and in nature.

Introduction

Marga Vicedo offers a different approach to evolutionary ethics, starting from the perspective of American geneticists of the early twentieth century. Davenport, East, Jennings, and Conklin made some interestingly different attempts to provide natural bases for ethics, attempts that led them to ethics, but also to genetics and issues of biological determinism. The ethical claims by those biologists were in some ways naive philosophically, yet they speak to a sincere impulse to provide a grounding for ethical behavior and to find that grounding in biology. East, for example, saw hope for the "new religion of science" that could give "us both an emotional inspiration and a practical procedure for enriching human life." The causes of ethical and social behaviors lay within one's biological nature, not outside. Yet there they diverged. Some held that this biological nature meant that actions were determined, in some cases genetically determined – thus the loss of a role for free will and responsibility. Others urged that the fact that ethical norms arise internally allows for free will and places responsibility in the individual behaviors of intelligent human beings. These men were not seeking to justify ethics through appeal to nature. Rather, they sought to root our nature in biological laws and then to provide, in their different interpretations, ways for humans to operate within those laws. Humans might either be relatively constrained or, alternatively, have opportunity to exercise free will and direct future evolution.

Diane Paul and Raphael Falk look at a slightly later period in which there were other efforts to ground ethical views in nature and then to act on what was found there. The Nazis embraced biological science and actively encouraged research. In their studies of behavior, of genetics, and of Darwinism, biologists found much that could contribute to the basis for action by the Nazi political movement. Discussions of racial hygiene and evolutionary considerations led easily to research in modern human genetics. Implicitly, that study then reinforced the assumptions of racial hygiene with which those calling for the research had begun. Thereby, Paul and Falk also introduce issues of ethical action by biologists. Is it ethical to engage in research that is used for undesirable purposes? "Sure, as long as it is good science," their biologists would have responded. Their essay plays two themes together: What science is being done, and is it ethical? Once the science is done, what does it tell us about human behavior and about what is ethical based on that interpretation?

Peter Woolcock rejects all such efforts. Though he focuses specifically on evolutionary ethics, it is clear that he believes that no naturalized ethics can avoid the naturalistic fallacy completely, and none can escape the problem of the "altruism guarantee." In fact, Woolcock believes that any version of evolutionary ethics that grounds ethical norms in nature will necessarily fail to explain why individual humans might act in any way other than that dictated

by their own egoistical self-interest. Once they realize that there are no objective normative truths, why would they ever act ethically? Notwithstanding Ruse's attempts to show that they might do so because of their membership in an evolved group that shares their values and reinforces them through
training, Woolcock rejects such claims. Thus, evolutionary ethics fails to
provide any guidance for normative ethics. He concludes that "evolutionary
theory, then, can't serve the moral reassurance role previously filled by religion. It looks as if we shall have to resolve our moral differences through the
hard grind of normative justification." Yet that will not, Woolcock suggests,
prove as difficult as skeptics such as our Professor Dingle would suggest.

Robert McShea and Daniel McShea offer an alternative interpretation.
Rejecting evolutionary or genetic accounts, they look instead to emotivist
interpretations of the development of ethics. Neither a god nor any other pur-
ported factor outside human nature can account for the development of ethical behavior and ethical choices – only the "human nature value theory" or
emotivism will do. Feeling, as exhibited in brain states, is separate from behavior and guides behavior in humans. Human nature, then, is the validation
of value judgments responding to the "human feeling profile" and produces
ethical behavior. The result is that humans must do what they will to do – what
they really will to do. The authors acknowledge that they do not have proof
for their point of view, but feel that the strength lies in the problems that be-
set all other alternatives and the relative strengths of their interpretation.

We see herein a rich diversity of viewpoints, all within the context of concern to ground ethics in nature. Some engage in metaethics, seeking to justify
and provide epistemic warrant for ethical norms in biological nature; these
authors generally hold to some version of ethical objectivity or truth. Others
provide descriptive accounts and explanations, even causal explanations of
ethical behaviors in nature, but stop short of demanding justification or even
objectivity. Still others rest content to outline what happens in nature. All the
arguments from nature to ethics are problematic in various ways, yet the appeal of the general effort is compelling. For the naturalist, who has rejected
the efficacy of religion and higher-order values, where else is there to look?
As Woolcock points out, if all we have are descriptions, it is difficult to see
why humans would behave ethically (implying socially rather than selfishly)
toward others. So the discussion continues.

With the diversity of views presented in these essays, and with the recognition that there are many different ways to approach the relationships between
biology and ethics, we have made real progress toward posing our questions
more cogently. And as Farber clearly shows, we also see ways in which we
can learn from history and from the failures of past arguments.

The studies presented here focus on biologists and their contributions to discussions of biology and ethics; other biologists, ranging from E. O. Wilson to Franz de Waal, have advanced further arguments and viewpoints. Yet these studies have been done by historians and philosophers of science reflecting on the past biological arguments. With this foundation and these interpretations, we hope that moral theorists will begin to take the biological contributions more seriously and that biologists can begin to make their arguments more persuasive philosophically.

REFERENCES

Dingle, H. 1996. 50 years ago. *Nature* 382 (August 8): ix.
Wilson, E. O. 1978. *On Human Nature.* Cambridge, MA: Harvard University Press.

1

Aristotle on the Biological Roots of Virtue
The Natural History of Natural Virtue

James G. Lennox

INTRODUCTION

Traditionally, attempts to relate Aristotle's well-known passion for biology to his ethical thought have focused on two points of intersection. In *Nicomachean Ethics* I 7, Aristotle relies heavily on the idea that human excellence (or virtue) must be the excellence of a specifically human functional capacity, and his approach to defining that capacity is reminiscent of *De Anima* II 2–3: Functions shared with plants and animals are ruled out in favor of those that are distinctively human, those associated with reason. This so-called function argument has generated a good deal of scholarly discussion, but a careful reading of it, especially in light of Aristotle's remarks at 1102a13–32 about the limited need for a scientific understanding of the soul for the purposes of ethics, must lead one to conclude that no detailed understanding of biology was needed to craft it.[1]

A second point of possible intersection on which scholars have focused is Aristotle's concept of a political animal, used often in the biological works and extending to many species other than man. Furthermore, Aristotle's *Politics,* much more than his *Nicomachean Ethics,* is a "biological" work, opening with an extended narrative about the natural growth of the *polis* out of essentially biological origins:

This essay has benefited at various stages from the helpful comments of a number of people. I want especially to single out the detailed written suggestions of David Depew, Mark Gifford, Aryeh Kosman, and Ursula Wolf, and discussions with John McDowell at the 4-Way Workshop on Human Nature in Berlin. I would also like to thank Mark Gifford, the organizer of a conference on *NE* VI at Virginia Polytechnic Institute and State University, for sharing my essay with the participants and for sharing the circulated papers with me. Because of prior commitments, I was unable to attend, but I have learned much from the papers presented, especially those of Professors Dahl and Gifford noted herein.

> Accordingly, if the first communities exist by nature, so too does every *polis;* for the *polis* is the goal of these communities, and nature is present in the goal. For we call that which each thing is when its generation is completed, whether horse, human, or household, the nature of each thing. Again, what a thing is *for,* and its goal, is best, and self-rule is the goal [of communities] and best. Hence it is apparent from these considerations that the *polis* is natural, and that humans are by nature political animals. [*Politics* I 1, 1252b28–33]

This account of the growth and development of the *polis* is simply another instantiation of Aristotle's natural teleology. The *polis* is the natural goal of human community; it is within the *polis* that human nature is truly expressed. Indeed, though we are *by nature* political – even preeminently so – this does not, in itself, set us apart from the other animals:

> It is clear why humans are political animals *to a greater degree* than are any of the bees or gregarious animals. . . . For it is a distinguishing feature of humans that only they are able to perceive good and bad, justice and injustice, and other such things, and it is the common awareness of these things that produces a household and a *polis.* [1253a8–18]

This connection between the biological and the political conceptions of 'political animal' has been fruitfully explored,[2] but it must be admitted that the connection to Aristotle's ethics is indirect, through the principle that happiness is fully achievable only in the *polis.*

In this essay we shall explore a very different intersection between Aristotle's biological and ethical interests, one that brings him into much closer proximity to Darwin: his rooting of the human virtues in what he refers to as 'natural virtues'. Just as Darwin, in the third chapter of *The Descent,* claimed to be approaching the topic of the moral sense "exclusively from the side of natural history,"[3] so Aristotle, the astute interpreter of animal behavior, provides, in *Historia Animalium* (*HA*) VII–VIII, the natural history, specifically the *ethology,* for his virtue ethics.

NATURAL VIRTUE IN THE *NICOMACHEAN ETHICS*

Nicomachean Ethics (*NE*) II–V provides an extensive discussion of the nature of virtue[4] and of specific virtues of character: courage, temperance, generosity, justice, and the like. In book VI, Aristotle turns to an extended taxonomy of the *intellectual* virtues, with the goal of isolating those most relevant to moral and political action. In chapter 13 he takes up the crucial question of how the intellectual virtue of practical intelligence (φρόνησις) is related to virtue of character (ἠθικὴ ἀρετή). It is a crucial question for him, because it has become clear, in his account of the virtues, that a life lived virtuously involves intelligence in the determination of the ends that constitute a good life and of the means of achieving those ends in the concrete settings of our daily lives. He begins by reminding us of the way in which virtues of character and reason are integrated in his very account of being human:

> Furthermore, our function is fulfilled [by living] in accordance with practical intelligence and virtue of character; for virtue makes the goal correct, and practical intelligence makes that which promotes the goal correct. [1144a6–9]

Now this may sound as if practical intelligence *isn't* involved in determining the correctness of the end, but we are immediately disabused of that misunderstanding by being shown our continuity with certain animals when these two distinct sorts of virtue are considered in isolation from each other. First, as to virtue of character:

> Let us consider virtue again; for as practical intelligence is related to cleverness – not the same, yet similar – so too is natural virtue related to virtue strictly speaking. In all of us each of the states of character are present somehow by nature; for we have just, temperate, courageous and other such states of character already from birth; but nevertheless we are seeking something else, that which is good in the strict sense, and [we are seeking] such states present in another manner. For the natural states are present even in children and beasts, but without reason they appear to be harmful. This much seems clear – that just as a powerful body moving without sight may well take a 'powerful' fall because of the lack of sight, so too here. If, however, a person acquires intelligence, it makes a difference to his actions; the disposition which is *like* virtue will at that point *be* virtue in the strict sense. So, just as in our doxastic nature there are two forms, cleverness and practical intelligence, so also in our character there are two forms, natural virtue and virtue in the strict sense, and virtue in the strict sense does not arise without practical wisdom. [1144b1–16][5]

Acting solely on the basis of natural virtue is "flying blind" in moral space. Yet when Aristotle remarks that these 'natural virtues' are present in children and beasts, yet are harmful without reason, it is natural to ask whether it is only in humans that natural virtue is harmful without reason, or in other animals as well. This turns out to be a difficult question to answer. Clearly, for us, *unqualified* virtue of character requires some connection to human reason. And yet Aristotle regularly characterizes the animals with natural virtues as practically intelligent and skilled thinkers, as we shall see. How, then, can he ascribe both natural virtue and practical intelligence to other animals and yet deny to them the unqualified virtue that comes to us when we have both? The answer, supported by Aristotle's account of animal character in *HA* VII–VIII, lies in the *independence* of the cognitive capacities and character traits in other animals, and thus in the fact that animals act 'in character' without deliberative choice. The other animals do not need to integrate practical intelligence with natural virtues to achieve excellence of character – in humans, however, it is this very integration that is the essence of both practical intelligence and virtuous character.

We began by worrying that Aristotle had made an utterly artificial distinction between end-determining virtues of character and the means-determining intellectual virtue of practical intelligence. In this passage, on the contrary, the move from natural to unqualified virtue seems simply to *be* the acquisition of practical intelligence by beings with natural virtue. And indeed Aristotle goes on from the foregoing passage to remind us that this is the appeal of the Socratic attempt to identify all of the virtues with knowledge: Without it we lack virtue, and once we have it we are virtuous (1144b26–30). The trick, then, is to keep moral virtue and practical intelligence distinct from, and yet properly intimate with, one another. Aristotle does so by reminding us that virtue involves correct reasoning (ὀρθός λόγος) and that "practical intelligence just is correct reasoning concerning such matters" (1144b27–28). Thus, when a bit later he concludes that "virtue makes us reach the right end in action, while practical intelligence makes us reach what promotes that end" (1145a5–6), it is clear that the virtuous character that consistently promotes the actions of a good life must be guided by practical intelligence; and it is equally clear that practical intelligence isn't mere cleverness at calculating the means to any given end, but is intelligence guided by a conception of the good life that allows us habitually to choose actions that promote that end. Cleverness (δεινότης) is therefore defined as a *capacity* (δύναμις) to act in ways that achieve any assumed goal (1144a24–26), whereas practical intelligence, though it *requires* cleverness, is defined as a *state* (ἕξις) that cannot exist without virtue, because the starting point of practical reasoning is the end to be achieved, and only if that starting point is good will it be appropriate to term the underlying state practical intelligence.[6]

The distinctions that frame this discussion are fundamental to Aristotle's ethical project. The opening of *NE* II carries over from the first book's conclusion the distinction between virtues of thought and of character. And when the argument turns to defining virtues of character, he frames that discussion in terms of whether we should view them as *capacities, feelings,* or *states.*[7] He concludes that because virtues are acquired by habituation (indeed, as he points out, this is suggested by the etymology of 'ethical' from 'ethos', habit), they must be states. Natural capacities come to be prior to their active realization, whereas states come to be as a result of training. You can throw a rock up a thousand times, he reminds us, but you will never habituate an 'upward' power in it (cf. 1103a20–b2). And because virtues dispose us to feel appropriately (e.g., to be angry to the appropriate extent, at the appropriate times, toward the appropriate people, etc.), they must *underlie* our feelings, just as they do our actions, and so must not be identified with them.[8] This dependence of appropriate feeling on virtuous character can be seen in the following defense of the view that virtue aims at what is intermediate between excess and deficiency:

> By virtue here I mean virtue of character; for this is concerned with feelings and actions, and these admit of excess, deficiency and an intermediate condition. Both fear and confidence and appetite and anger and pity, and in general pleasure and pain may be felt too much and too little, and in both cases not well; but to feel them at the right times, with reference to the right objects, toward the right people, with the right consequences in the right way, is what is both intermediate and best, and this is characteristic of virtue. Similarly too with regard to actions – there is excess, deficiency, and the mean. [*NE* II 6, 1106b16–24][9]

Virtues, then, are states rather than natural capacities. Yet, though the virtues of character in *NE* II are said *not* to arise by nature (because they are acquired by the habitual performance of correct actions), humans are said to be *naturally suited* to receive them. This natural suitability allows *NE* VI 13 to treat the virtues of character as based on *natural* virtues – as we saw, just, temperate, and courageous states of character we possess 'in a way *by nature*' and 'already from birth'.[10] In fact, at one point these are referred to as *natural states,* though his constant qualifications show him to be aware that this strictly violates his usual way of talking. How, then, do such character states emerge from the natural capacities that are 'like' them and, in properly educated children, develop into them? Because we are repeatedly told that these natural states of character are found in other animals as well as in human children, an understanding of the process of virtue acquisition may benefit from examining Aristotle's zoological account of character.[11]

As noted previously, the *NE* II distinction between natural capacities and acquired states of character is also mirrored in the account of cleverness in *NE* VI 13. There, cleverness is pointedly defined as a *capacity* precisely in order to distinguish it from practical intelligence, a *state* (and thus, presumably, acquired by habituation and teaching). But again, he begins by stressing the essential role that moral virtue plays in the transition from the natural capacity to the acquired state: "this eye of the soul doesn't arise without virtue" (1144a29–30). On the other hand, although practical intelligence is not *simply* the capacity of cleverness, cleverness is its sine qua non (1144a28–29).

Thus there is a suggestive parallel between the cleverness/practical-intelligence distinction and the natural-virtue/full-virtue distinction. In particular, this parallel suggests that *both* aspects of the virtuous personality emerge from natural roots shared with 'the beasts'. This suggestion is elevated to a plausible hypothesis by Aristotle's account of one of the key elements connecting the virtues of character to intelligence, namely, choice (προαίρεσις).[12] When, near the end of *NE* II, Aristotle gathers up his results into a definition of virtue of character, it is the ability to make intelligent choices that is given pride of place:

> Therefore virtue is a state expressed in choice (ἕξις προαιρετική), present in a mean relative to us, defined by an account such as the person of practical intelligence would define it. [*NE* II 6, 1106b36–1107a2]

His account of choice rests on his account of voluntary action, which is a necessary condition of choice:

> Choice, then, is apparently voluntary, but not the same as what is voluntary, which extends more widely. For children and the other animals share in what is voluntary, but not in choice; and actions done on the spur of the moment are voluntary, but do not express a choice. [*NE* III 2, 1111b6–10]

Once again, then, human beings are said to *begin* in precisely the same state that other animals *remain* in, capable of voluntary action, but not in possession of that excellence of character that is a state expressed in the kinds of *reasoned choices* and *appropriate emotions* found in the person of practical intelligence. In a long discussion of the relationship between bravery and emotional impulse, for example, Aristotle insists that someone brave in the strict sense acts out of a conception of what is noble, with his emotions cooperating, whereas beasts, and humans who have not acquired bravery, act purely out of anger and pain. But he concludes:

> Still they have something similar. The bravery due to emotional impulse would seem to be the most natural sort, and *is* bravery once choice and that for the sake of which are added. [1117a4–5]

Moreover, in a passage contrasting reasoning in the mathematical sphere and reasoning in the moral sphere, Aristotle notes that it is reason that teaches about the starting points in mathematics, whereas in the moral sphere "it is virtues, either *natural* or *habituated,* that teaches the right opinion about the starting point" (*NE* VII 8, 1151a15–19; emphasis added). The stress, I take it, should be on 'opinion' here, because Aristotle goes on to discuss the person who acts contrary to right *reason*. But I take it that there is also an important point about the person who acts on the basis of natural virtue implied by these two passages. Natural virtue is *similar* to virtue strictly speaking, in that it is a natural capacity to fix on the goal of action, a capacity we are able to train into a state expressed in practically intelligent choices.[13]

Aristotle, it seems, is prepared to endorse a counterfactual very like that expressed by Darwin in *The Descent of Man,* "that any animal whatever, endowed with well-marked social instincts, would inevitably acquire a moral sense or conscience, as soon as its intellectual powers had become as well developed, or nearly as well developed, as in man."[14] If one takes Darwin's 'social instincts' to be among the 'natural virtues', then this is the view expressed in the *Nicomachean Ethics* – with Aristotle adding the claim, non-counterfactually, that in fact man begins in this very state, and gradually the intellectual virtues transform the natural virtues into the Aristotelian equivalent of a moral sense, virtuous character.

The theory of the virtues that is at the core of Aristotle's ethical theory holds that there is a natural capacity, cleverness, that is like practical intelligence, absent its focus on the good life, and a natural virtue, which may develop into virtue in the strict sense when habituated by actions under the guidance of practical intelligence. The natural virtues are said to be present in animals other than man, and children are held to possess them in a manner virtually indistinguishable from that found in some other animals. Moreover, Aristotle acknowledges the soundness of the claim that when animals seem to act for their own good based on 'foresight', they are practically intelligent.

This may be, as I suspect most readers of the *Ethics* assume, a mere acknowledgment of generally accepted opinion.[15] Yet its fit with his account of animal character and behavior in *Historia Animalium* suggests that there is a general theoretical account of differences in animal behaviors that underwrites the concept of natural virtue.

HA VII–VIII:[16] A NATURAL HISTORY OF THE VIRTUES

Two books of the *Historia Animalium* are devoted to differences in the activities (πράξεις) and ways of life (βίοι) of animals. An animal's way of life differs according to its character and mode of feeding (κατὰ τὰ ἤθη καὶ τὰς τροφάς) (cf. *HA* VII, 588a17–19).[17] Most other animals have traces of these 'manners connected with soul', which differences "are more apparent in humans" (588b19–21), a claim repeated in introducing his discussion of character in *HA* VIII (cf. 608b4–7). He includes bravery and cowardice, as well as 'resemblances of that understanding connected with thinking' (τῆς περὶ τὴν διάνοιαν συνέσεως), in a list of examples, and he applies the machinery of similarity and difference by degree (or more and less) versus by analogy defended in *HA* I's introduction [and in *Parts of Animals* (*PA*) I 2–4] to these behavioral differences.[18] That introduction will structure the rest of our discussion, and we should have it before us in full:

16

> For some of these traits differ by the more and the less in relation to man, as does man in relation to many of the animals (for certain of them are more present in man, some more in the other animals), while others differ by analogy; for as art, wisdom and understanding are present in man, so in some of the animals there is some other such natural capacity (τις ἑτέρα τοιαύτη φυσικὴ δύναμις). Just such a thing is most apparent when we look at children when they are young; for in them it is possible to see 'traces' and 'seeds' of the states (τῶν ἕξεων) that will be present later in life, though at that time their soul hardly differs at all from that of the beasts (τῆς τῶν θηρίων ψυχῆς), so there is nothing unreasonable if some traits are the same in the other animals, some similar, and some analogous. [588a25–b3][19]

It is clear that Aristotle has in mind that some psychological characteristics differ in degree (by more and less) between humans and animals, whereas others are sufficiently different that they should be described as analogous; but the opening sentence of this passage also suggests that there are two ways in which animals and humans can differ in degree. David Balme's note on this passage[20] points out that the second stresses the ways in which man differs from *most* animals, and he offers the helpful suggestion that Aristotle means to distinguish two quite different sorts of contrast. Certain traits don't distinguish humans from the other animals as a group, even in degree. Other traits, by contrast, do distinguish us, if only in degree, from most of the earth's other creatures: Although other animals may be tame or political, humans are more so than any of the other animals. Thus, whereas the trait itself doesn't distinguish us from the rest of the animal kingdom, the degree to which we exemplify it does.

Three claims in this passage evidence a studied coherence with Aristotle's theory of ethical virtue:

1. The characteristics in other animals that are said to bear comparison to art, wisdom, and understanding are identified carefully as sorts of *natural capacities.*
2. The traces and seeds that children have develop into *states of character* when they are adults.
3. Nevertheless, when they are children, their souls differ 'hardly at all' (διαφέρει οὐδεν ὡς εἰπεῖν) from the souls of beasts.

As we have seen, these three claims all have interesting connections with the passages in the *Nicomachean Ethics* discussing the relationships between the virtues of intellect and those of character. There, recall, the position that Aristotle defended seemed to be that unqualified ethical virtue developed out of a natural virtue shared with certain other animals. This development paralleled that of practical intelligence out of cleverness. Indeed, Aristotle's view appeared to be that one becomes practically intelligent and morally virtuous in virtue of the gradual integration of the virtues of intellect and of character – that this integration was part and parcel with the development of natural virtue and cleverness into ethical virtue and practical intelligence.[21] It is thus significant that Aristotle is here concerned with traits that bear a likeness to *both* virtues of character and intellectual virtues. The discussion in these books should then provide further insight into the relationship between these two sorts of traits in other animals, and thus in young children.

In what follows, the case will be made that, for Aristotle, these characteristics, likenesses of which are found in other animals, are present (a) in seed in children and (b) in those very natural sources of (mature) human character and intelligence referred to in *Nicomachean Ethics*. It will be argued that in Aristotle's view humans begin life with the very *natural capacities* that are the beast's *likenesses* of bravery, temperance, understanding, or intelligence, yet end up with quite different *learned* and *acquired states,* namely *true* bravery or intelligence. The argument will be in part direct, providing further textual evidence from Aristotle's 'cognitive ethology' for this view, and partly indirect, by showing how this fits with, and accounts for, his views about the connection of natural virtue to ethical virtue that we have already surveyed in his ethics.

Developing this interpretation requires that Aristotle tell us more about 'this other natural power' that is found in other animals and is alike, somehow, to the human states of wisdom, art, and reasoned understanding. This he does, in the concrete accounts of behavior and character differences in *HA* VII–VIII.[22]

The argument of *HA* VII begins with the assertion that beyond generation, which is common to all living things, differences in the ways of life of animals begin with the ways in which they raise their offspring and feed:

> . . . as soon as perception is added their lives differ both in regard to mating, because of the pleasure involved, and in regard to the bearing and rearing of offspring. Now some, like plants, simply accomplish their own reproduction according to the seasons; others also take trouble to complete the nourishing of their young, but once that is accomplished they separate from them and have no further association; but those that have more understanding (τὰ συνετώτερα) and possess some memory continue the association, and have a more social relationship (πολιτικώτερον) with their offspring.[23] Thus, while one part of living consists for them in the activities to do with the producing of young, a further and different part consists in those to do with food; for these two objects in fact engage the efforts and lives of all animals. [588b28–589a5]

The claim is not that all animal activity can be reduced to these two pursuits, but rather that all animals partake in these two. Differences associated with manner of cooling (respiration, as we would say) are quickly added, as are movements related to climate change – migration, hibernation, and emergence from the larval stage of development (596b20–601a22). And finally, consideration of behavior associated with change in climate leads to a discussion of seasonal patterns of health and disease (601a23–605b22) and differences that seem to be local adaptations. Near the end of *HA* VII, Aristotle looks forward to the discussion of character traits by citing evidence that "locations also produce differences in the characters (τὰ ἤθη) of animals" (607a9–10).

As recent scholarship would lead us to suspect, these books are concerned with locating patterns of relationships among *differences* in activities, ways of life and character. Insofar as animal groups are referred to, they serve as a general way of ordering the discussion. Animal groups below the level of Aristotle's 'great kinds' (birds, fish, oviparous quadrupeds, crustaceans, cephalopods, etc.) are referred to only to illustrate a certain differentia or correlation between differentiae.[24]

In the foregoing passage, greater investment in the rearing of young is correlated with two behavioral traits: greater *understanding* (the comparative form of the word used in our framing passage, translated earlier) and a more complex 'political' life. Greater understanding appears, in turn, to be associated with increased *memory capacity,* as we would suspect from the familiar opening page of the *Metaphysics,* where we are told that those animals with memory are more practically intelligent (φρονιμώτερα) than those without (*Metaph.* A 1, 980b1).

These passages are certainly consistent with the discussion of the virtues of intellect in *NE* VI. Aristotle there weaves a complex web of relationships among the dispositions of practical intelligence (φρόνησις), understanding (σύνεσις), and political skill (πολίτικη). Although practical intelligence and understanding are *concerned with* the same things, he tells us, they are nevertheless not the same (1143a5–7). Practical intelligence is focused on determining what must be done in a particular situation, whereas understanding simply is a faculty of judging about such things and may be exercised independently of a concern about action. Still, it is clearly Aristotle's view that typically the person of understanding is also a person of practical intelligence, because the person of understanding judges well about what the person of practical intelligence recognizes he must achieve (cf. 1143a25–32).[25]

The connection between being political and being practically intelligent is even tighter: "Political skill and practical intelligence are the same state, though their being is not the same" (1141b23–24). Aristotle acknowledges that common usage tends to suggest that the former term refers to ability to deal with particular legislative decrees, and the latter to ability to manage one's personal affairs well. But he argues that in fact political skill is a particular application of practical intelligence, in recognition of the fact that human good cannot be fully realized outside of a properly organized community; and hence, conversely, practical intelligence must in part be focused on how to achieve proper community organization.[26]

Nevertheless, as Jean-Louis Labarrière argues,[27] Aristotle's regular and unqualified use of these terms in application to other species raises a conceptual puzzle, for it shows that Aristotle is perfectly at ease ranking animals as more or less intelligent, using (as we do, of course) the same language that is used in discussing human cognition and behavior, while asserting, at the beginning of *HA* VII, that animals other than man have natural powers that are different to varying extents from understanding, wisdom, and art in human beings. It is natural to read this passage, in fact, as saying that the natural capacities in question are only analogues of the human ones. But if animals have cognitive capacities that are only *analogues* of ours, that will make the claim that animals are more or less intelligent, or brave, problematic. On Aristotle's understanding of analogy, the fact that feathers in birds are analogous to scales in fish does not permit one to say that one fish is more feathered than another. This way of talking is reserved for differences in degree, which he *opposes* conceptually to analogous difference.

The self-consciously philosophical, framing characteristic of these two discussions tells strongly against writing the problem off as a result of traditional and colloquial use of language.[28] It is the central concern of all of these passages to determine precisely the right way to think about human intelligence in relation to the cognitive capacities of other animals.

Nor can the problem be finessed by appeal to Aristotle's philosophical development, for in the very books we are discussing Aristotle asserts that the deer seem to be φρόνιμος (611a16), that other quadrupeds, and particularly the weasel, act φρόνιμῶς (612a3, 612b1), and that the διανοία (the thinking) of birds is resourceful in promoting their ways of life (616b20–32). These passages are found in the very discussion of differences of activity, way of life, and character that our problematic passage introduces. And the same use of this language can be found throughout the biology, not to mention the *Metaphysics* and the ethical treatises.[29]

The solution to this puzzle comes from attending somewhat more closely to Aristotle's words. Doing so provides, in addition to a resolution of our puzzle, a deeper understanding of Aristotle's concept of a natural virtue.

Let us begin by considering the case for interpreting our passage to be saying that the other animals have a natural capacity that is only *analogous* to 'art, wisdom, and understanding'. The case, in fact, is not very strong. The passage begins by stressing that some animals have, along with the same 'emotional' and character traits as humans, *likenesses* to the understanding connected with thinking. Next, Aristotle introduces the distinction between more-and-less likeness and analogous likeness – apparently in order to place the discussion of differences in behavior and character within the conceptual framework of sameness and difference established in *HA* I 1, 486a15–b17. It is then asserted that whereas in humans there are art, wisdom, and understanding, in *some* animals there is another *such* natural capacity. Analogy is not mentioned, but because this statement follows his general claim that some characters differ by more and less between man and some or most animals, whereas others differ by analogy, it is easy to assume that the natural capacity referred to is an analogue to the three human capacities named. Yet there is no compelling reason to make this assumption. It is more naturally taken to be leaving the *sort* of likeness dependent on which trait is being compared in which animals.[30]

In favor of taking the claim this way is the fact that the remark that immediately follows, which insists that children are hardly different at all from the beasts, is clearly intended to provide grounds for the claim about natural likenesses to these human states. But if the souls of children are initially scarcely different at all from those of various beasts (and likewise bear only traces of the traits to be), then humans must begin life with souls *very* similar to those of certain animals, presumably differing from members of certain other species by, at most, more and less differences. From this fact about us, Aristotle concludes that it is not unreasonable "that some things present in the other animals are the same, others quite similar, and others analogous [to those found in humans]." Surely the point made clear by appeal to our children is that humans can be compared to other animals in all three ways, depending on the nature of the comparison.

Suppose, however, that we accept the idea that the 'other natural capacity' is to be viewed as a 'beastly analogue' of certain human capacities. What may we conclude from that about this capacity? As noted earlier, two features of an organism can differ in degree from one another – as bone and cartilage do, when considered as materials – and yet be analogous by virtue of playing similar roles in different kinds of organisms. Thus features that, considered in themselves, vary only in degree may nevertheless be analogous, because of their association with different kinds of animals. Were we to over-interpret the passage to be saying that other animals have a natural analogue to all three of the cognitive states mentioned, that still would provide us no license for denying that it differs only in degree from them.

A careful reading, then, doesn't provide evidence that the human intellectual capacities mentioned must *all* have *only* analogues in 'the beasts'. On the other hand, the use of the language of the natural traits of intelligence in

HA VIII is *never* used to compare the intelligence of other animals directly to their human counterparts, but always to distinguish one sort of animal from another. And, for our purposes, it is especially important that this language is scrupulously restricted to those capacities that *NE* VI links tightly together in their role as focused on achieving the good life and that it regularly claims to be present in other animals as well as in humans.[31] And indeed, this very focus is pointedly stressed in every case, via repeated use of adjectives such as εὐβίοτος and βιομήχανος (roughly, well-disposed and re-sourceful with respect to their way of life).[32]

One extended example will illustrate his approach. As Aristotle turns from character differences among viviparous quadrupeds to those of the birds, he begins with a general claim:

> On the whole, with regard to the ways of life of the other animals, one may study many imitations of human life, and more especially in the smaller than in the larger animals one may see the precision of their thinking (τὴν τῆς διανοίας ἀκρίβειαν) a first example being, among the birds, the nest-building of the swallow. For in the mixing of straw into mud she keeps the same order. She interweaves mud with the stalks; and if she lacks mud she moistens herself and rolls her feathers into the dust. Further, she builds the nest just as men build, putting stiff materials underneath first, and making its size commensurate with her own. Both parents labor over the feeding of their offspring; they give food to each nestling, carefully watching the one who has already taken food so that it doesn't take it twice. And initially they themselves throw out the dung, but when the nestlings have grown they teach them to turn around and discharge it themselves. [612b18–32]

I have quoted this passage in full so that readers unfamiliar with this part of Aristotle's thought may taste the careful descriptions upon which the general remarks rest. The general remark of interest here is that in making these observations, what one is studying are 'imitations of human life', and the observations are intended to show that, supposing the goal of their behavior is nest-building and feeding suitable to the successful rearing of young, their behavior displays 'precision thinking' with respect to that goal.

Now if we instinctively adopt a common post-Cartesian (yet pre-Darwinian) view of this passage, we might say this: "Aristotle here is saying that these birds behave *as if* they were actually thinking about these things – that must be the force of 'imitation'." But if we adopt a Darwinian view, the one that best coheres with Aristotle's overall attitude to cognition, we shall read it differently. In their behavior, the swallows imitate human life in many ways; such behavior, we know, requires a process of thought – consideration of goals, of the organization of actions needed to achieve them, and of how to adapt one's plans to difficult situations (such as the unavailability of needed materials). Aristotle never says that 'thinking'[33] requires self-consciousness or deliberation, and indeed he insists, in an argument for the validity of attributing goal-directed behavior to spiders and ants, that artisans can do what they do without deliberation (*Physics* II 8, 199b26–33). He claims, and we should accept it as his view, that we can *see,* in the behavior of swallows, thinking.

The term 'thinking' (διανοία) in these books is clearly not the term for a character trait, but for an intelligent activity that is conjoined with differences in character, depending on the sort of animal being discussed. In each passage where it is used, Aristotle uses contrastive particles (μὲν . . . δὲ), first characterizing the animal's character (τὸ ἦθος) and then its thinking, as in the following:

> The Krex is, in its character (τὸ μὲν ἦθος), pugnacious, and in its thinking (τὴν δὲ διανοίαν) resourceful in relation to its way of life. [616b20–21]

This should bring to mind the language used to distinguish the two sorts of virtue that structure the overall plan of the *Nicomachean Ethics:*

> There are, then, two sorts of virtue, virtue of thought (τῆς μὲν διανοητικῆς) and virtue of character (τῆς δὲ ἠθικῆς). Virtue of thought takes its genesis and growth more from teaching, and accordingly needs experience and time. Virtue of character results from habituation (ἐξ ἔθους), hence its name is a slight variation on habit (ἀπὸ τοῦ ἔθους). [*NE* II 1, 1103a14–18]

The intellectual capacities Aristotle mentions when discussing animal behavior are those associated with action in relation to the maintenance of life, and once again this reflects the views expressed in the ethical treatises. At *NE* VI 7, 1141a20–b2, for example, Aristotle begins with the following conditional:

> If what is healthy and good is different for human beings and for fish, while the white and the straight are always the same, then we should agree that wisdom is always the same, but practical intelligence is different. [1141a20–25]

This passage would seem to imply that it is appropriate to attribute practical intelligence, but not wisdom, to other species, and Aristotle indeed goes on to remark that this is why "even some of the beasts are said to be practically intelligent (φρόνιμον), namely all those that appear to have the capacity of forethought about their own life" (1141a26–28); and he later argues that there is no more one 'political skill' (πολίτικη) for all animals than there is one 'medicine', because the good for each animal is different (1141a31–33). True, the phrases 'are said to be' and 'appear to have' express caution; but the statements that other kinds of animals have other sorts of practical intelligence and political skill are not so qualified.

23

Again, this passage is fully consistent with the use of the language of the intellectual virtues in *HA* VIII. It requires no more, in the attribution of practical intelligence, political skill, or understanding to other animals, than that they display a capacity to organize their behavior for their own good, and/or a capacity to include the role of community in that good, and/or a capacity to make judgments about such matters (as in the swallow's ability to determine that no mud is available and its ability to determine how to produce mud at the moment when practical intelligence determines that it is necessary, because a nest must be built so that eggs can be laid in a protected place).

There is, however, one puzzling feature of Aristotle's discussion of thought and character in other animals, one shared by the *Nicomachean Ethics* and the *Historia Animalium.* In the *Historia Animalium* he says that other animals have a different natural capacity that is comparable to such traits as 'understanding', 'practical intelligence', and 'thinking' in adult human beings. In the *NE* VI 12–13, Aristotle distinguishes cleverness from practical intelligence in a way that parallels the distinction between natural virtue and unqualified virtue. But he regularly ignores these distinctions in practice. In both works, animals are referred to as practically intelligent, more or less understanding, thoughtful, courageous, temperate, and so on, without any qualifications or reservations.

This should not come as any surprise to contemporary readers – one simply needs to ask, What option did he (or do we) have? Indeed, I think that the extended discussions of the ways in which animals are similar to and different from us that open *HA* VII and VIII are essentially justifications for extending language applied in its strict sense to virtuous human beings to characterize the behavior and character of other animals.

Nevertheless, the extension remains deeply problematic. Virtue, or excellence, of character in the human realm is a state expressed in *reasoned choice,* and Aristotle is as clear as he can be that other animals have no such state. Nor do they have, as our children do, the natural capacity to *acquire* the ability to choose according to 'right reason'. And the intellectual virtues relevant to virtuous activity depend on a conception of the good life, which, again, Aristotle clearly thinks is lacking in these other animals. I shall conclude with a suggestion as to how Aristotle might respond to the problem, put in this way.

WHAT IS MISSING IN THE CHILDREN AND THE BEASTS?

Aristotle's discussion of differences in the ways of life and the underlying traits of character and intelligence in other animals provides the basis for two crucial elements in the virtue theory of the *Nicomachean Ethics:*

1. It grounds, in the actual study of animal behavior, the claim that the two crucial ingredients in the psychological makeup of the virtuous person – virtuous character and practical intelligence – have natural likenesses in the other animals.
2. It helps us to make sense of the claim that humans begin, as young children, with natural (innate) capacities that are needed if the virtues of character and practical intelligence are to develop from proper education and training.

But if the other animals – and young children – are regularly said to be more or less courageous, practically intelligent, gentle, and so on, why, in Aristotle's ethical theory, must we stop short of ascribing unqualified virtue of character or practical intelligence to children and other animals?

The answer is to be seen in the crucial role, in our moral development, of the *integration* of the relevant intellectual virtues with the virtues of character. It is telling that in Aristotle's discussions of the traits of the other animals, he regularly *contrasts* the cognitive traits and the traits of character. Further, he never speaks, in the *Historia Animalium,* of these traits as *virtues,* even though the very same words are used that denote states of *virtuous* character in humans: Animals are said to have a courageous character, but this is not referred to as a virtue. Nor does he ever suggest that the practical intelligence of animals plays a role in the determination of the actions that emanate from their character traits.

I want to suggest that these two omissions are connected: What is missing in children and in the other animals is that *integration* of practical intelligence with dispositions to feel and act, such that one's feelings and actions tend to be *appropriate* expressions of the life of a rational and political animal. For humans, this requires that the child's native 'virtues' and cleverness be trained and educated according to the standards with which the virtuous, that is, the practically intelligent, adults are familiar. This process of education and training is a matter of being encouraged to perform the actions that their caretakers know to be the just, temperate, or benevolent actions on each specific occasion. Gradually, children learn to use their own developing practical intelligence to determine (now for themselves) the appropriate actions and reactions to the concrete situations in which their lives consist and thus learn to integrate, as their caretakers have, practical intelligence and virtue of character. They now have a state expressed in *deliberate choices* to act and react in the manner defined by the person of practical intelligence – that is, they now have *complete* virtue. It is the *ability* to be in such a state that differentiates the souls of young children from the souls of the other animals, souls that are otherwise so much like our own.

Two related objections can be raised to the attribution of this solution to Aristotle.[34] The first points out, as my account has, that what most obviously appears to be missing in animals and young children is the ability to reason, and the ability to choose the good on the basis of deliberation. The second questions that the like states in other animals are not integrated, and thus whether or not such integration can be the crucial difference.

To the first objection, it simply need be pointed out that the account offered here does not deny that the virtuous human adults deliberate and choose their actions on the basis of a reasoned account. But Aristotle's account of deliberation, choice, and practical intelligence depends, as I have argued, on the idea that cleverness divorced from a properly educated character will not be able to deliberate or choose well and that a natural tendency to choose the virtuous course of action, in the absence of practical intelligence, will "hit the mean" only by chance. So the issue raised by this objection comes down to which of these admitted features of the virtuous person is more fundamental.

A thought experiment can be used to differentiate the two positions. Suppose Aristotle encountered creatures with the ability to think only at the level of 'experience'. Suppose further that they could be educated to use this thinking to transform their natural dispositions to act on the basis of desire. Would Aristotle consider them to have the rudiments of virtuous character? Or, because they lacked reason, defined as the ability to form abstract universals and reason discursively, would he deny that they had any virtue at all? I am arguing that he would take the former position.

The second objection points to an ambiguity in the notion of 'integration' as I have been using it, for it is true that Aristotle thinks of all animal functions as naturally tending toward the preservation of the animal's life. Insofar as their thinking and their character are naturally directed to the same end, they *are,* in one sense, integrated. But these creatures do not undergo a process of education aimed at *uniting* certain intellectual traits with certain character traits that are naturally distinct. As I have pointed out, Aristotle talks of the traits of intellect and character in other animals as if they remain naturally separate, whereas it is the essence of his discussion in *NE* VI 13 (so I have argued) that practical intelligence and the virtues of character become, in the virtuous human being, two aspects of one complex state.

NOTES

1. For a recent contribution to discussion of the function argument, and references to earlier discussion, see Whiting (1988).
2. Good recent discussions of the biological basis of this concept are available (Kullmann 1980; Cooper 1990; Depew 1995).

3. Darwin (1871, p. 71).
4. The Greek term ἀρετὴ is sometimes translated 'excellence', and sometimes 'virtue'. Greeks of the classical period were comfortable discussing the ἀρετὴ of instruments and animals, as well as human character, and thus using 'excellence' as a translation has much in its favor. Because the uses of the term to be discussed in this essay are all in ethical contexts, and because the very question at issue is how the human life of virtue is connected to the lives of other animals, I shall stick with the traditional translation, but its conceptual link to the idea of human excellence should be kept in mind.
5. See *Eudemian Ethics* (*EE*) III 7, 1233b24–32, which gives a taxonomy of natural affections that "tend toward the natural virtues" and looks forward to a distinction (presumably in the foregoing passage, which is common to both *EE* and *NE*) between the natural and unqualified virtues, the latter involving thought. There is an illuminating discussion of this passage in an unpublished manuscript by Norman O. Dahl entitled "Phronesis as an intellectual and practical virtue" (esp. pp. 21–5 and note 22). This is a manuscript of a lecture presented at a conference on *Nicomachean Ethics* (VI (see note 35).
6. Mark Gifford has pointed out (personal correspondence) that at 1143a28 Aristotle refers to φρόνησις as one of a number of capacities (δυνάμεις). But it should be noted that the paragraph begins, at 1143a25, by referring to the same cognitive traits as states (ἕξεις). As I note later, and as is clear from *Metaphysics* Δ 12, 1019b8–12, there is a wide sense of 'capacity' that includes 'states'.
7. He takes this to exhaust the "things found in the soul" (see *NE* II 5, 1105b19–28).
8. The especially intimate connection between the virtues of character and our feelings or emotions (πάθη) has been stressed by Kosman (1980) and Broadie (1991, ch. 2).
9. Again compare *Eudemian Ethics* III 7, 1233b24–32.
10. Some wiggle room is provided by his qualification that humans are nevertheless *able to acquire* the virtues by nature (*NE* II 2, 1103a24–26).
11. A number of authors (e.g., Burnyeat 1980; Broadie 1991) properly stress Aristotle's complex account of moral education. But none of those authors pay any attention to Aristotle's repeated claims that young children begin with natural dispositions virtually identical with those of certain other animals.
12. I am resisting the temptation to translate this term in such a way as to build rationality into the concept; see Broadie's 'rational choice' (1991). If rationality were conceptually built into the concept, Aristotle would not feel the need to argue, as he does in book III, that it is neither a sort of appetite nor emotion. He then goes on (at 1112a13ff.) to argue that it involves reason and thought.
13. Compare Nicias' exchange with Socrates over the common practice of attributing courage to beasts and children at *Laches* 196c1–c4.
14. Charles Darwin (1871, pp. 71–2). I should stress that Darwin treats the self-regarding virtues as the more advanced ones; but he is providing an evolutionary account of the development of human moral psychology, and it is his view that it is from the social instincts shared with other higher animals that morality originates (as evidenced by the emphasis on the 'tribal' virtues of primitive cultures) (Darwin 1871, pp. 95–7).
15. Indeed, in a stimulating and thoughtful commentary on this essay given at the 4-Way Workshop on Human Nature in Berlin (May 4–7, 1997), Dr. Ursula Wolf argued for this point of view. But she did not discuss any of the many passages

James G. Lennox

I use as evidence for a pervasive appeal to what we share with certain other animals as the natural starting point for our intellectual and moral virtues, and she raised only one text against my argument (*NE* VII 6, 1149b31–1150a10). But in fact this passage seems to argue for my thesis, for it is a long comparison of the human virtues and vices of temperance and intemperance with the similar 'natural' traits in certain animals. Aristotle says that these terms can be applied only metaphorically to animals (1149b31–32) and he says this is so because they lack choice and reasoning (b34–35). This seems fully consistent with the arguments of book VI that I have been discussing and surely invites the question of what is the basis of the metaphoric extension to other animals.

16. As David Balme explains (Balme 1991, pp. 30, 48–9, 56, note a) the modern numbering of the last four books of *Historia Animalium* is due to the Latin translation of Theodorus Gaza, who took book IX in the manuscripts and made it book VII (thus making books VII and VIII in the manuscripts books VIII and IX in his Latin translation). In the Balme rendition of *HA* VII–X for the Loeb Classical Library, edited for publication after Balme's death by Allan Gotthelf, the manuscript order, which should have precedence for both historical and philosophical reasons, has been restored, with the Gaza numbers placed in parentheses. However, because the Bekker page and line numbers are a universal form of reference, these continue to reflect the Gaza order. For our purposes, this will not cause any problems, because I shall be working almost exclusively with books VII (Gaza VIII) and VIII (Gaza IX), which have sequential page and line numbers. This restoration will also be reflected in David Balme's magisterial edition, near completion at the time of death, of the entire *Historia Animalium,* currently being prepared by Gotthelf for publication by Cambridge University Press.

17. In *HA* I's introduction to the methodology of the whole work, Aristotle states that animals differ in accordance with ways of life, activities, characters, and parts (487a11–15), and in the flow of the discussion there seems to be a rough distinction between a discussion of activities and ways of life (487a15–488b11), on the one hand, and of character on the other (488b12–28). Reflecting the sense that these are in fact three parallel sets of differentiae, *HA* VIII introduces itself as a discussion of character differences, distinct from the preceding discussion of differences in activity and way of life. Yet Aristotle will occasionally use phrases such as 'the activities of their way of life' (τάς τοῦ βίου πράξεις), suggesting that ways of life are constituted of the animals' activities (cf. 588b23). Clearly, however, the fact that activities and ways of life differ 'according to character' doesn't preclude differences in character being discussed separately from differences in activities and ways of life.

18. As I have discussed elsewhere (Lennox 1987), the many *forms* of animals that make up a kind for Aristotle are said to be one in kind but to differ by 'the more and the less', or by 'excess and deficiency'; whereas members of different *kinds* are said to be alike only by analogy. But because the concepts 'form' and 'kind' are relative to each other, can refer to a wide range of extensions, and to parts as well as animals, the same parts can be said to differ by more and less in one context and to be analogical in others. For example, cartilage and bone are said to be the same in nature and to differ by more and less when compared with each other (*PA* II 9, 655a32–34, *HA* III 8, 516b32–33), because they are quite similar both functionally and in their sensible properties; but they are also analogues, in

28

that they play functionally similar roles in *different kinds* of animals (*HA* III 8, 517a1–2). See also Pellegrin (1982/1986, pp. 84–90).

19. The passage immediately following this (588b4–589a9) is more famous, being one of two statements (the other being *PA* IV 5, 681a10–b13) that historians of biology identify as the sources of the idea of a continuous *scala naturae.* It takes on added philosophical significance when read in its wider context, however, because its connection with the passage we are currently discussing is to make the point that all animal differentiae vary by small degrees from one animal to the next.

20. Balme (1991, p. 59, note d).

21. In the comments mentioned in the earlier note 13, Dr. Wolf has properly asked for more clarification of this concept of 'integration'. What I have in mind is this: Moral education involves training the natural capacity of 'cleverness' on the goal of living a good and noble life, and developing the natural virtues into states of acting according to reasoned choice, rather than according to impulse and un-reasoned emotion. But this means, not that two separate capacities develop along parallel paths into two separate states, but that natural virtue develops into intelligent virtue, and cleverness develops into virtuous intelligence – the same state, Aristotle might say, but different in account.

22. Again, Dr. Wolf has countered that in this opening passage, after we are told about a different natural capacity, we are told nothing else about it. My claim, however, is that we learn a great deal about it by looking at the actual descriptions of animal behavior and character in these books. I take it Aristotle is claiming, in the theoretical passages that introduce these books, that we are sanctioned in relying on our understanding of these terms in human contexts, provided we are warned at the outset that it will be 'similarities' and 'traces' we are looking for, not the traits referred to by the same terms in properly educated virtuous adults.

23. This phrase literally translates 'they deal with their offspring in a more political manner'. *Generation of Animals* III 2 also stresses that there is a strong correla-tion between the degree of long-term parental commitment and being practically intelligent (φρονιμώτερα): "And those that share most of all in practical intelli-gence (φρόνησις) develop intimacy and parental love (συνήθεια και φιλία) even after their young have matured, as in humans and some of the quadrupeds" (753a11–14).

24. See Balme (1987), Lennox (1987), and Pellegrin (1982/1986).

25. I am merely scratching the surface of a difficult topic here. For a fuller discus-sion of the connections among 'understanding', 'political skill', and 'practical intelligence', see Gifford (1995, pp. 51–60).

26. Again, see the discussions of this issue by Kullmann (1980/1991), Depew (1995), and Cooper (1990).

27. Labarrière (1990, pp. 405–28).

28. As in G. E. R. Lloyd (1983, pp. 18–29). For a critical response, see Parker (1984, esp. pp. 181–2) and Lennox (1985).

29. *PA* II tells us that bees show more practical intelligence in their nature (φρον-ιμώτερα τὴν φύσιν) than some blooded animals (648a6–10), that they along with the ants show more understanding in their soul (συνετώτερον τὴν ψυχὴν) than some blooded animals (650b24–25), and that generally animals with thin,

pure blood have greater practical intelligence (φρονιμώτερα) and thought (διανοία) than those with thick clotting blood (648a10–11, 650b19–20).
30. Once again Labarrière (1990) is on the right track here; cf. his discussion of this
    passage on pages 411–12.
31. Thus Aristotle never says that any animal other than humans has scientific knowl-
    edge (ἐπιστήμη) or wisdom (σοφία), and he uses 'reason' (νοῦς) only once in a
    general introductory passage saying that animals can differ in reason or lack of
    it – it isn't clear, in fact, that humans are excluded from the scope of the remark,
    which is at *HA* VIII 3, 610b22. He goes on to comment that sheep are εὔηθες
    καὶ ἀνόητον – David Balme translates 'simple-minded and stupid' – followed by
    a good deal of convincing shepherd lore as evidence. But when he turns to the in-
    telligent ones, such as deer, he refers to them as φρόνιμος, not νόητον (at 611a16).
32. E.g., at 614b32–35, 615a19, and 615a34 and the repeated uses from 616b10–33.
33. That is, διανοία: Of course, deliberation is crucial to human moral action, but
    that is not relevant here.
34. Both objections were raised, in slightly different ways, by Mark Gifford and Ur-
    sula Wolf.

REFERENCES

Balme, D. M. 1987. Aristotle's use of division and differentiae. In: *Philosophical Is-
    sues in Aristotle's Biology,* ed. A. Gotthelf and J. G. Lennox, pp. 69–89. Cam-
    bridge University Press.
    (ed. & trans.) 1991. *Aristotle: History of Animals,* books VII–X. Cambridge, MA:
    Harvard University Press.
Broadie, S. 1991. *Ethics with Aristotle.* Oxford University Press.
Burnyeat, M. 1980. Aristotle on learning to be good. In: *Essays on Aristotle's Ethics,*
    ed. A. O. Rorty, pp. 69–92. Berkeley: University of California Press.
Cooper, J. 1990. Political animals and civic friendship. In: *Aristotles Politik: Akten
    des XI. Symposium Aristotelicum,* ed. G. Patzig, pp. 221–41. Göttingen: Vander-
    hoek und Ruprecht.
Darwin, C. 1871. *The Descent of Man and Selection in Relation to Sex.* London:
    J. Murray.
Depew, D. J. 1995. Humans and other political animals in Aristotle's history of ani-
    mals. *Phronesis* 40 (2): 156–81.
Gifford, Mark. 1995. Nobility of mind: the political dimension of Aristotle's theory
    of intellectual virtue. In: *Aristotelian Political Philosophy,* ed. K. I. Bourdouis,
    pp. 51–60. Athens: Kardamitsa.
Kosman, L. A. 1980. Being properly affected: virtues and feelings in Aristotle's
    ethics. In: *Essays on Aristotle's Ethics,* ed. A. O. Rorty, pp. 103–16. Berkeley:
    University of California Press.
Kullmann, W. 1980. Der Mensch als politisches Lebeswesen bei Aristoteles. *Hermes*
    108: 419–43. English translation (1991): Man as a political animal in Aristotle.
    In: *A Companion to Aristotle's Politics,* ed. D. Keyt and F. D. Miller, Jr.,
    pp. 94–117. Oxford: Blackwell.

Labarrière, J.-L. 1990. De la phronesis animale. In: *Biologie, Logique et Méta-physique chez Aristote,* ed. D. Devereux and P. Pellegrin, pp. 405–28. Paris: CNRS.

Lennox, J. G. 1985. Demarcating ancient science. A discussion of G. E. R. Lloyd, *Science, Folklore, and Ideology. Oxford Studies in Ancient Philosophy* 3: 307–24.
1987. Kinds, forms of kinds, and the more and the less in Aristotle's biology. In: *Philosophical Issues in Aristotle's Biology,* ed. A. Gotthelf and J. G. Lennox, pp. 339–59. Cambridge University Press.

Lloyd, G. E. R. 1983. *Science, Folklore, and Ideology.* Cambridge University Press.

Parker, R. 1984. Sex, women, and ambiguous animals. *Phronesis* 34 (2): 174–87.

Pellegrin, P. 1982. *La classification des animaux chez Aristote.* Paris: Les Belles Lettres. Translation by T. Preuss (1986): *Aristotle's Classification of Animals.* Berkeley: University of California Press.

Whiting, J. 1988. Aristotle's function argument: a defense. *Ancient Philosophy* 8: 33–48.

2

The Moral Status of Animals in Eighteenth-Century British Philosophy

Michael Bradie

INTRODUCTION

The contemporary debate over the moral status of animals reflects a mixture of traditions. Utilitarianism, which measures moral standing in terms of the ability to suffer, has been used to defend the widening-circle conception of morality. The difference between humans and other animals vis-à-vis moral standing diminishes in its light. Focusing on questions of agency, conscience, and reflective powers, the differences between humans and nonhumans seem greater. Darwinism has been invoked to bridge the gaps between the intellectual and moral capacities of humans and those of other animals. This has led some to argue either that differences in agency, conscience, and the power to reflect are overrated (e.g., Rodd 1990) or that they may not be absolute indicators of moral consideration (e.g., Rachels 1990).

Debates over the moral status of nonhuman animals have had a long and stormy history. The considerations that drive the current debate – questions of the intellectual capacities of animals, questions about their moral sensibilities, and questions about their sensitive capacities – are ancient (Sorabji 1993). Nonetheless, they first began to assume their modern form in the writings of the eighteenth-century British moral philosophers. Those individuals were struggling with the moral implications of (1) discoveries in the natural sciences and (2) an increasingly secularized approach to philosophical issues. Vivisection experiments and rudimentary anatomical studies not only showed that the bodies of animals were remarkably similar to the bodies of human beings but also showed that the brutes were capable of experiencing pain and suffering. The physical and behavioral similarities between humans and the brutes meant that those who wished to argue for a moral difference were hard pressed to find where and with respect to what their argument could gain any leverage.

The mechanical approach to questions of human nature ushered in by the impact of Newton's worldview and its appropriation as a model by pro-topsychologists and moral theorists alike, coupled with those empirical studies, had profound effects on the development of moral philosophy in the eighteenth century. The moral gap between humans and the brutes could no longer be sustained by an appeal to divine revelation.

This essay is an examination of some of the central arguments in the eighteenth-century debate. The issues are both conceptual and empirical. What counts as moral standing of a given kind is a conceptual issue. Whether or not animals (or *all* humans, for that matter) have the relevant factors needed to enjoy that moral standing is an empirical issue. The eighteenth-century discussions concerning the moral status of animals exhibited a broad shift from one conceptual foundation at the beginning of the century to a different foundation at the end. At the beginning of the century the question was whether or not animals had souls. As the relevance of having a soul declined, it was replaced by questions about the intellectual capacities of animals. At the end of the century, the utilitarian tradition redefined the relevant conceptual base as the capacity to suffer. The path from the Cartesian position, which denied that screaming animals were suffering, to Bentham's position, which was that they could suffer, and so deserved proper moral consideration, wound through a rising accumulation of evidence that supported the view that animals could suffer. That animals can suffer, of course, is in itself no reason to adopt the conceptual stance that the capacity for suffering should be the basis of moral standing. But the irrelevance of divine revelation suggests that the choice of a proper conceptual base for determining moral standing is not something that can ignore empirical results. The proper foundation of moral standing has not yet been resolved. This essay is an exploration of the development of some of the themes from the eighteenth century that have helped to shape the modern debate.

The issues in moral philosophy that were debated in the seventeenth and eighteenth centuries revolved around three questions: (1) Is the source of moral order external to us or internal to us? (2) Is knowledge of that order accessible to everyone or only to a few? (3) Need we be compelled to be moral, or is there something in our nature that leads us to a moral life? (Schneewind 1990, vol. 1, p. 18). The brutes served as convenient models against which the integrity or divinity of human beings could be contrasted. They were, in effect, mere foils for establishing the moral condition of human beings.

In the following, I offer a broad sketch of some of the important themes relating to the status of animals in eighteenth-century British moral philosophy. I begin with a theme inherited from the past: man's dominion over the brutes. I then turn to a discussion of the implications of the gradual realization that animals had an "inner life" and perhaps not only were worthy of moral consideration but also were capable of being moral agents in their own right. I conclude with some remarks on the contemporary implications of the eighteenth-century discussion.

MAN'S DOMINION OVER THE BRUTES

In the early modern period, English Christians tended to view nature and animals as opportunities for exploitation, not stewardship (Thomas 1983, pp. 24f.). Part of that pervasive anthropocentrism was the view that human beings had dominion over the brutes. In the eighteenth century, Bishop Butler argued that men were superior to the brutes in virtue of their reason: "There are several brute creatures of equal and several of superior strength to that of men; and possibly the sum of the whole strength of brutes may be greater than that of mankind; but reason gives us the advantage and superiority over them, and thus man is the acknowledged governing animal upon the earth" (Crooks 1855, p. 128). It does not follow, Butler went on to say, that rational beings will always prevail over irrational beings or brutes. But reasoning beings can form cooperative and coordinate ventures, so that over time they will prevail. So, too, he thought, would virtue prevail over vice. Later in the century, Adam Ferguson echoed those sentiments: "Animals have power, consisting in muscular strength; and, in this respect, Man is inferior to many of the brutes: But his dominion in nature is derived from a different source; – from his superior skill, and the authority of mind over-ruling and wise" (Ferguson 1978, vol. 1, p. 2). And, later, "to this peculiarity of aspect and form, on the part of man, is joined a decided superiority of condition and power. Compared to the other animals, he is every where the Lord among his vassals, and the master among his slaves; or, where any species remains untamed, and disposed to dispute his ascendant, the contest in fact is unequal, or the balance, by some evident advantage of superior resource and contrivance, ever inclines to his side" (Ferguson 1978, vol. 1, p. 51).

Francis Hutcheson offered an elaborate defense of that traditional point of view from within a theistic framework. It can be found in Hutcheson's *A System of Moral Philosophy,* which appeared posthumously in 1755. A chapter entitled "The adventitious Rights, Real and Personal, Property or Dominion" contains the arguments for the claims to man's dominion over what Hutcheson calls "the inferior animals" (Hutcheson 1968, vol. 1, pp. 309–24).

According to Hutcheson, "real rights" are rights to goods or things. Among these goods and things are animals, and their labor (Hutcheson 1968, vol. 1, p. 309). Inferior animals are led by their appetites and instincts. They do not have the capacity for distinguishing between right and wrong. As such, they use "such fruits of the earth as their senses recommend and their appetites crave for their support." They do so as a matter of instinct. So did human beings before they appreciated that a wise God had created all "these curious forms [of inanimate objects]" as well as the notions of right (Hutcheson 1968, vol. 1, p. 310). With that knowledge, they discovered that "it was the will of *God* that they should use the inanimate products of the earth for their support of more comfortable subsistence." In addition, humans came to see that they had a right to use them. Hutcheson then offered the following "considerations" in support of that right:

1. Humans come to see themselves as the most deserving species and to see that they would perish without the use of these inanimate objects.
2. They come to see that "their instincts and senses were plainly destined to lead them into the use of them."
3. Humans come to see that the instincts of lower animals clearly show that inanimate forms were destined to support animals.
4. Inanimate forms are destined to perish, and unless utilized to promote the happiness of animal life, they would serve no valuable purpose.
5. The inanimate objects are indifferent to their states – changing them affects happiness only to the extent that they are used to further the ends of animate beings.

Hutcheson's final argument is this:

P1. The happiness of animals will be greatly increased through the use of inanimate forms.
P2. The use of inanimate forms by animals does not result in any decrease in happiness for the forms, for they have none in themselves.
C1. Therefore it is right that animals should use inanimate forms to further their own ends.
C2. Furthermore, it is the intention of God that they should be so used (Hutcheson 1968, vol. 1, p. 310).

Hutcheson observed that

[a] new created pair [of humans or any animals] indeed could scarce subsist even in the finest climates, without a place cultivated for them artificially, and stored with fruits ready for their subsistence. Their first days must be anxious and dangerous, unless they were instructed about the fruits proper for their use, the natures of animals around them, the changes of seasons, and the arts of shelter and storing up for the future. They would not need a revelation to teach them their right, but would need one to teach them how to use it. [Hutcheson 1968, vol. 1, pp. 310–11]

The right of human beings to use "inferior animals," in addition to their right to exploit their environment, was not so obvious. But the power of reason soon presented a case. A rational being with an understanding of right and wrong would, should the occasion arise, be conscious of his "natural power by means of reason to make such creatures subservient to his support and happiness." So apprised, the rational being would "readily presume upon his right, and a little further reflection would confirm his presumption" (Hutcheson 1968, vol. 1, p. 311).

The problem is that inferior animals, however inferior, have the capacity to suffer – to experience happiness or misery. Our use of inferior animals, unlike our use of inanimate nature, can result in suffering or misery. To the extent that it does, such actions contribute a negative value to the sum total of happiness in the world. For Hutcheson, who was a proto-Utilitarian, that created a problem. The saving grace was that the *necessary* uses of animals that contributed to their suffering could be balanced by the amount of happiness for us that such uses provided. Hutcheson allowed that we "must condemn all unnecessary cruelty toward them as showing an inhuman temper" (Hutcheson 1968, vol. 1, p. 311). His emphasis was on *us; they* suffer, but *we* are diminished in the process. Is the inflicting of unnecessary cruelty on inferior creatures bad because of their suffering or because it makes us inhuman? As it turns out, it is a little bit of both.

"Could we subsist sufficiently happy without diminishing the ease or pleasure of inferior animals, it would be cruel and unjust to create to them any needless toil or suffering, or to diminish their happiness" (Hutcheson 1968, vol. 1, p. 311). But, we can't. Why not? Our capacity for happiness and misery is greater than that of the beasts. Their "external senses" may be as acute or more acute than ours, but their "internal senses" are not:

> . . . men have superior senses or powers of enjoyment or suffering; they have sublimer pleasures by the imagination, by knowledge, by more extensive and lasting social affections, and sympathy, by their *moral sense,* and that of honour. [Their] reason and reflection collect joys and sorrows, glory and shame, from events past and future, affecting others as well as themselves; whereas brutes are much confined to what at present affects their senses. [Hutcheson 1968, vol. 1, p. 312]

Given that mix of animate beings (it is not clear whether plants were to be included or, as the earlier passage about "fruits" suggests, belonged rather to the realm of the inanimate forms), each with, as Hutcheson put it, its appropriate "dignity," how would an "impartial governor" [shades of Adam Smith, who was a subscriber to Hutcheson's volume (two sets)] aim for the "best state"? Presumably the best state was that in which each kind of creature would enjoy that degree of happiness, all things considered, that its dignity allowed it. Suppose, Hutcheson said, that human beings multiplied so fast that they became "oppressed with immoderate toil and anxiety" without the use of animals. In such a case, all would be worse off; the tamer animals would be at the mercy of the wild and would tend to perish. Some of them could be harnessed to do labor to increase the happiness of human beings, at a cost to them far less than they would incur should they be allowed to perish in the wild as a consequence of their labor not being expropriated by human beings. The benefits accruing to mankind would far outweigh the evils imposed upon the beasts. So liberated, humans could care for the animals, and the numbers of the animals would increase (Hutcheson 1968, vol. 1, pp. 312–13). Hutcheson concluded:

> Here is plainly a well ordered complex system, with a proper connexion and subordination of parts for the common good of all. It tends to the good of the whole system that as great a part as possible of the severer labors useful to the whole be cast upon that part of the system to which it is a smaller evil, and which is incapable of higher offices requiring art and reason: while the higher part, relieved from such toil, gains leisure for nobler offices and enjoyments of which it alone is capable; and can give the necessary support and defence to the inferior. Thus by human dominion over the brutes, when prudently and mercifully exercised, the tameable kinds are much happier, and human life exceedingly improved. And this sufficiently shows it to be just. [Hutcheson 1968, vol. 1, p. 313][1]

That happy state, however, would create another potential problem. Suppose human beings and other animals multiplied so fast that there were greater and greater demands for fewer and fewer resources. Problems of scarcity and conflicts of interest between humans and brutes would arise. How to adjudicate between them? Those who had forethought (humans) would suffer more loss in a given situation than those that did not (the brutes). Hutcheson concluded that "the brutes therefore can have no right or property valid against mankind, in any thing necessary for human support" (Hutcheson 1968, vol. 1, p. 313).

That created a problem of application. Just what was *necessary* for human support? Hutcheson did not say. Those arguments have reemerged in the contemporary debate over the uses of animals (as food, as laboring animals, in research, etc.). Because that system is best that apportions rewards to those most deserving in proportion to their desert, any suggestion that the brutes had claims against humans in such situations was absurd, for such a right would be "opposite to the greatest good of the system" (Hutcheson 1968, vol. 1, p. 314). Had God intended that animals have such rights, he would have given them the wherewithal to make them known, to exercise them, and to deal with humans where conflicts arose (Hutcheson 1968, vol. 1, p. 314). So the nub of it is that the brutes get short shrift in view of their incapacity to reason and reflect.

That being said, animals were seen to have the right "that no useless pain or misery should be inflicted" on them; our sense of pity is an intimation of that right (Hutcheson 1968, vol. 1, p. 314). This is so despite the fact that brutes "have no notion of rights or of moral qualities" (Hutcheson 1968, vol. 1, p. 314). Infants, Hutcheson reminds us, are in the same state, yet we feel obligated to respect their rights not to be abused or mistreated. So, on Hutcheson's view, one can have a right without being aware of it or without the capacity for being aware of it. These rights of brutes are predicated upon their ability to "needlessly suffer." In addition, there is a further consideration: "frequent cruelty to brutes may produce such a bad habit of mind as may break out in like treatment of our fellows" (Hutcheson 1968, vol. 1, p. 314).

But when all is said and done, humans care for the tamer animals for what can be gotten out of them; those that are not useful to the promotion of human happiness are or should be left to fend for themselves. Barring that, they should be banished and "be left to perish miserably in desarts [*sic*] and mountains by savage beasts" (Hutcheson 1968, vol. 1, p. 315).

> Those of the inferior species thus destined for food to the superior, enjoy life and sense and pleasure for some time, and at last perish as easily as by old age, winter-cold, or famine. The earth and animals must have had quite different constitutions, otherwise these seeming evils could not have been prevented. [The] superior orders must have had some food provided: 'tis better this food be animated for some time, and have some low sense and enjoyment, than be wholly insensible, and only subservient to nourish animals. [In other words, better to have lived and lost than never to have lived at all.] These lower orders also during their lives may do considerable service in the world, as naturalists observe that the smaller insects, the ordinary prey of birds and fishes, by feeding on all putrefaction, prevent the corruption of the air, and thus are useful to the whole system. [Hutcheson 1968, vol. 1, p. 316]

> It would be the interest of an animal system [systems have interests!] that the nobler kinds should be increased, though it diminished the numbers of the lower. A violent death by the hands of men may be a much less evil to the brutes than they must otherwise have endured, and that much earlier too, had they been excluded from human care. [Thus the old refrain: But for us, they never would have lived at all.] By this use of them for food men are engaged to make their lives easier and to encourage their propagation. They are defended and fed by human art, their numbers increased, and their deaths may

be easier; and human life made agreeable in those countries which otherwise must have been desolate. Thus the intention of nature to subject the brute animals to men for food is abundantly manifest, and its tendency to the general good of the system shows that men have a right to make this use of them. [Hutcheson 1968, vol. 1, p. 316]

Those considerations would occur to anyone who had the need to use brutes either for work or for their flesh. The right of human beings to exploit the brutes was the consequence. Hutcheson concluded: "And yet this right is so opposite to the natural compassion of the human heart that one cannot think an express grant of it by revelation was superfluous" (Hutcheson 1968, vol. 1, p. 316). In other words, although reason leads us to the conclusion that man has dominion over the beasts, this conclusion is so contrary to our natural compassion that it is a good thing that such dominion was revealed to us as our right in the Bible.

Hutcheson acknowledged that such a call for reasons on this issue was not a live topic of debate in Scotland. But in other places, he noted, people had rejected the right of dominion except on God's revelation. If their reasons were sound (and the foregoing considerations were supposed to show that they were not), then Hutcheson allowed that "any grant of it by revelation [would] appear incredible" (Hutcheson 1968, vol. 1, p. 317).

Despite Hutcheson's defense of the dominion of man over the brutes, there were alternative voices as well. Commenting on G. M. Trevelyan's sense of the need to preserve nature in the 1930s, Keith Thomas noted that until the beginning of the eighteenth century the march of civilization meant the cultivation and clearing of nature. As an inheritance of man's dominion over nature, "agriculture stood to land as did cooking to raw meat. It converted nature into culture. Uncultivated land meant uncultivated men" (Thomas 1983, p. 15). In his "inquiry concerning virtue or merit," Lord Shaftesbury developed a contrasting theme of nature as an organic whole, whose parts mutually interact to produce a viable system.

For Shaftesbury, nature was an ordered and balanced economy. If that balance was thrown out of kilter, the system as a whole suffered. One might call this a homeopathic view of the state of nature. Shaftesbury, were he alive today, no doubt would find comfort in the Gaia hypothesis.

The important point is that the several species of animals were parts of this whole, and their interactions as individuals and as groups were parts of the system. So the males and females of a species were both parts of a "particular race or species" with "one common nature" and cooperated in the conservation of the species and in its support (Shaftesbury 1963, vol. 1, p. 245). "In the same manner, if a whole species of animals contribute to the existence or well-being of some other, then is that whole species, in general a part only of some other system" (Shaftesbury 1963, vol. 1, p. 245). So spiders need flies, and in some sense flies are to spiders as limbs are to organs or as branches of a tree are to the trunk. "There is a system of all animals: an animal-order or economy, according to which the animal affairs are regulated and disposed" (Shaftesbury 1963, vol. 1, pp. 245f.). In considering nature as this interconnected system, we can see that one organism's suffering is often another organism's sustenance. The pain suffered by some is required for others to survive. The system as a whole prospers. Private pains are no ills in themselves, no more, says Shaftesbury, than "the pain of breeding teeth is ill in a system or body which is so constituted that, without this occasion of pain, it would suffer worse by being defective" (Shaftesbury 1963, vol. 1, p. 246).

> So that we cannot say of any being that it is wholly and absolutely ill, unless we can positively show and ascertain that what we call ill is nowhere good besides, in any other system, or with respect to any other order or economy whatsoever.
>
> But were there in the world any entire species of animals destructive to every other, it may be justly called an ill species, as being ill in the animal system. And if in any species of animals (as in men, for example) one man is of a nature pernicious to the rest, he is in this respect justly styled an ill man. [Shaftesbury 1963, vol. 1, p. 246]

But what makes a man ill is not something that befalls him, as plague spots, but rather in affections that move him – and he is then deemed good or ill only when "the good or ill of the system to which he has relation is the immediate object of some passion or affection moving him" (Shaftesbury 1963, vol. 1, p. 246).

We can now wonder about the following:

1. Do nonhuman animals have these affections, and, if so, are they moral agents or patients?
2. Given the complex network of interconnections that constitute nature, it will be the case that individuals are often and for the most part relating to a number of different systems. This will generate a conflict of assessments with respect to whether a given action is good or ill depending on which system one takes to be overriding. If so, how is one to determine which overrides which? Shaftesbury does not give us any clues in this regard. All he says is that each animal or kind has an "economy of passions" that, when observed, leads to right or moral conduct, and when out of balance leads to wrong.

THE INNER LIFE OF ANIMALS:
REASON, REFLECTION, AND AGENCY

At the beginning of the eighteenth century, the moral status of animals, for the most part, turned on the question whether or not animals had souls and could reason. Keith Thomas noted that "the Cartesian view of animal souls generated a vast learned literature, and it is no exaggeration to describe it as a central preoccupation of seventeenth- and eighteenth-century European intellectuals" (Thomas 1983, p. 35). Indeed, "the suggestion that a beast could feel or possess an immortal soul, commented John Locke, had so worried some men that they 'had rather thought fit to conclude all beasts perfect machines rather than allow their souls immortality'" (Thomas 1983, p. 34). In an attempt to defuse that notion, Butler argued that the limbs and organs of sense were mere instruments that human beings used to navigate in the world and to make their way about in it. As such, our bodies stood to ourselves as mere instruments. Their destruction and dissolution no more meant the destruction of our souls than did the destruction of a microscope mean the destruction of the scientist who used it. Butler then continued: "But it is said, these observations are equally applicable to brutes; and it is thought an insuperable difficulty, that they should be immortal, and by consequence, capable of everlasting happiness" (Crooks 1855, p. 100).

Butler accepted the natural immortality of brutes, but rejected that line of reasoning on two grounds: First, even if the natural immortality of brutes entailed that they would become moral and rational agents, that was no objection, because, for all he knew, they did possess such latent powers. Second, even if brutes were naturally immortal, that would not mean that they were endowed with latent powers of reason and morals. For all he knew, "the economy of the universe might require that there should be living creatures without any capacities of this kind." So, Butler urged, we cannot reject the immortality of the human soul based on unknown considerations pertaining to the souls of brutes (Crooks 1855).

The eighteenth-century British moralists staked the moral divide between men and the brutes on the quality of their respective inner lives. Much of that contrast had to do with the human capacity for reason and its associated abilities. Humans could reflect; the brutes could not. Humans had a conscience or moral sense; the brutes did not. Humans could form contracts; the brutes could not (Erasmus Darwin notwithstanding). Humans had interests; the brutes – well, if they had any, they were subservient to human interests. Humans could improve their conditions; the brutes could not. Humans could speak; the brutes could not. Humans could suffer; the brutes – well, again, they could suffer, but, at best, that made them moral patients. That is, they were due some consideration, but their suffering was not taken to be as great or as significant as human suffering. Hume, who more than anyone at the time emphasized the continuity between the capacities of brutes and the capacities of men, found room in the difference in degree to distinguish mankind from the beasts. Even Bentham, who turned the question "Can they suffer?" into a moral litmus test, saw nothing wrong with killing animals and eating them. The general drift of the argument was that even though animals suffer, they suffer less because of their limited mental capacities to anticipate what might have been. In short, their lesser capacity for reflection and reason serves them ill even when the moral focus shifts from "Can they reason?" to "Can they suffer?" So, in rejecting the Cartesian picture, which suggested that brutes had no "inner life" at all, the eighteenth-century British moral theorists took pains to emphasize either the qualitative or quantitative differences between the inner lives of animals and the inner lives of human beings. The secularization of the foundations of ethics meant that none but the theists could comfortably take refuge in scriptural revelations as the basis for the moral superiority of human beings.

Reason, Reflection, and Agency

According to Bishop Butler, both brutes and humans "obey their instincts or principles of action, according to certain rules; suppose the constitution of their body, and the objects around them" (Butler 1970, p. 7). Brutes who are so moved act in accordance with their "whole nature." They do not pause to reflect or reason. How do we know? Butler claimed that we are led to this conclusion by the fact that we do not observe anything that would lead us to think that there is anything more to their natures (Butler 1970, p. 7). It is otherwise with men. We have reason to suspect that there is more to their natures than the sum of the principles that lead them to act the way they do. There is a power of reflection. But how can we be so sure on Butler's view, that this power is lacking in animals? Do we not observe them to hesitate, to tentatively assess and reassess situations in which they find themselves, just as we observe in human beings? On what grounds do we not ascribe to the brutes at least some sense of reflection and approbation (or judgment as to the relative worthiness of one course of action as opposed to another)? Butler did not indicate the answer, although he was clear that failing to always give our conscience its due is to act in a way that is contrary to *our* nature or constitution (Butler 1970, p. 8).

On the other hand, Butler rejected the view that humans might live by reason alone – to do away with our natural affections and passions would be as bad as if we were to be "entirely governed by them" (as are the brutes) (Butler 1970, sermon V, sect. 3, p. 51). Were we to be entirely governed by our passions, we would have sunk almost to "the condition of brutes; and that would leave us without a sufficient principle of action." Midgley argued that for Butler, what distinguished humans from the brutes was that humans had an integrated life-style that the brutes, driven by immediate contingencies, did not (Midgley 1995, ch. 11). But brutes have natures too; they have their own agendas and presumably their own sufficient principles of actions. They may not be ours, but they must suffice, or else the brutes could not survive. This raises the question: Could Midgley's reading of Butler be right? How could Butler, or indeed any of the eighteenth-century writers, fail to see that organisms that were at the mercy of whatever was their immediate passion could not long survive? One possibility is that the then-current theories of the psychology of animal behavior were extremely crude. What we might see as a complex pattern of behaviors (e.g., stalking prey) was conceptualized by them as a single indivisible unit of behavior. Another possibility is that the details of animal psychology were of no great concern to Butler and his contemporaries. They used animals as contrasts to humans, and stereotypical characterizations were sufficient for their purposes (cf. Wilson 1995).

This theme is echoed by other writers. Adam Ferguson, for example, wrote that what distinguishes man from the animals is intelligence or mind, "intimately conscious of itself, as it exists in thought, discernment, and will" (Ferguson 1978, vol. 1, p. 48). The proofs of this distinctiveness lie in external appearances and internal evidence. "The animals, for ought we know, might be supposed to partake in the intelligence of man, if the external effect did not serve to evince his distinction" (Ferguson 1978, vol. 1, p. 49).[2]

Hume, on the other hand, argued that it is evident that animals *are* endowed with thought and reason. They behave in ways that are so much like our own, when we reason from ends to means, that "all our principles of reason and probability carry us with an invincible force to believe the existence of a like cause [in them]." That was so obvious that Hume said that examples would be superfluous (Hume 1978, p. 176).

Hume argued that the causal structure of the dog's reasoning and inferring was the same as for human reasoning – involving sensation, judgment, memory, and inference (Hume 1978, pp. 177f.). Hume's point seems to have been that human reasoning, for the most part, is no more than the result of custom on experience and the imagination (Hume 1978, p. 178). We behave like the beasts, but "beasts certainly never perceive any real connection among objects [and, therefore, neither do we]." Experience operates on animals through custom alone. They never suspect mistakes. (But humans do?) "Reason is nothing but a wonderful and unintelligible instinct in our souls, which carries us along a certain train of ideas, and endows them with particular qualities, according to their particular situations and relations" (Hume 1978, p. 179). So, wherein lies the difference between human and beast? "Men are superior to beasts principally by the superiority of their reason; and they are the degrees of the same faculty, which set such an infinite difference betwixt one man and another. All the advantages of art are owing to human reason" (Hume 1978, p. 610). There Hume was a lone voice against the general tide. Whereas the majority of his contemporaries used animals as contrast cases to distinguish humans from the brutes, Hume focused on the similarities and concluded that human beings were less distinguished from the brutes than they might like to believe. That did not, however, lead him to confer the mantle of agency upon them.

Thomas Reid was somewhat more cautious. He acknowledged that we share with the brutes certain faculties that precede our use of reason and powers of reflection. These powers in humans are gradually developed in individuals. Because they are not fully operational from the beginning, their development is difficult to trace. But "some operations of brute-animals look so like reason that they are not easily distinguished from it. Whether brutes have anything that can properly be called belief, I cannot say; but their actions shew something that looks very like it" (Reid 1967, vol. 2, p. 548). For Reid, appetites (those inclinations that have an "uneasy sensation associated with them") and desires are principles acting on the will and intention common to both brute animals and human beings (Reid 1967, vol. 2, p. 551).

> When appetite is opposed by some principle drawing a contrary way, there must be a determination of the will, which shall prevail, and this determination may be, in a moral sense, right or wrong. Appetite, even in a brute-animal, may be restrained by a stronger principle opposed to it. A dog, when he is hungry and has meat set before him, may be kept from touching it by the fear of immediate punishment. In this case his fear operates more strongly than his desire. Do we attribute any [moral] virtue to the dog on this account? I think not. Nor should we ascribe any virtue to a man in a like case. [The] animal is carried by the strongest moving force. [This] requires no exertion, no self-government, but passively to yield to the strongest impulse. [This], I think, brutes always do [a Butlerian point!]; therefore we attribute to them neither virtue nor vice. We consider them as being neither objects of moral approbation, nor disapprobation. But it may happen that, when appetite draws one way, it may be opposed, not by any appetite or passion, but by some cool principle of action, which has authority without any impulsive force – for example, by some interest which is too distant to raise any passion or emotion, or by some consideration of decency or of duty. [Reid 1967, vol. 2, p. 554]

But such is not the lot for brutes who "have no power of self-government. From their constitution, they must be led by the appetite or passion which is strongest for the time. On this account, they have, in all ages, and among all nations, been thought incapable of being governed by laws, though some of them may be subjects of discipline" (Reid 1967, vol. 2, p. 554).

For Reid, our good, on the whole, is "that which, taken with all its discoverable connections and consequences, brings more good than ill" (Reid 1967, vol. 2, p. 581). A concern for this is beyond the capacity of brute animals, presumably because they cannot discern the necessary discoverable connections and consequences. To do so would require the ability to reflect upon the past and consider prospects for the future. Such an ability is beyond the ken of brutes. Reid goes on to contrast reason with the passions. The animal principles, or blind desires, are "hot" desires that render us inattentive, so that we do not see the consequences of what we do, nor wherein lies our interest and happiness. Heat leads to "blindness," whereas the "cool" light of reason allows us to attend to and contemplate at our leisure where the best course of action lies (Reid 1967, vol. 2, p. 581).

What were the external circumstances that persuaded the philosophers that animals were not possessed of the higher powers of reason? The apparent absence of any linguistic ability was the key for most. Ferguson articulated the majority view that what distinguished man from the animals was the power of speech or language: "Although other animals learn from human beings to articulate sounds, and thereby show that there is not any absolute inability of their organs for this purpose; yet, they have not the meaning affixed to the sounds they articulate. And, if some animals, without being able to articulate, take the meaning of words, as the dog or the horse knows his name, and obeys the command of his master; yet we cannot, by any means, admit that they are fitted to partake with man, in the formation or use of language" (Ferguson 1978, vol. 1, p. 40).

Conscience

For Butler, what precluded the brutes from being moral agents was that they lacked a conscience. Exactly what Butler meant by conscience – whether it was a separate principle or a manifestation of the integrated reflective equilibrium that is characteristic of human nature – is a matter of some dispute. We find a similar analysis by Reid.

Conscience, for Reid, was an active and intellectual power that, like all human powers, matured with age (Reid 1967, vol. 2, p. 595). Conscience, which was the faculty of distinguishing right from wrong, was a universal trait that developed from "seeds . . . planted in the mind by him who made us." It was a power peculiar to human beings. "We see not a vestige of it in brute animals. It is one of those prerogatives by which we are raised above them" (Reid 1967, vol. 2, p. 596).

We share many faculties with the brutes. Reid listed among these shared faculties perception, sensibility to pleasure and pain, parental affection, sympathy, anger, emulation, pride, shame, and the ability to be trained. Were conscience to be resolved into some combination of these, we would have to acknowledge that

> some brutes are moral agents, and accountable for their conduct. But common sense revolts against this conclusion. A man who seriously charged a brute with a crime, would be laughed at. They may do actions hurtful to themselves, or to man. They may have qualities, or acquire habits, that lead to such actions; and this is all we mean when we call them vicious. But they cannot be im-moral; nor can they be virtuous. They are not capable of self-government; and when they act according to the passion or habit which is strongest at the time, they act according to the nature that God has given them, and no more can be required of them. They cannot lay down a rule to themselves, which they are not to transgress, though prompted by appetite, or ruffled by passion. We see no reason to think they can form the conception of a general rule, or of obli-gation to adhere to it. [They] have no conception of a promise or contract; nor can you enter into any treaty with them. They can neither affirm nor deny, nor resolve, nor plight their faith. If nature had made them capable of these oper-ations, we should see the signs of them in their motions and gestures. The most sagacious brutes never invented a language, nor learned the use of one before invented. They never formed a plan of government, nor transmitted inventions to their posterity. [Reid 1967, vol. 2, p. 595]

Reid concluded: "These things, and many others that are obvious to com-mon observation, shew that there is just reason why mankind have always considered the brute-creation as destitute of the noblest faculties with which God hath endowed man, and particularly of that faculty which makes us moral and accountable beings" (Reid 1967, vol. 2, p. 597).

Finally, with respect to conscience, Reid noted that conscience had au-thority over the other active powers of the mind and was "evidently intended by nature to be the immediate guide and director of our conduct, after we ar-rive at the years of understanding" (Reid 1967, vol. 2, p. 597). The last clause is necessary because the principle of conscience is notably stunted in the young. That most of us develop at least a modicum of it cannot be denied. But, how, under the circumstances, can Reid conclude that it is evident that nature so intended? Indeed, he later remarked that "the authority of con-science over the other active principles of the mind, I do not consider as a point that requires proof by argument, but as self-evident" (Reid 1967, vol. 2, p. 597). And Reid takes that to mean no more than that "in all cases a man ought to do his duty" (Reid 1967, vol. 2, p. 598).

46

The last point concerns the conflicts between the dictates of conscience and the other active powers. Certainly such can arise. When they do, conscience is supposed to tell us where our duty lies. For Reid, there could not be any conflict between the principle that we ought to act for our good as a whole and the principle that urges us to duty: "While the world is under a wise and benevolent administration, it is impossible that any man should, in the issue, be a loser by doing his duty. Every man, therefore, who believes in God [!], while he is careful to do his duty, may safely leave the care of his happiness to Him who made him. He is conscious that he consults the last most effectually by attending to the first" (Reid 1967, vol. 2, p. 598).

What of the nonbeliever, the atheist? An atheist who believes that, for himself, virtue and happiness, on the whole, are opposed is faced with a dilemma, well noted by Shaftesbury: "It will be impossible for the man to act so as not to contradict a leading principle of his nature. He must either sacrifice his happiness to virtue, or virtue to happiness; and is reduced to this miserable dilemma, whether it be best to be a fool or a knave" (Reid 1967, vol. 2, p. 598) – at least, so he must be convinced, because if the world is under a wise and benevolent administration, he will be virtuous in striving to be happy on the whole or happy on the whole while striving to be virtuous – unless being happy on the whole requires that one believe oneself to be so. In that case, the atheist who chooses virtue must count himself to be a fool, although in fact, given Reid's supposition, he is not.

Can They Suffer?

Having granted that animals possessed an inner life of some sort, it seemed to follow that they could suffer. The moral relevance of that fact was another question. For Bentham, it was *the* question:

> The day *may* come, when the rest of the animal creation may acquire those rights which never could have been withholden from them but by the hand of tyranny. The French have already discovered that the blackness of the skin is no reason why a human being should be abandoned without redress to the caprice of a tormentor. It may come one day to be recognized, that the number of legs, the villosity of the skin, or the termination of the *os sacrum,* are reasons equally sufficient for abandoning a sensitive being to the same fate. What else is it that should trace the insuperable line? Is it the faculty of reason, or, perhaps, the faculty of discourse? But a full-grown horse or dog is beyond comparison a more rational, as well as a more conversable animal, than an infant of a day, or a week, or even a month, old. But suppose the case were otherwise, what would it avail? [The] question is not, Can they *reason?* not, Can they *talk?* but, Can they *suffer?* [Bentham 1948, pp. 310f.]

Michael Bradie

CONCLUSION

The eighteenth century marked an important philosophical shift concerning the moral status of animals. The general scientific climate suggested that animals were more like human beings than people were generally prepared to admit. Despite the theological frameworks within which Butler and Hutcheson propounded their views, epistemic and cognitive considerations were replacing theological ones. The question shifted from "Do the brutes have souls?" to "Do the brutes have the requisite epistemic and cognitive capacities to have moral standing?" The focus on cognitive capacities and eventually on the capacity for suffering helped to undermine any categorical difference between brutes and human beings. The advent of Darwinism in the next century furthered that trend. But even without the Darwinian contribution, the legacy of the eighteenth-century discussion raises vexing issues for contemporary moral theorists.

Just how are human beings superior to the brutes? At the beginning of the eighteenth century the emphasis was on differences in the capacity to reason or to speak. Butler introduced the idea that human beings, in contrast to brutes, had consciences, which allowed them to reflect upon their actions. The brutes, who lived for the moment, were incapable of reflecting on what they did and, properly speaking, did not *act* at all. Hume's analysis, which cast the inner life of beasts in such a way as to make the differences between us and them a matter of degree, threatened to undermine the conviction that humans were superior. Although, as a man of his times, Hume was an essentialist, there are grounds for thinking that he would have welcomed Darwin's ideas with open arms. Nevertheless, although he made the differences between humans and brutes smaller than most, he never took the final step of including them fully within the moral sphere. Animals might have some moral standing and thereby be due consideration in that they ought not to be made to suffer unduly, but that they had rights or that their interests ought to figure in our moral calculations was a step that Hume did not make. That step was taken by Bentham, who took suffering, rather than reason, to be the criterion of moral standing.

The problem is that neither an appeal to cognitive considerations nor an appeal to the capacity to suffer serves to mark a sharp distinction between humans and the brutes. In emphasizing the physical and behavioral similarities between humans and brute animals, we are led to the conclusion that animals are analogous to human beings who have severe defects. If the brutes have lesser moral standing, then defective or incomplete human beings should as well. The eighteenth-century philosophers did not fully come to grips with this issue. For us, however, blurring the distinction between hu-

48

mans and other animals raises a number of thorny issues with respect to the treatment of animals and humans with diminished capacities (Rachels 1990; Rodd 1990; Pluhar 1995; Bradie 1997). Questions of vegetarianism and animal rights aside, consider, for example, the propriety of animal experimentation. If it is proper to experiment on animals, then why not on human fetuses or defective children and adults? Or consider the moral issues involving health care, diminished capacity, and quality-of-life considerations. If human beings with diminished capacities are no different from brutes, then how does this affect their rights to life and our duties to care for and nourish them? Our contemporary struggle with these and other difficult practical moral issues is part of the legacy of the eighteenth-century turn in moral philosophy.

NOTES

1. Note that Hutcheson's concerns were for the tamer animals that toiled in the service of humankind, but no word about the wild beasts or vermin or insects or the like. Do they have some rights too?
2. This inability to reflect has its positive side. Regardless of the caution or recklessness with which animals may live their lives, no plants or animals are immortal. None can "resist the violence of that general stream on which he is borne to his end. This itself is the order of things in which we must revere the arm of power that removes the fleeting generations of plants and animals, no less than the creative hand that provides a continual supply of new generations to perpetuate the race. In mere animals, incapable of reflection, this destination is not any cause of distress. In such as do, or may reflect on their lot, it is an admonition that the value of life is to be estimated from the good it contains, not from the length of its period" (Ferguson 1978, vol. 1, pp. 16f.).

REFERENCES

Bentham, Jeremy. 1948. *The Principles of Morals and Legislation.* New York: Hafner. (Originally published 1789.)

——— 1962. *The Works of Jeremy Bentham,* 11 vols. New York: Russell & Russell.

Bradie, Michael. 1994. *The Secret Chain: Evolution and Ethics.* Albany: State University of New York Press.

——— 1997. Darwin and the animals. *Biology & Philosophy* 12: 73–88.

Butler, Joseph. 1970. *Fifteen Sermons Preached at the Rolls Chapel and A Dissertation of the Nature of Virtue.* London: Society for Promoting Christian Knowledge.

Cottingham, John. 1986. *Descartes.* Oxford: Blackwell.

Crooks, G. R. (ed.). 1855. *Bishop Butler's Analogy of Religion, Natural and Revealed, to the Constitution and Course of Nature with an analysis, left unfinished by the*

late Rev. Robert Emory, D.D., completed and edited, with a life of Bishop But-
ler, notes and index. New York: Harper & Brothers.

Descartes, René. 1984. *The Philosophical Writings of Descartes,* vol. 1, trans. John
Cottingham, Robert Stoothoff, and Dugald Murdoch. Cambridge University
Press.

    1985. *The Philosophical Writings of Descartes,* vol. 2, trans. John Cottingham,
Robert Stoothoff, and Dugald Murdoch. Cambridge University Press.

    1991. *The Philosophical Writings of Descartes,* vol. 3, trans. John Cottingham,
Robert Stoothoff, and Dugald Murdoch. Cambridge University Press.

Fearing, Franklin. 1964. *Reflex Action: A Study in the History of Physiological Psy-*
chology. New York: Hafner.

Ferguson, Adam. 1978. *Principles of Moral and Political Science,* 2 vols., ed.
R. Wellek. New York: Garland. (Originally published 1792.)

Fowler, Thomas. 1883. *English Philosophers: Shaftesbury and Hutcheson,* ed. I. Muller. New York: Putnam.

Gray, P. H. 1968–9. The early animal behaviorists: prolegomena to ethology. *Isis* 59:
372–83.

Hobbes, Thomas. 1958. *Leviathan,* 2 vols. Indianapolis: Bobbs-Merrill.

Hume, David. 1978. *A Treatise of Human Nature: Analytical Index by L. A. Selby-*
Bigge, 2nd ed., with text revised and notes by P. H. Nidditch. Oxford: Clarendon Press. (Originally published 1739–40.)

Hutcheson, Francis. 1968. *A System of Moral Philosophy,* 2 vols. (with separate pag-
ination, bound as one). New York: Augustus M. Kelley. (Originally published
1755.)

Huxley, Thomas Henry. 1978. (1874). On the hypothesis that animals are automata,
and its history. In: *Significant Contributions to the History of Psychology 1750–*
1920, ed. D. Robinson, pp. 555–80. Washington, DC: University Publications
of America.

Leroy, Charles Georges. 1870. *The Intelligence and Perfectibility of Animals from a*
Philosophic Point of View with a Few Letters on Man. London: Chapman & Hall.
(Originally published 1751.)

Mautner, Thomas (ed.). 1993. *Francis Hutcheson: On Human Nature.* Cambridge University Press.

Miall, L. C. 1969. *The Early Naturalists: Their Lives and Work (1530–1789).* New
York: Hafner. (Originally published 1912.)

Midgley, Mary. 1984. *Animals and Why They Matter.* Athens: University of Georgia
Press.

    1995. *Beast and Man: The Roots of Human Nature,* rev. ed. London: Routledge.

Pluhar, Evelyn B. 1995. *Beyond Prejudice: The Moral Significance of Human and*
Nonhuman Animals. Durham, NC: Duke University Press.

Rachels, James. 1990. *Created from Animals: The Moral Implications of Darwinism.*
Oxford University Press.

Raphael, D. D. (ed.). 1991. *British Moralists 1650–1800,* 2 vols. Indianapolis:
Hackett.

Reid, Thomas. 1967. *Philosophical Works, with notes and supplementary dissertations by Sir William Hamilton,* 2 vols. Hildesheim: Georg Olms.

1969. *Essays on the Active Powers of the Human Mind.* Cambridge, MA: MIT Press.

1990. *Practical Ethics, Being Lectures and Papers on Natural Religion, Self-government, Natural Jurisprudence, and the Law of Nations,* ed. E. t. K. Haakonssen. Princeton, NJ: Princeton University Press.

Robinson, Daniel (ed.). 1978. *Significant Contributions to the History of Psychology: Physiological Psychology,* vol. 1. Washington, DC: University Publications of America.

Rodd, Rosemary. 1990. *Biology, Ethics and Animals.* Oxford: Clarendon Press.

Ryder, Richard D. 1989. *Animal Revolution: Changing Attitudes towards Speciesism.* Oxford: Blackwell.

Schneewind, J. B. (ed.). 1990. *Moral Philosophy from Montaigne to Kant,* 2 vols. Cambridge University Press.

Selby-Bigge, L. A. (ed.). 1897. *British Moralists,* 2 vols. Oxford: Clarendon Press.

Shaftesbury, Anthony, Earl of. 1963. *Characteristics of Men, Manners, Opinions, Times, etc,* 2 vols., ed. John M. Robertson.

Shaftesbury, Lord. 1806. *A Philosophical Inquiry into the Origin of our ideas of the sublime and beautiful with an introductory discourse concerning taste, and several additions.* Philadelphia: Samuel F. Bradford.

Singer, Bernard. 1982. History of the study of animal behavior. In: *The Oxford Companion to Animal Behavior,* ed. D. MacFarland. Oxford University Press.

Smith, Adam. 1982. *The Theory of Moral Sentiments.* Indianapolis: Liberty Classics.

Sorabji, Richard. 1993. *Animal Minds and Human Morals: The Origins of the Western Debate.* Ithaca, NY: Cornell University Press.

Stephen, Sir Leslie. 1949. *History of English Thought in the Eighteenth Century.* New York: Peter Smith.

Thomas, Keith. 1983. *Man and the Natural World: A History of the Modern Sensibility.* New York: Pantheon Books.

Turner, E. S. 1964. *All Heaven in a Rage.* London: Quality Book Club.

Walker, Stephen. 1983. *Animal Thought.* London: Routledge & Kegan Paul.

Warden, C. J. 1927a. The historical development of comparative psychology. *Psychological Review* 34: 57–85, 135–68.

1927b. *A Short Outline of Comparative Psychology.* New York: Norton.

Willey, Basil. 1964. *The English Moralists.* London: Chatto & Windus.

Wilm, E. C. 1925. *The Theories of Instinct: A Study in the History of Psychology.* New Haven, CT: Yale University Press.

Wilson, Margaret Dauler. 1995. Animal ideas. In: *Proceedings and Addresses of the American Philosophical Association* 69 (2): 7–25.

3

From Natural Law to Evolutionary Ethics in Enlightenment French Natural History

Phillip R. Sloan

The relationship between ethics and biology within the French Enlightenment tradition can only loosely be associated with the Darwinian meaning of "evolutionary ethics." There was no notion of "evolution" in our modern sense until the writings of Jean-Baptiste Lamarck (1744–1826) in 1800, nor was there a unified understanding of how moral reasoning relates to the biological constitution of human beings. As a feature of the general Enlightenment philosophical project, one must consider individual ethical vignettes united in a family resemblance of relations. The unity of these diverse developments can best be characterized as efforts to supplant the authority of revealed religion and tradition by naturalistic foundations for morality, society, and economic order. Reflections on the "biological" foundation for ethical action provide a component of this more general enterprise.

Darwin's connections of ethics and evolutionary biology, developed primarily in chapter five of the *Descent of Man* (1871), associate him with the eighteenth-century Scottish moral-sense tradition of David Hume, Adam Smith, Dugald Stewart, and especially James Mackintosh (Manier 1977, pp. 96–101).[1] Nonetheless, indirect connections from Darwin to French traditions can be specified. His membership in the student Plinian Society during his medical studies in Edinburgh (1825–7) led him into contact with Robert Edmond Grant, the foremost advocate of Lamarckianism in the British Isles. That may have acquainted him with the contemporary French discussions of transformist ethics.[2] With him on HMS *Beagle* were four important French works that he seems to have consulted: Lamarck's *Histoire naturelle des animaux sans vertèbres* (1817–22); an important compendium of Lamarck-inspired natural history and natural philosophy, Bory de Saint-Vincent's *Dictionnaire classique d'histoire naturelle* (1822–31), which contained important articles on the developmental natural history of human beings (Bory de Saint-Vincent 1825; Corsi 1988, pp. 222–9); the English translation of Bougainville's account of his voyage to Tahiti; and Bernardin de Saint-Pierre's *Voyage à l'Ile de France,* a work that will figure in our subsequent discussion.[3] Through those works Darwin could have become more familiar with French ethical discussions relating ethics to biology.

The goal of this essay is to illuminate that preceding French ethical tradition, allowing a comparative analysis of the French treatment of issues that later would be of interest to Darwin's project. The essay will begin with an examination of the background of French ethical inquiry in the prior natural-law tradition. I shall then narrow the focus of the discussion by examining the ethical views of three French authorities on natural history, Buffon, Bernardin de Saint-Pierre, and Lamarck, all associated at some point with the primary French institution for investigation in natural history: the Paris Muséum national d'histoire naturelle and its ancien régime predecessor, the Jardin du Roi. These three authors will provide contrasts among a hybrid natural-law dualistic position, an ethical system grounded on *sentiment* owing considerable debt to the view of Jean-Jacques Rousseau, and a view that combined elements of those previous traditions with biological transformism. In narrowing the inquiry to the confines of a group of individuals associated with one institution, it will be possible to discern the ways in which diverse French Enlightenment ethical discussions were refracted within a context of inquiry that was explicitly engaged in pursuing empirical investigations in zoology, physiology, medicine, botany, and physical ethnography.

NATURAL-LAW TRADITION AND FRENCH MORAL DISCOURSE

Variants of non-Utilitarian ethical theories in the eighteenth century can be loosely classified into four categories: (1) derivatives of traditional natural-law theories; (2) theories relying on some kind of special ethical sense or property, represented in their most highly developed form by the Scottish moral-sense tradition of Hume, Adam Smith, William Ferguson, and James Mackintosh and by the ethics based on the *sentiments du coeur* of Rousseau in a French context; (3) theories of rational duty that were best developed by Kant; and (4) variants of materialistic theories that sought to derive ethics from more elementary properties of *vital* matter, best represented by Denis Diderot in his *Rêve d'Alembert* and by the writings of members of the French *Idéologie* school, illustrated by Pierre-Jean-Georges Cabanis (1757–1808). Historical studies of French ethical theory during the Enlightenment suggest that categories (1), (2), and (4) will be most relevant to this inquiry.

The Enlightenment ethical project in its many variants did not necessarily challenge traditional ethical standards, at least not in its expressions by the most widely read authors, but rather sought to alter the *grounds of obligation* for sustaining those traditional norms (MacIntyre 1981, pp. 50ff.). To accomplish that, ethical thinkers sought to discover grounds for ethical obligation that did not rely on revealed religion, authority, Scholastic philosophy, or mere custom. It was the search for those grounds of obligation that defined both the urgency and the richness of the ethical inquiry of the Enlightenment, whatever might have been the variety of its expressions. In the traditions we shall explore, that project can be characterized as developing those grounds of obligation on the assumption of a universal human nature that provided objective foundations for ethical norms.

An ethics founded on the assumption of a universal "human nature," incorporating a strong biological component, was not a novelty of the Enlightenment. It had been a primary premise of the preceding natural-law tradition, a tradition that historians of French ethical thought have described as the primary backdrop for early modern French moral theory (Crocker 1963, ch. 1). But that was also a tradition that underwent some unusual transformations within the context of French natural history, and through those transformations it was able to provide an important foundation for transformist ethics as developed by Lamarck.

The natural-law tradition, particularly in its Aristotelian-based formulations, rested on the assumption of an intimate connection between a dynamic biological conception of human nature and an objective normative ground for ethical reasoning. The relationships between ethics and biology integral to that tradition were developed initially for the Classical tradition by Aristotle in his biological and ethical writings, and they became, through Stoic and Christian transformations, the basis of the received Scholastic natural-law tradition, most commonly identified in the western tradition with the teachings of Thomas Aquinas, but by no means exhausted by that high Scholastic expression.

A definitive principle of that ethical tradition was the notion that 'nature' (*phusis, natura*) supplies, through the realization of its inherent teleological ends, a normative pattern of moral action that is in harmony with the essential realization of a human level of existence. Furthermore – this point is crucial – that 'nature' was primarily the individuated principle of action, closely related to *eidos* and individual soul-as-form, rather than referring to some superintending, providential, and demiurgic Nature as developed particularly by the Stoic tradition.

A second important principle in that ethical tradition was the notion of a functional hierarchy of properties displayed by living things: Humans, governed by the functions of reflection, display a more complex and more nearly perfect set of functions than those manifest in animals, but these still incorporate and build upon the vegetative and animal soul-as-form that governs plants and animals.

A third key precept of that tradition was the assumption that the full realization of our natural ethical end required the intervention of reason, language, education, and socialization. Humans in some alleged "state of nature" would not spontaneously develop into fully ethical beings without the presence of those necessary conditions.

These general features serve to differentiate ethical systems within the natural-law tradition from their successors and eventual competitors, such as the moral-sense theories and *sentiment* theories. Human actions could deliberately violate the dictates of natural law. Ethics also was not grounded purely on instinctual properties within human nature that manifest themselves without the intervention of reason. Only under certain conditions could ethical behavior be fully realized.

The writings of the Spanish Jesuit Francisco Suarez (1548–1617) provide an important and influential late Scholastic expression of those principles, as codified in his synthesis *Tractatus de legibus, ac Deo legislatore* (1612). Following Aquinas, Suarez distinguished among four categories: (1) divine positive law; (2) human positive law, expressing the will of a superior; (3) divine eternal law; and its reflection in (4) natural law. However, in Suarez's formulation there was also an important revision of some central features of the Aquinian tradition that seem significant for subsequent French discussions. Whereas Aquinas had clearly denied that natural law could be likened to a habit or instinct, and must involve human reason (Aquinas 1952, par. 2, Q94, p. 425),[4] Suarez was more willing to grant this an instinctual aspect, manifest by principles known intuitively within the 'heart':

> . . . I have no doubt but that, it is in the actual judgment of the mind that natural law, in the strictest sense, exists. I must add, however, that the natural light of the intellect – which is inherently to prescribe what must be done – may be called the natural law, since men retain that law in their hearts, although they may be engaged in no [specific] act of reflection or judgment. It must be taken into consideration, then, that natural law, as we are now using the term, is looked upon as existing not in the Lawgiver, but in men, in whose hearts that Lawgiver Himself has written it. . . . [In] the case of natural law, which exists in the lawgiver as none other than the eternal law, there is, in the subjects, not only an active judgment, or command, but also the [mental] illumination itself in which that law is (as it were) permanently written, and which the law is always capable of incorporating in action. [Suarez 1944, pp. 186–7]

On this rendering, ethical action in accord with natural law does not require deliberate reflection or education to be realized, even though Suarez does not deny that reason plays a role in understanding the precepts of natural law. Conscience, for example, is the innate expression of this engraved natural law and acts even without reflection, and it

> bears witness to and reveals the work of the [natural] law written in the hearts of men, since it testifies that a man does ill or well, when he resists or obeys the natural dictates of right reason, revealing also, in consequence, the fact that such dictates have the force of law over man, even though they may not be externally clothed in the form of written law. Therefore, these dictates constitute natural law; and, accordingly, the man who is guided by them is said to be a law unto himself, since he bears law written within himself through the medium of the dictates of natural reason. [Suarez 1944, p. 184]

Enlightenment French moral theorists received the natural-law tradition both from Catholic Scholastic sources such as Suarez and more immediately from Protestant reinterpreters, such as Hugo Grotius (1583–1645) and particularly Samuel Pufendorf (1632–94).[5] Furthermore, in the Protestant formulations of natural-law theory, the Scholastic natural-law tradition was altered more in the direction of a theory of natural rights, supporting the rights of conscience against authority (Haakonssen 1996, ch. 1). Those formulations had particular appeal to the *philosophes.*

It is important for understanding the French discussions of natural law to be fully aware that natural-law ethics could be, and were, embraced without appeal to revealed religious traditions or Scholastic philosophical systems. The normative character of that ethical system did *not,* it should be emphasized, rest on authority, but on some normative conception of nature, and that appeal to nature often was explicitly accepted by many of the *philosophes* in their efforts to return to foundations for ethics and society that were either prior to the Christian tradition (typically Roman antiquity) or to foundations in natural religion discoverable by natural reason that did not entail acceptance of revealed religious authority. Roman moralists such as Cicero could therefore serve as sources for natural-law theory alternative to those from the Judeo-Christian and Scholastic traditions.[6]

French Enlightenment discussions of natural law also reveal some important revisions of the tradition. Those were taking place in the wake of the strong dualism of soul and body, mind and matter, that followed upon the Cartesian revolution in philosophy, and that dualism was reinforced in the eighteenth century by Jansenist traditions within French thought.[7] That dualism severed the intimate connection of matter and biology from the rational soul-as-form-of-the-body integral to Aristotelian interpretations of that moral theory.

The intervention of Cartesian mechanism into natural-law discussions had also deeply affected the concept of 'nature', either in the sense of an individual principle related to the soul-as-form or in the sense of a larger teleological order established by creation. Within the Aristotelian tradition, *phusis* could function as an ethical norm precisely because it was an integral, developing, and dynamic principle intimately related to *psuche*. But once 'nature' was reduced by the mechanical philosophy to simply "God himself, or the ordered system of created things established by God" (Descartes 1984, vol. 2, p. 56), the grounds in nature for moral obligation became much less clear. The resultant tendency in eighteenth-century discussions was to make "natural law" equivalent to commanded "laws of nature" imposed by God on creation. That tended to collapse traditional understanding of natural law into dictates of divine positive law.

We find the evidence for those alterations in the formulations of natural-law ethics that were immediately important for some of the French Enlightenment discussions. In the influential *Principes du droit naturel* (1747), by the Swiss Protestant professor of natural and civil law at the University of Geneva, Jean-Jacques Burlamaqui (1694–1748), a work strongly influenced by Pufendorf's discussions, we find an explicit acknowledgment of a natural and biological foundation for ethical principles. The treatise begins with a discussion of human nature, its interaction with social structure, and the foundation it supplies for ethical norms. But Burlamaqui also makes a sharp distinction between the material and the spiritual and gives only to the latter feature of human beings a role in moral action:

> Man is an animal, endowed with understanding and reason; a being, composed of an organized body and a rational soul.
>
> With regard to his body, he is pretty similar to other animals, having the same organs, properties, and wants. This is a living body, organized and composed of several parts. . . .
>
> But man, besides the marvellous disposition of his body, has likewise a rational soul, which eminently discriminates him from brutes. It is by this noble part of himself that he thinks, and is capable of forming just ideas of the different objects, that occur to him; of comparing them together; of inferring from known principles unknown truths; of passing a solid judgment on the mutual fitness or agreement of things, as well as on the relations they bear to us; of deliberating on what is proper or improper to be done; and of determining consequently to act one way or other. . . .
>
> Such is the general idea, [*sic*] we are to form of the nature of man. What results from it is, that there are several sorts of human actions; some are purely spiritual, as to think, to reflect, to doubt, &c. others [*sic*] are merely corporeal, as to breathe, to grow, &c. and some there are that may be called mixt, in which the soul and body have both a share . . . such as to speak, to work, &c. [Burlamaqui 1823, vol. 1, pp. 2–3][8]

57

It is through his reason and reflection, argues Burlamaqui, that "[the human being] is able to see and know his situation, as also to discover his ultimate end, and in consequence thereof to take the right measures to attain it." It is also definitive of his "natural" (as opposed to merely "primitive") state that one has freedom "conformable to his nature, constitution, and reason, as well as the good use of his faculties, considered in their full maturity and perfection" (Burlamaqui 1823, vol. 1, pp. 30–1). Burlamaqui's discussions also display clearly the conflation of "natural law" with imposed "laws of nature" commanded by God, defining natural law as "the system or assemblage of [the rules which nature alone prescribes to man], considered as so many laws, imposed by God on man" (Burlamaqui 1823, vol. 1, p. 1).

These changes have important implications for the subsequent discussion we shall follow. Accompanying the strong dualism we encounter in many French ethical discussions of natural law is the sharp distinction between instinct, which governs animals and some human actions purely automatically and mechanistically in response to natural laws, and reason, which operates exclusively in the realm of human action and alone has genuine freedom. 'Nature' also loses much of its individuated meaning as a teleological principle of action intimately associated with life.

The implications of these alterations for an understanding of "natural" ethical action is strikingly illustrated by Burlamaqui. Drawing explicitly on the early moral-sense theory of Francis Hutcheson, Burlamaqui attributes many moral actions, as well as the ability to make an instantaneous attribution of good or evil to other actions, directly to the action of a "moral instinct," "a faculty of the mind, which instantly discerns, in certain cases, moral good and evil, by a kind of sensation and taste independent of reason and reflection" (Burlamaqui 1823, vol. 1, p. 101).[9] This moral instinct acts for the good and is imposed upon us by "the author of our being." It is also present even among savage peoples, even though it may be obscured by customs and prejudices (Burlamaqui 1823, vol. 1, pp. 103–4). Reason, implanted by God, is to "enable us the better to discern and comprehend the true rules of conduct" (Burlamaqui 1823, vol. 1, p. 105). It strengthens, expands, and develops the promptings of natural moral instinct and deduces subsequent rational moral principles from natural law. But the basis of moral action itself is generally common to both animals and humans and resides in the nonrational component of our being.

French eighteenth-century moralists who may have accepted the claims of the natural-law tradition, but who did not follow Burlamaqui and Pufendorf's grounding of that tradition in revealed religion or traditional creationist metaphysics, could nonetheless still embrace the main features of natural-law ethics through the assumption of its grounding in a common human nature. The normative character of that natural foundation derived from the purposive ordering of nature by a deistic creator. In that form it was able to supply a framework of realistic moral reference to which even free-thinking *philosophes* could appeal. Writing in the *Encyclopédie*, Diderot thus defined *loi naturelle* as "the immutable and eternal order which must serve as the rule of our actions. It is grounded on the essential difference which is found between good and evil." Furthermore, in spite of varieties of customs or disputes over competing moral precepts, Diderot still affirmed that

> the eternal distinction of good and evil, the inviolable rule of justice without difficulty gains the approval of all men who reflect and who reason. Because there is no man who voluntarily comes to transgress this rule in important occasions, who does not feel that he acts against his own principles, and against the light of his reason, and who would not thereby make secret reproaches against himself. To the contrary, there is no man who, after having acted in conformity to this rule, would not be pleased with himself, and who would not commend himself for having had the strength to resist this temptation, and having done only that which his conscience had dictated to him to be good and just. ([Diderot] 1765, p. 665, unsigned article)

With the claim that "natural law is written on our hearts in characters so fair, with such strong expression, and with characteristics so lucid, that it is impossible not to recognize it" (Diderot 1765, p. 666), Diderot in many respects was only a short remove from the views of Suarez.[10]

ETHICS AS NATURAL HISTORY

The incorporation of human beings into the framework of a "natural history" of man marked one of the most important developments in the human sciences during the Enlightenment (Wokler 1993; Sloan 1995, pp. 112–51). That natural historical approach went beyond the Aristotelian tradition in collapsing the human into the purely natural and animal. We can plausibly date that event to 1735, when Linnaeus included human beings within the taxonomic order of the other animals (Broberg 1983; Sloan 1995). Because that tradition experienced such dramatic development through the lengthy examination of the natural history of the human species by Georges-Louis Leclerc, comte de Buffon (1707–88), in the second and third volumes of his *Histoire naturelle, générale et particulière, avec la description du cabinet du roi,* published by the royal press in 1749, his reflections on the relationships of humans and animals and the relationship of ethics to the new "natural history" provide the logical starting point for detailed inquiry.[11] Buffon's treatments of ethical questions, like most of his views on philosophical topics, are to be found embedded in various articles in the *Histoire naturelle* (1749–67) and in its seven volumes of *Suppléments* (1774–89), rather than being collected together to provide the subject of a specific treatise.

Buffon's general positions on moral theory and the relationships of human and animal behaviors are best revealed by reflections contained in the 97-page "Discours sur la nature des animaux" that opened the fourth volume of the *Histoire naturelle* of 1753.[12] Coming only two years after publication of the first volumes of Diderot and d'Alembert's *Encyclopédie,* five years after Montesquieu's *Esprit des lois,* and two years before the publication of Rousseau's landmark *Discours sur l'inégalité,* Buffon's treatise was situated historically within a remarkable midcentury French discussion of social, ethical, and political questions. It was located, in Buffon's massive survey, at the point of transition between the general biological and anthropological discussions of volumes two and three of the *Histoire naturelle* (1749) and the survey of the quadruped animals that would occupy twelve volumes published between 1753 and 1767. In that discussion, Buffon set forth a set of principles that remained consistently expressed through the remainder of his highly influential work, even though important developments were to occur in his thought after 1753.

Running as a constant theme through the lengthy "Discours" of 1753 is a Cartesian metaphysical distinction between animals, composed of a machine-like body, functioning deterministically in relation to the organization of its parts, and human beings, who have conjoined to their bodies rational and reflective (but not necessarily immortal) souls. Much of the article was specifically intended to define the resemblances and differences between human existence and animal existence on the basis of that metaphysical dualism.

Following a specification of many physical and physiological resemblances between humans and animals, Buffon nonetheless unequivocally affirmed that the primary differences depended on the possession by human beings of reason and a *substance spirituelle:*

> . . . this spiritual substance has only been conferred on man, and it is only by this that he thinks and reflects. The animal, to the contrary, is a purely material being, which neither thinks nor reflects, and which, while acting and determining itself, we cannot doubt but that the principle determining movement in the animal is a purely mechanical effect, absolutely dependent on its organization. [Buffon 1954a, p. 323a][13]

That immediately served to undermine any inferences drawn from the apparently similar behaviors of animals and man. In claiming that, Buffon was adopting a position on the ongoing debate over "animal soul" that was still current in general French intellectual circles (Rosenfield 1941). Buffon was also aware of the recent fascination in French circles created by travelers' reports of "wild, hairy men . . . midway between apes and us" (Maupertuis 1756, vol. 2, pp. 350–1; Sloan 1995). Like Darwin a century later, he was also a keeper and admirer of dogs, fully aware of their apparent intelligent behavior and their ability *combiner des ideés* that seemed to suggest strong analogies between humans and animals in terms of consciousness and motivation. Nonetheless, Buffon rejected any genuine identity of primary human properties with those of animals:

> . . . in order for this analogy [between humans and dogs] to be well-grounded, would it not be necessary that there be something more, that at least nothing would contradict it, that it would be necessary that animals could do, and would do on some occasions, all that we can do? But the contrary is evidently demonstrated. They do not invent, they perfect nothing, and as a consequence, they reflect on nothing. They never do anything except the same things in the same way. We can thereby already reduce the force of this analogy, and we can even doubt its reality. We must inquire whether there is not another principle different from our own by which they are governed, and whether their senses are not sufficient to produce their actions, without according knowledge by reflection to them. [Buffon 1954a, p. 328a]

The specifically ethical issues are explored in the second part of that essay, entitled "Homo duplex." There Buffon articulated at least some of the traditional theses of the natural-law tradition, including the claim that ethical principles had a firm foundation in human nature. He also argued that the realization of that ethical potential was not automatic, but involved a gradual moral development that took place through education, socialization, intercourse with others, and the development of reflection:

> The animal principle is developed first, since it is purely material and consists in the duration of the vibrations (*ébranlemens*) and the renewal of the impressions formed in our internal material sense by the objects analogous or contrary to our appetites. . . . The spiritual principle is manifest later; it is developed and perfected by means of education. It is by the communication of thoughts from others that the child acquires it and becomes himself thinking and rational, and without this communication he would be only stupid or capricious, according to the degree of inactivity or activity in his material internal sense. [Buffon 1954a, p. 338a]

Buffon saw moral deficiency as resulting from inadequate conditions that prevent the full development of those moral features. Other moral conflicts within human beings were products of the tension of the "material" and "spiritual" principles that reside within, creating the conflict of reason and passion. Happiness consisted either in the sole dominance of one of those principles (e.g., the material principle in the newborn infant) or in the difficult harmonious balancing of the two principles in the ethically mature adult (Buffon 1954a, pp. 338–9).

Buffon's dualism was complex and showed some tendencies to return in places to a form–matter relationship, similar in some respects to that of Aristotle, rather than to the radical dualism of Descartes. He attributed to organisms many properties found in humans, including sensation, consciousness of their own existence, and *sentiment.* Only reason and reflection were denied to animals. He also assumed a graded hierarchy of functional activity leading from animals, such as polyps, that merely exhibit a kind of "sleep" in their existence to the higher quadrupeds that display many analogies with human behavior (Buffon 1954a, pp. 328–9, 318bff., 347b). Some forms of behavior (e.g., fear reactions) were properly viewed as identical in humans and animals (Buffon 1954a, p. 341). Buffon acknowledged that dogs, for example, displayed forms of friendship, ambition, and pride, along with fidelity, humility, and educability. But those latter developments of the *sentiments naturelles* were only superficially similar to the human manifestations and were not considered to be the source of their human expressions. The example of friendship was particularly revealing. Although there were analogues to friendship that might be seen in the behavior of infants or animals, true friendship was possible only at a human level and had to involve the intervention of reflection:

> Since friendship supposes this power of reflection, it is of all the attributes (*attachemens*) the most dignified of man and the only one which does not degrade him. Friendship only emanates from reason; sense impression has nothing to do with it. It is the soul (*âme*) of his friend that one loves, and in order to love a soul, it is necessary to have one, to have made use of it, to have known it, to have compared it and found the parity at which one can know that of another. Friendship thus supposes not only the principle of knowledge, but the active and reflective exercise of this principle. [Buffon 1954a, p. 342b]

Buffon's distinction between the reflective activities of humans and the basically instinctual activities of animals led him to deny the efforts of René Réaumur and Bernard Mandeville to draw strong analogies between the social organizations of humans and insects and to ascribe moral dimensions to animal associations. All human society, even at the level of the family, implied the intervention of rational reflection and in that way was fundamentally distinct from animal society (Buffon 1954a, p. 346b).

Buffon came closest to the natural-law tradition in his claim that the good "is within ourselves; it has been given to us" (Buffon 1954a, p. 330a), supplying a moral foundation that is perfected by reason.[14] But though many features associated his reflections with the preceding natural-law tradition, his natural-history approach to these questions raised important complexities that would reverberate through his writings after 1753.

THE CHALLENGE OF THE NEW ETHNOGRAPHY

The new wave of eighteenth-century encounters of European explorers and navigators with the flora and fauna of the South Seas, Hawaii, and the East Indies, made possible by solution of the problem of longitudinal navigation in the middle decades of the eighteenth century (Beaglehole 1969; Frost 1976; Marshall and Williams 1982, ch. 9; Duchet 1995), brought with it a significant challenge to the assumptions of a universal human nature that had underpinned the main lines of the Enlightenment ethical project in all its variants, including Buffon's own conclusions. For the French tradition in particular, the expedition of Louis de Bougainville to Tahiti (1766–9), described in his idyllic account of unspoiled Tahitians living a life of sexual freedom and natural gentility, provided a convenient model with which to challenge the ethical principles of the received western tradition, displaying at last the true "noble savage" in the state of nature that, disappointingly, had not been encountered among the American aborigines. Exploited for its ethical significance most effectively by Diderot in his posthumous *Supplément au voyage de Bougainville* that had circulated among the *philosophes* in manuscript in the late 1770s,[15] the ethical significance of those discoveries had been first explored in print by Bernardin de Saint-Pierre in his popular *Voyage à l'Ile de France* of 1773 (Saint-Pierre 1773).[16] The initial conclusion drawn from the Tahitian encounters – that "natural man" had finally been discovered and had been found to be naturally good, generous, gentle, and loving – could be exploited in ways that accentuated the instinctual ethical character of human nature.

The existence of exotic peoples on islands like Tahiti, thousands of miles from the nearest landmass, also posed a troubling problem concerning how such peoples had gotten to those locations in the first place. It was not too difficult to assume some kind of migration of a single human species by land to the New World, particularly after the discovery of the interconnections in the North Pacific. That theory was not available to explain the populating of the South Seas. Theories of origin had to develop new dimensions in response to those challenges, and in the French tradition the revival of polygenetic theories formed one solution, with the implication that human beings comprised more than one species (Blanckaert 1981, 1988). Buffon, as a firm monogeneticist, insisted that all human beings formed a single species with a single origin. To account for the complexities of human differences and distribution, he supplied an influential model for understanding those issues, although his most important writings on those topics predated the main impact of the new discoveries.[17]

Buffon's engagement with the ethical problem created by that expanding biogeographical and anthropological knowledge developed incrementally, forced upon him by his plan to survey the entire natural history of the four-footed animals of the globe. As his survey of the animals moved in the 1750s from a European focus to an African and New World discussion, he was forced to deal with the problem of the geographical relationships of forms by an expansion of his conception of organic species. Buffon never abandoned his view that the identity of a biological species, including the human species, was assured by the ability to reproduce fertile offspring. But as creatures subject to all the natural contingencies of geography and climate, human beings had "degenerated" into distinct races (Sloan 1973, 1995; Eddy 1984).

The implication was that aboriginal peoples could acquire semipermanent deficiencies within their biological nature, directly caused by geographical and climatic differences. Furthermore, those could not immediately be overcome by education and socialization, the two factors assumed on traditional natural-law theory to be the necessary conditions for the full realization of natural ethical goodness.[18] Buffon thus allowed for moral as well as physical degeneracy. The issue emerged dramatically in his discussion of New World aborigines in the 1761 article on the relationships of New and Old World animals (Buffon 1954b). Under the influences of the degenerative conditions operative in the New World, the human being had become less controlled by reason, more governed purely by the *substance matériel*, rendered "less sensitive and more fearful," without "vivacity, nor activity in his soul." Such peoples, leading a "dissipated life," reflected increasing losses of the *substance spirituel* to the point that they had almost become animals (Buffon 1954b, pp. 103–4).

As should be evident, Buffon never joined those who admired the "noble savage," nor did he idealize a morally superior primitive state of humankind as that thesis had been introduced by Rousseau in 1755.[19] Rather, his emphasis on reason and socialization as necessary for developing the human potential exalted civilized society as the perfection of nature, wherein reason had achieved its greatest domination over the natural sentiments.[20]

In Buffon's reflections we can see one way in which a metaphysical dualism of thought and matter could be combined with some of the traditional assumptions of natural-law theory, to which was added an effort to deal with biological information on human diversity and the concept of semipermanent historical change within the human species.

In the writings of Bernardin de Saint-Pierre we can observe an alternative solution to the same issues, one deeply influenced by the new anthropological discoveries of the latter part of the century and a field-observer natural-history approach to human nature.

FRENCH NATURAL HISTORY AND THE ETHICS OF *SENSIBILITÉ*

Succeeding Buffon within the tradition of the Paris Jardin as an ethical theorist of some importance was Jacques-Henri Bernardin de Saint-Pierre (1737–1814).[21] Although today he is known primarily as a novelist, poet, and man of letters, rather than as a natural scientist and ethicist, his popular voyager's account of the Isle of Mauritius (*Voyage à l'Ile de France,* 1773) and his writings on nature (*L'Arcadie,* 1781; *Études de la nature,* 1784–8) established him as a major author on natural-history topics in the last decades of the century, with sufficient reputation in that domain that he was appointed by the weakened post-Varennes Louis XVI in 1792 as Buffon's successor in the important post of *intendant* of the Jardin in its last year of existence in that institutional form.[22] Saint-Pierre also achieved sufficient reputation as an ethical theorist during the Revolutionary period that he was named, for a brief period, *professeur de morale* at the Collège de France in 1794, following the abolition of his intendency in the Convention reforms of the Muséum of 1793 (Morel 1867, p. 90). His ethical writings also led to controversies with the ethical theories of the *idéologues,* whose writings will be of brief concern subsequently in this essay (Saint-Pierre 1798, vol. 7, pp. 423–41).[23]

Drawing on the natural-history and anthropological observations he recorded during a four-year journey to Mauritius and the Cape of Good Hope (1767–71), Saint-Pierre extended his naturalistic approach to ethical questions in portions of the *Voyage à l'Ile de France,* one of the first works to take account of Bougainville's Tahitian discoveries. They were also explored in sections of the *Études de la nature* (1784–8) and in his immensely popular moral-education novel *Paul et Virginie* (1788) that originally formed the fourth volume of the *Études.*[24] In general, Saint-Pierre's ethical reflections were a popularized version of Rousseau's moral theory without the complications introduced by Rousseau's radical "conjectural history" of humanity that postulated the development of civilized human beings from prehuman beginnings. His ethical project, as he tells us in *Paul et Virginie,* was to expound an account of ethical development displaying that "our happiness consists in living according to nature and virtue" (Saint-Pierre 1900, p. 13).

The notion that virtue was founded on 'nature' was, of course, not new and would not have appeared so to his French readers. The novelty of Saint-Pierre's account lay in his claim that the natural foundation for ethics consisted in an instinctual *sentiment du coeur* that he placed in opposition to the conclusions of reason and, by extension, to the claims of revealed religion, systematic philosophy, and tradition.[25] Hence, rather than requiring education, socialization, and habituation to bring that ethical nature to completion, the human being, left to 'nature', would develop noble ethical sentiments spontaneously. That was the point of the Paul and Virginia story: two transplanted European infants, growing up with a minimum of education and formal religious instruction in a small French settlement on the island of Mauritius, the scene of the novel, developed into naturally good, devout human beings imbued with a religion of nature, tenderness, and love:

> Their life seems attached to that of the trees, like that of fauns and dryads. They knew no other historical epochs than those of the lives of their mothers, no other chronology that than of their orchards, [and] no other philosophy than to do good to all the world and to be resigned to the will of God.
>
> After all, what need would these young people have to be rich and learned in our fashion? Their needs and their ignorance added further to their happiness. . . . Thus grew these two children of nature. No anxiety troubled their countenance; no intemperance corrupted their blood; no unfortunate passion depraved their heart. Love, innocence, piety, developed each day the beauty of their souls with unspeakable gracefulness in their traits, their attitudes and their movements. They had all the freshness of the morning of life, such as our first parents appeared in the garden of Eden when leaving the hands of God. [Saint-Pierre 1900, p. 49]

As Saint-Pierre developed these views more systematically in his pre-Revolutionary writings, he expounded in more detail on the opposition between the natural sentiment for the good that develops naturally in all humans and the distortions of that sentiment produced by reason and human artifice. Rational science and its offspring were therefore generally seen as having distorted and defaced the true aspect of nature open to the sympathetic observer. When the abstractions of science had been applied to human beings, they had distorted and defaced the natural human goodness:

> The History of Man has been disfigured in a very different manner. If we except the interest which religion, or humanity, has prompted some good men to take, in favour of their fellow-creatures, the rest of Historians have written under the impulse of a thousand different passions. The Politician represents Man, as divided into nobility and commonalty, into papists and huguenots, into soldiers and slaves; the Moralist, into the avaricious, the hypocritical [*sic*], the debauched, the proud; the Tragic Poet, into tyrants and their victims; the

Comic, into drolls and buffoons; the Physician, into the pituitous, the bilious, the phlegmatic. They are universally exhibted [*sic*] as subjects of aversion, of hatred, or of contempt: Man has been universally dissected, and now nothing is shewn of him by the carcase. Thus the master-piece of Creation, like every thing else in Nature, has been degraded by our learning. [Saint-Pierre 1797, vol. 1, pp. 21–2]

The artificiality and moral corruption of current society were to be overcome by a rediscovery of the ethics of *sentiment,* revealed in its purity through the study of natural history, meaning not the inquiries of the museum collector, but those of the student of plants, animals, and human beings in their natural habitat. That alone could reveal nature's beneficence and inner harmonies, and the benevolent and providential order of things. That alone could disclose the key to the Rousseauist *cogito: Je sent, je suis:*

> We can know that only which Nature makes us feel; and we can form no judgment of her Works but in the place, and at the time, she is pleased to display them. All that we imagine, beyond this, presents only contradiction, doubt, error, or absurdity. [Saint-Pierre 1797, vol. 1, p. 57]

The twelfth contribution to the *Studies of Nature,* devoted to ethical topics, bears the revealing title "Of Some Moral Laws of Nature: Weakness of Reason, of Feeling; Proofs of the Divinity and of the Immortality of the Soul from Feeling." Here we see that ethical behavior, for Saint-Pierre, was basically the product of a natural instinct that humans shared with the animals. Whereas Buffon had drawn an absolute metaphysical distinction between the two domains, with reason governing truly human action, and instinct and *substance matériel* defining that of animals, Saint-Pierre saw strong analogies and continuities in human behavior and animal behavior, both deriving from the primacy of *sentiment:*

> What, then, is that versatile faculty, called *reason,* which I employ in observing Nature? It is, say the Schools, a perception of correspondencies, which essentially distinguishes Man from the beast; Man enjoys reason, and the beast is governed merely by instinct. But if this instinct always points out to the animal what is best adapted to it, it is, therefore, likewise a reason, and a reason more precious than ours, in as much as it is invariable, and is acquired without the aid of long and painful experience. [Saint-Pierre 1797, vol. 2, p. 333]

Those "philosophers of the last age" who denied reason to animals therefore were making "a direct attack on the Supreme Intelligence itself, which is invariable in its plans, as animals are in their instinct" (Saint-Pierre 1797, vol. 2, pp. 333–4). Though not claiming that human reason and animal reason were identical, much of the argument that Saint-Pierre developed was an attack on reason and on the artificial situations it had created in human beings. Reason could not be the best guide to moral action. Rather, one could best trust the innate sentiments of the heart that humans possessed in common with animals:

> Good cause, then, we have to mistrust reason, as, from the very first step, it misleads us in our researches after truth and happiness. Let us enquire, whether there is not in man some faculty more noble, more invariable, and of greater extent. . . .
>
> I substitute, therefore, in place of the argument of *Descartes,* that which follows, as it appears to me both more simple and more general: *I feel, therefore I exist.* It extends to all our physical sensations, which admonish us much more frequently of our existence than thought does. It has for its moving principle an unknown faculty of the soul, which I call *sentiment,* or mental feeling, to which thought itself must refer; for the evidence to which we attempt to subject all the operations of our reason, is itself simply sentiment.
>
> I shall first, make it appear, that this mysterious faculty differs essentially from physical sensations, and from the relations presented to us by reason, and that it blends itself in a manner constant and invariable in every thing that we do; so that it is, if I may be allowed the expression, human instinct. [Saint-Pierre 1797, vol. 2, pp. 337–8]

To the workings of that innate *sentiment* Saint-Pierre attributed virtually all goodness, love, and social bonding. It was also through the promptings of that nonrational, instinctual sentiment that humans contacted "the ineffable sentiments of the Deity. It is to this last instinct, much more than to his reflective powers, that he is indebted for the conviction which he has of the existence of God" (Saint-Pierre 1797, vol. 2, p. 343).

Although Saint-Pierre downplayed the rational and elevated the instinctual in behavior to such a degree, he had not simply identified human and animal motivations. He still maintained that "man . . . is not formed of a simple nature, like other animals . . . but of two opposite natures, each of which is itself farther subdivided into several passions, which form a contrast" (Saint-Pierre 1797, vol. 2, p. 343). That duality, however, was a duality of natural sentiments rather than of metaphysical principles. Humans shared a natural sentiment in common with animals, and from that developed the natural affections and virtues. The other was the *sentiment divine,* discovered by the maturing of the human person, that displayed to humans their connection to the divine providence of nature. The moral challenge was to bring those two sentiments into harmony:

> When these two instincts unite, in the same place, they confer upon us the highest pleasure of which our nature is susceptible; for, in that case, our two natures, if I may thus express myself, enjoy at once. [Saint-Pierre 1797, vol. 2, p. 353]

Saint-Pierre's eclectic moral theory illustrates an interesting set of transitions taking place within ethical theory in pre-Revolutionary France. As a natural historian of an early romantic cast, he joined together his empirical observations of human and animal existence in the natural state (much of which he claimed to have conducted personally in his travels to Mauritius) with an ethics of innate feeling. His empirical approach relied not on abstract moral theory or the reflections of the preceding moral tradition, but on observations made "in nature." Although there was no questioning of the claim that the foundation and objectivity of moral reasoning lay in its grounding in a universal human nature, that foundation had become primarily a nonrational instinct. Reason had virtually nothing to do with true moral action. It mainly served to corrupt the natural morality within human beings.

Not to be found in Saint-Pierre's work is any treatment of three primary issues that were already under discussion within the French ethical tradition. The first was the problem of racialization and human diversity that had led Buffon, for example, into a theory of historical degeneration. There was no effort to deal with Rousseau's revolutionary idea that behavior based on reason could have arisen from the instinctual natural *sentiment* of animals by a *historical* process. Nor was there an effort to reduce the foundation of the ethics of sentiment to an even more fundamental biological level. It is in the work of Saint-Pierre's younger contemporary, the new professor of invertebrate zoology at the newly constituted Muséum national d'histoire naturelle, Jean-Baptiste Lamarck, that we encounter a novel exploration of those three issues.

FROM A NATURAL HISTORY OF ETHICS
TO ETHICAL TRANSFORMISM

The introduction of biological transformism took place in French circles against the backdrop we have been following. Rousseau's fertile suggestions in the *Discours sur l'inégalité* of 1755 provided the more general framework in moral philosophy. In that work he had offered a conjectural history of how a prerational, solitary *Homo sylvestris,* inspired by descriptions of the orangutan (Wokler 1988, 1993; Moran 1995), had developed into the alienated individual of contemporary civil society over an extended period of time. Against that general framework, Lamarck and other early transformists (e.g., Julien-Joseph Virey), at the end of the century, could propose a realistic theory of species transformism that was also extended into an account of the origins of human society and morality. The Scottish school would most systematically pursue the implications of that conjectural history for ethical and social issues in the latter decades of the century, particularly through the writings of Lord Kames and Monboddo (Wokler 1988, 1995, pp. 44–5; Wood, 1996; Haakonssen 1996, pp. 177–9). But French inquirers associated with the Paris Muséum and Institut national were situated to pursue such inquiries in a more explicitly biological direction, with more intimate connections to empirical zoology, physiology, anthropology, and psychology. The short-lived Société des observateurs de l'homme, founded in 1799 in close association with the Classe des sciences morales et politiques of the Institut national, included among its members Pierre-Jean-Georges Cabanis (1757–1808), Julien-Joseph Virey (1775–1846), author of the *Histoire naturelle du genre humain* (1801), and Jean-Baptiste Lamarck. That society drew together workers interested in the *histoire naturelle de l'homme,* where anthropological discussions and their relations to philosophical questions were pursued (Stocking 1982; Copans and Jamin 1994, pp. 7–51). As a chaired professor of invertebrate zoology at the Muséum national and a member of both the Société des observateurs and the Institut, Lamarck was ideally suited to relate issues being discussed in the Société and Institut with his researches into natural history at the Muséum.

Lamarck's main theses concerning the historical transformation of species were first expounded in his Muséum lectures of 1800, with their first extended appearance in print in his *Recherches sur les corps organisés* of 1802. However, extension of his transformist theory explicitly to human beings was first made public in his *Philosophie zoologique* of 1809, followed by remarks in the "Discours préliminaire" to the opening volume of the *Histoire naturelle des animaux sans vertèbres* (1815). To these were added important articles in the second edition of Virey's *Nouveau dictionnaire d'histoire naturelle* (1816–19) and the philosophical treatise *Système analytique des connaissances positives de l'homme* (1820).

As is well known, Lamarck's original transformist theory envisioned species as changing in response to changes in circumstances. But, more fundamentally, species changes resulted from an inner dynamism of matter itself. That explanation of an inner cause for change drew upon a metaphysical monism that manifested a heavy debt to aspects of French medical vitalism, as developed by physicians such as Paul-Joseph Barthez, Théophile de Bordeu, and other workers within the Montpellier medical tradition.[26] More immediately, Lamarck's writings show many debts to the reflections of the vitalist physician Cabanis, a fellow member of the Société des observateurs de l'homme and the author of the *Rapports du physique et du moral de l'homme* (1802), a collection of discourses given to the Institut national between 1796 and 1800.[27]

In the original model of his dynamic transformism, Lamarck envisioned present forms to have "ascended" from more primordial organisms in a continuous and linear *séries* that eventually terminated in human beings. In embracing a developmental and progressive view of life, Lamarck's biological reflections were in marked contrast to the "degenerational" model of change advocated by Buffon.

By arguing for an ascending, rather than degenerating, transformation of living forms as part of a realistic history of nature, Lamarck was also equipped with a model of life that enabled him to postulate the historical emergence of novel higher properties from more primitive states of other organisms. Extending the theories of the French medical vitalist tradition that had explained life activity and its related processes of generation and sensation on the basis of the constitutive fibers of an organism possessing an inherent property of *sensibilité*, Lamarck, similar to Cabanis, proceeded to draw from that a naturalistic foundation for ethics and consciousness. That involved the derivation of both the "moral" and the "physical" from a common physiological source.[28] Referring directly to Cabanis's *Rapports,* Lamarck argued that

> in their source, the *physical* and the *moral* are, without doubt, one and the same thing. And it is in studying the organization of the different orders of known animals that it is possible to make this truth most evident. [Lamarck 1994, pt. 1, pp. 66–7][29]

As a transformist and taxonomist, however, Lamarck envisioned the relationship between the physical and moral to be revealed by a developmental natural history that displayed the *continuity* of le morale and le physique as a historical relationship, beginning from the dynamic properties given to living matter by caloric and electrical fluids at the origin of life. That provided a series of stages in the emergence of higher functions and eventually the *facultés éminentes* that humans possess. Discernment of that historical process could best be revealed by study of the organization and function of lower organisms:

> If one had compared these different objects between themselves and with that which is known with reference to man; if one had considered it from the organization of the most simple animal up to that of man, who is the most complex and most perfect, [and] the *progression* which is displayed in the composition of organization, as well as the successive acquisition of different special organs, and following upon that as many new faculties as new organs are obtained; then one would have been able to perceive how the *needs,* at first reduced to nothing, and of which the number following is gradually acquired, have drawn out the appropriate inclinations to satisfy them; how the habitual and dynamic (*énergiques*) actions have occasioned the development of the organs which

71

enact them . . . and finally, how it has become the source of sensibility, and in the end the acts of intelligence. [Lamarck 1994, pt. 1, p. 68]

Lamarck's original progressive model of evolution was based on an ascending series of complexifying forms, moving from the infusoria, created by spontaneous generation from the entry of caloric and electrical fluids into nonliving matter, through the sponges, coelenterates, flatworms, and arthropods and on through the remainder of the fourteen groups that formed the great *séries* of organizational plans. The series terminated in the mammals, with the highest expression of that group being the human being. The primary axis of that hierarchy was to be seen in the development of nervous tissue, manifest as elementary *sensibilité* in the lowest forms of life and by complex brain and sense organs in the highest.[30]

In the third part of the *Philosophie zoologique,* Lamarck then extended his thesis of a gradual ascent of forms to cover human actions and eventually the development of ethical sensibility. In view of that history, it was the natural historian who had privileged access to the roots of ethical behavior:

If the *physical* and the *moral* have a common source; if the ideas, thought, and imagination itself are only phenomena of nature, and consequently only true facts of organization, it then belongs primarily to the zoologist, who is engaged in the study of organic phenomena to investigate what are the ideas, how they are produced, how they are preserved. [Lamarck 1994, pt. 3, p. 463]

Animal actions were not different in kind from those of human beings. They arose from the same biological source and differed from human actions in that only certain levels of animal organization were accompanied by an *inner* sentiment (*sentiment intérieure*) that emerged when sufficient complexity of the nervous system was achieved. It was that interior sentiment that immediately gave rise to *sensibilité morale,* forming in its highest development in man "the source of humanity, goodness, friendship, honor etc." (Lamarck 1994, pt. 3, p. 537). With many similarities to the arguments of Saint-Pierre, but within the framework of a developmental monistic metaphysic, the appearance of that faculty in man was spontaneous and natural, impelled by the inner dynamism of life itself.

This developmental, transformist framework provided Lamarck with the means to relate animal and human properties, including moral behavior. The emergence of higher degrees of function with increased organization was accompanied by increasing degrees of inner sentiment, and with that the greater control of actions of *sensibilité physique* by *sensibilité morale:*

This [moral] *sensibility,* considered in the developments that a perfected intelligence can enable it to acquire . . . appears to me a product and even a blessing (*bienfait*) of nature. It then forms one of the most beautiful qualities of man; because it is the source of humanity, of goodness, friendship, honor etc. Sometimes, however, certain circumstances render this quality as fatal for us as it can be advantageous in others, and to prevent these inconveniences that it can produce, it is only necessary to moderate the impulses (*elans*) by means of the principles to which a good education can alone direct us. [Lamarck 1994, pt. 3, pp. 536–7]

As Lamarck explored those issues in the latter sections of the *Philosophie zoologique* and in more specific detail in important articles in the second edition of Virey's *Nouveau dictionnaire d'histoire naturelle,* we see that he envisioned the relationships between human and animal properties and behaviors to be complex. All organic functions, including moral and rational ones, must eventually be derived causally from properties originating in material organization. The distinction between animal "instinct" and human "intelligence" could not, therefore, be based on a metaphysical difference in two kinds of substances. Animal actions ultimately were derived from the impelling action of an innate *sentiment intérieure.* That force, directly related to the elementary vital properties of life – sensibility and irritability – was manifest in various degrees as mere sensitivity, as instinct, as habit, and eventually as rational *judgment* and *intelligence* (Lamarck, "Instinct," in Roger and Laurent 1991, p. 162).[31] In that final stage, action was effectively governed by two different sources of action (Lamarck, "Homme," in Roger and Laurent 1991, p. 119). Some highly organized animals, along with humans, would act with foresight in response to rational *judgment* and *intelligence.* Others would respond to the input of sensations only on the basis of their *sentiment intérieure* and would be governed by instinct and habit, lacking the active power of *attention* that was manifest only in higher life (Lamarck, "Instinct," in Roger and Laurent 1991, pp. 130–3). The human being underwent a development from one to the other:

> At birth, man is not supplied with any acquired idea, and has not yet exercised any judgment. He then possesses only a single source of action, that constituting *instinct*. . . . But soon afterward, he acquires a second; because, among the diverse objects which then strike his senses, his attention, excited by the sensations he receives, begins to exercise itself. He fixes it, actually, on certain of these objects, compares them one to another, and finally *judges.* [Lamarck, "Jugement," in Roger and Laurent 1991, p. 191]

Although Lamarck did not expound at length on the issue, it is evident that he found similar states of inner awareness, and even similar acts of judgment and reason, in *animaux intelligens* that were directly "homologous" to those in humans (Lamarck, "Instinct," in Roger and Laurent 1991, p. 158).

Lamarck's derivation of all action from a primordial *sentiment intérieure* resulted in a sequence that led from that internal sentiment to more specific sentiments that governed specific human actions: those "that man forms internally, that he preserves or alters according to circumstances relative to his personal interest, and that have a great influence on his actions" (Lamarck, "Homme," in Roger and Laurent 1991, p. 121). For Lamarck, all human action ultimately must follow from the interplay of the environment and the four innate natural tendencies that derived from a more primordial natural inclination for self-preservation. Those tendencies motivated all human beings as essential properties that were developed in response to circumstances. They then could be "mastered, modified or directed" by judgment (Lamarck, "Homme," in Roger and Laurent 1991, p. 120):

1. a natural tendency to well-being
2. self-love
3. the tendency to dominate
4. aversion to one's destruction

Fundamental to Lamarck's evolutionary ethics was the assumption of a superintending 'nature' that acted for ordained ends. 'Nature', as Lamarck defined it, was an intermediate agency, "being neither an intelligence, nor even a creature, but an order of things constituting a universal power (*puissance partout*) subjected to laws," acting in accord with the will of Lamarck's deistic God (Lamarck, "Nature," in Roger and Laurent 1991, p. 305). Study of nature and natural history was the means by which one could determine those natural laws and also gain an understanding of how the "physical" and the "moral" could be seen to reside "entirely in the domain of nature" (Lamarck, "Nature," in Roger and Laurent 1991, p. 316).

The difficulty for Lamarck's ethics was to explain evil and the morally reprehensible dimensions of humanity, particularly because no "natural evil" was attributed to the actions of nonhuman organisms in the tradition of Scottish moralists like David Hume and, later, Darwin.[32] The similarities to aspects of Saint-Pierre's ethics of *sentiment* were again strong. Lamarck readily acknowledged the deep moral duality of human beings, who could be the most noble of creatures, but could also display "hardheartedness, malice, cruelty and barbarity that even the most fierce wild animals cannot equal" (Lamarck, "Homme," in Roger and Laurent 1991, p. 117). Those reprehensible moral properties were not attributed to nature or natural instincts, however. With Rousseau, he considered the primary explanation for evil to lie in the consequences of civilization and the results of the historical progression from a more primitive state of humanity. Socialization, arising originally out of the need for mutual protection, also established the means by which important disparities could become instituted as heritable traits. Those included hereditary differences in intelligence that resulted in relations of domination and finally in the moral squalor of modern industrial society, where masses of people were "tormented by diverse passions, seeing, without noticing, their health being altered, their blood being tainted in a thousand ways, forming, a quantity of diverse disorders in their organization and in the end the germ of a considerable and always intersecting number of endemic illnesses transmitting and perpetuating themselves by generation" (Lamarck, "Homme," in Roger and Laurent 1991, p. 124). Such moral conflict resulted ultimately from the undesirable interactions that civilization brought about between the *sentiment physique* and the *sentiment morale,* and the reforming role of education and proper reason was required to correct that. But to the degree that human nature itself had been altered by the historical process, resulting in a natural hereditary gradation of intelligence, such correctives would necessarily take a long time and would require reform of the social environment.

CONCLUSION

In the transformist ethical naturalism of Lamarck we can perceive remnants of the prior natural-law theory, with an overlay of several novel developments. Ethical principles certainly had obtained their objectivity from a grounding in biological nature. Humans displayed emergent properties that were related to, but not simply identical with, features displayed by animals. Lamarck did not follow Rousseau and Saint-Pierre in opposing a natural ethics of *sentiment* to reason and reflection. Education, reason, and a suitable social environment were needed to bring ethical sensibility to its natural fulfillment.

Furthermore, although it has been commonplace in the literature to view Lamarck as a "materialist," his materialism was of a very complex kind that in no sense reduced organic life and functions to the actions of inert matter and purely physical forces. His "matter" was possessed of self-realizing powers and contained within itself the sources of both the physical and rational aspects of human beings. Because of that dynamic conception of organic matter, he did not rely on dualistic theories of soul and body or mind and matter as alien substances in order to establish a natural ethical foundation. That was all closer to the matter–form substance theory of the Aristotelian and Scholastic traditions than might first appear. Furthermore, the ethical dimension of human existence drew its objective merit from the relationship of human action to a more comprehensive conception of 'nature' that itself was realizing some larger ordained ends determined by an external agency distinct from nature (Lamarck, "Nature," in Roger and Laurent 1991, pp. 303–5).[33]

75

Even though many of the features we have seen in Lamarck's ethical theory were not remarkably alien to the traditional assumptions of the natural-law heritage, particularly in its Aristotelian-Scholastic forms, the novelty introduced by Lamarck's historical transformism raised a new set of issues. The traditional hierarchy of functional manifestations of *psuche* assumed within the Aristotelian tradition had become, to borrow Charles Gillispie's apt phrase, an escalator of historical derivation. Any moral meaning in that historical progressionism of life depended, for Lamarck, intimately on the larger directedness of nature as a whole. This Lamarck accepted, at least in the sense that 'nature' was viewed as endowed with inherent laws and purposes by its creator, even though he denied that nature in itself acts intentionally or for an end of its own. To reject these deistic assumptions, and to render nature "the aggregate action and product of many natural laws" (Darwin 1934, p. 64), would seriously undermine, at least on the Lamarckian grounds we have outlined, the degree to which the "natural" could serve as the basis of the "moral" in a transformist framework.

NOTES

1. Darwin's reading of the primary moral-sense Scottish moralists can be dated at least to his reading of the *Ethical Philosophy* of his family relative James Mackintosh in May of 1839 (Darwin's reading notebooks, Darwin Papers, Cambridge University Library, hereafter cited as DAR 119–20), and it probably dates to much earlier. See the letter to W. D. Fox, 3 January 1830, in Burkhardt and Smith (1985, p. 97). His reading of French moralists, other than Bernardin de Saint-Pierre, seems to have been confined to a reading of Montaigne's *Essays* in August of 1838 and October of 1843 (DAR 119; Darwin, "Notebook M," in Barrett et al. 1987, p. 540).
2. At meetings of the Plinian Society as a student, he was present at discussions that explored the identity of animal and human characteristics, including discussions of the continuity of instinct and reason (Desmond and Moore 1991, pp. 32–9).
3. The catalogue of the *Beagle* library is given in the Darwin correspondence (Burkhardt and Smith 1985, app. iv). Darwin was familiar with Bernardin de Saint-Pierre's moral-education novel, *Paul et Virginie* (1788), during the *Beagle* period. See the letter of C. D. to W. D. Fox, 15 February 1836 (Burkhardt and Smith 1985, p. 493).
4. The opening article of that question explicitly rejects the notion that natural law was a "habit" or passion. For translation, see Aquinas (1967).
5. On the history of natural law in France, see Crocker (1963, ch. 1). Schneewind claimed, without documentation, that Pufendorf was "the most widely read of these [natural-law] theorists, and undoubtedly the most widely read of anyone who wrote on moral philosophy during the entire period" (Schneewind 1990, p. 21).
6. See "Loi naturelle (morale)" (Diderot 1765, p. 665). I am following Lester Crocker

(1963) in attributing this unsigned article to Diderot. Diderot quoted Cicero's *De Legibus,* book 2. See also Cicero's *Res Publica,* book 3.xxii, which was quoted for non-Christian authority by Suarez (1944, p. 185).

7. On the political importance of Jansenism in that period, see Van Kley (1996).

8. All subsequent quotations are from this sixth American edition (the first London edition was in 1748). On Burlamaqui's importance for discussions of eighteenth-century natural law, see Haakonssen (1995, pp. 336–41) and Crocker (1963, pp. 25–7).

9. The English translation mistakenly attributed this to "Mr. Hutchinson." The moral-sense theory of ethics, clearly familiar to Burlamaqui through Hutcheson's formulations, only slowly filtered into French discussions in the eighteenth century, and its full impact on French ethical thought is not clear. Historians of French ethics have suggested that it played little role in French ethical discussions, although the tradition was available directly to French readers through Diderot's 1745 translation of such important works as Shaftesbury's *Inquiry Concerning Virtue.* Crocker identified Jean-Baptiste Robinet's *De la Nature* (1766) as an important exposition of the theory, and Adam Smith's important treatise was translated by Sophie Condorcet in 1798: *Théorie des sentimens moraux ou essay* [*sic*] *analytique sur les principes des jugemens que portent naturellement les hommes* (1st ed. 1798; 2nd ed. 1830).

10. On Diderot and natural law, see also Crocker (1963, pp. 52ff.). The complex developments within Diderot's ethical theory lie beyond the range of this essay.

11. For important historiographic comments on the overemphasis on Buffon in the history of anthropology, see Blanckaert (1993).

12. The natural history of man had been treated previously in volumes two and three of the *Histoire naturelle* (1749). The 1753 article made Buffon's views on the differences between humans and animals more explicit and contained his clearest statement of a dualistic metaphysics.

13. All quotations and references to the *Histoire naturelle* will be to the Piveteau edition, except where lacking in that volume. Others will be cited according to the original Imprimerie royale edition.

14. Buffon's readings in the natural-law tradition can in part be documented by his interest in the writings of Samuel Pufendorf in 1737; see the fragmentary letter of 1737 (Lanessan 1884–5, vol. 13, pp. 29–30).

15. First published in 1796 in his *Opuscules philosophiques et littéraires* (Paris: Chevet).

16. Saint-Pierre met Bougainville and his crew when they stopped at Mauritius in 1768, and he was able to study personally the Tahitian man being transported by the expedition to Paris (Cook 1994).

17. The initial series of the *Histoire naturelle* was completed in 1767, prior to the return of Bougainville's expedition. It was only in the *Suppléments* to the anthropological articles of the *Histoire naturelle* that he was in a position to take partial account of the extensive new anthropological information gathered in the latter decades of the century (Duchet 1995, pp. 229ff.; Sloan 1995, pp. 138–9).

18. Buffon allowed that with transplantation to temperate climates and education, aboriginals could slowly be improved.

19. Buffon's circumspect comments on the original state of man, addressed briefly in the *Epoques de la nature* of 1778, described original human beings as "both morally and physically in the pure state of nature, with neither clothes, nor

religion, nor society" (Buffon 1954d, p. 183b). Such beings evidently had been created directly and endowed with reason, and had not emerged from other animals. Buffon's negative reply to Rousseau's thesis of a gradual emergence of human beings from a prehuman state was delivered in his "Les animaux sauvages," *Histoire naturelle,* vol. 6 (1756). For a recent discussion of the views of Buffon and Rousseau on these issues, see Cherni (1992).

20. In discussing the relationship of human beings to the higher primates, treated in the articles of 1766 and 1767, Buffon considered the similarities in behavior to have been the products of extreme degeneration of members of the human species, leading to a virtual extinction of reason (Buffon 1954c, p. 32).

21. Biographical details have been drawn principally from Morel (1867). For remarks on Saint-Pierre's place in the history of French ethical thought, see Crocker (1963, pp. 206–10).

22. Buffon had been immediately succeeded at his death in April of 1788 by Charles-César de Flauhaut, the marquis de la Billarderie, who held the post until 1792 (Barthélemy 1979).

23. Saint-Pierre's survival through the violence and ethical conflict of the Revolutionary period renders aspects of his pre- and post-Revolutionary ethical writings difficult to reconcile on major points. His post-Thermidor ethical treatise (Saint-Pierre 1798, vol. 7, pp. 423–41) emphasized the importance of a transcendant foundation for ethics and contrasted a *morale des passions* to a *morale de la raison.* As he developed the meaning of his *morale de la raison,* it was still based on a "sentiment des lois établies par la nature de l'homme à l'homme, que sont dérivées toutes les vertus fondamentales des sociétées" (Saint-Pierre 1798, vol. 7, p. 427). That treatise was interpreted by his posthumous editor as an attack on the ethical naturalism and anti-religion of the *Idéologues* of the Institut; see editor's remarks in Saint-Pierre (1798, vol. 7, p. 412). My concern is with his pre-Revolutionary sentiments.

24. The first separate edition of *Paul et Virginie* was published as an unauthorized version in Lausanne in 1788. The first authorized version was published in Paris in 1789.

25. He reported a conversation with Rousseau, approvingly, in which Rousseau was said to claim that "when man begins to reason, he ceases to feel" (Saint-Pierre 1797, vol. 1, p. 15). Saint-Pierre's more extensive account of his relations with Rousseau was contained in several essays (Saint-Pierre 1818, vol. 12).

26. For remarks on the revival of vitalism in that period, see Staum (1980, ch. 3) and Rey (1988, pp. 31–59). It should be evident that I include Lamarck in that tradition, while acknowledging that Lamarck went beyond the tradition in supplying a materialistic interpretation for the causes of the traditional vital properties of sensibility and irritability by reducing them to the actions of caloric and electrical fluids.

27. Martin Staum has argued that Lamarck's reflections may be equally important for understanding the development of Cabanis's ideas (Staum 1980, pp. 182–6).

28. On the background and importance of the distinction between *le physique* and *le morale* in French ethical thought, see Wokler (1995, pp. 34–7). Wokler noted the tendency in earlier conjectural history to make one the outcome of the other. Cabanis, and after him Lamarck, drew more radical conclusions by deriving both from properties of material fibers and vital *sensibilité*. The impact of Cabanis's *Rapports* of 1802 may explain the important differences in the treatment of

human beings between Lamarck's *Recherches sur les corps organisés* of 1802 and the *Philosophie zoologique* of 1809. Lamarck's indebtedness to Cabanis's *Rapports* is evident throughout the discussion of ethics and intelligence in the *Philosophie zoologique,* even when he is critical of Cabanis. Working out the full relationship between Lamarck to Cabanis would exceed the bounds of this essay.

29. All Lamarck quotations from the *Philosophie zoologique* are translated from the 1994 reprint edition.

30. I am ignoring the complexity introduced in the "Additions" to the final part of the *Philosophie zoologique,* most likely reflecting the impact of Cuvier's critiques of linear transformism. There Lamarck embraced an alternative model of a branching derivation of groups from more than one origin. The model was further expanded in the "Discours préliminaire" to the first volume of the *Histoire naturelle des animaux sans vertèbres* of 1815. I have failed to see evidence that Lamarck ever adequately integrated those revisions into his more general philosophical conclusions drawn from the linear model of the *Recherches* and the *Philosophie zoologique.* It seems to have had no impact on the Virey *Nouveau dictionnaire* articles, nor on the *Système analytique des connaissances positives de l'homme* of 1820.

31. All references to the Virey *Nouveau dictionnaire d'histoire naturelle,* 2nd ed. (1816–19), articles by Lamarck are taken from this reprint edition. Lamarck's articles appeared only in the second edition of Virey's compendium.

32. E.g., the parasitism of the *Ichneumonidae,* appealed to by Philo in Hume's *Dialogues Concerning Natural Religion,* section x.

33. That crucial article clarified many of the issues raised in the *Philosophie zoologique* and in Lamarck's thought more generally, for there Lamarck clearly rejected two "errors" he saw commonly accepted. The first was the identification of God and nature, and the second was the reduction of 'nature' to the physical world (Roger and Laurent 1991, p. 303). His conception of nature was, to the contrary, dynamic; it was the ultimate source of life and vitality; it was monistic; and it was the ultimate source of the *morale* and the *physique.* In most of those respects his conception of nature was alien to that of the mature Darwin of the later editions of the *Origin.* On changes in Darwin's conception of nature, see Richards (Chapter 6, this volume) and Sloan (1981).

REFERENCES

Aquinas, Thomas. 1952. *Summa Theologiae,* ed. P. Caramello. Rome: Marietti.
    1967. *Summa Theologiae,* pt. 2, questions 90–97. In: *Treatise on Law,* ed. S. Parry. Chicago: Regnery.
Barrett, P. H., Gautrey, P. J., Herbert, S., Kohn, D., and Smith, S. (eds.). 1987. *Charles Darwin's Notebooks, 1836–1844.* London: British Museum of Natural History, Cambridge University Press.
Barthélemy, Guy. 1979. *Les Jardiniers du Roy.* Clercy: Pélican.
Beaglehole, J. C. 1969. Eighteenth century science and the voyages of discovery. *New Zealand Journal of History* 3: 107–23.

Phillip R. Sloan

Phillip R. Sloan

Phillip R. Sloan

Phillip R. Sloan

Phillip R. Sloan

Phillip R. Sloan

Phillip R. Sloan

Phillip R. Sloan

Phillip R. Sloan

Phillip R. Sloan

Phillip R. Sloan

Beaune, J.-C., Benoit, S., Gayon, J., Roger, J., and Woronoff, D. (eds.). 1992. *Buffon 88.* Paris: Vrin.

Darwin, Charles. 1934. *Origin of Species,* 6th ed. New York: Random House. (Originally published 1872.)

——— 1981. *Descent of Man.* Princeton, NJ: Princeton University Press. (Originally published 1871.)

Descartes, René. 1984. *Meditations on First Philosophy,* trans. J. Cottingham, R. Stoothoff, and D. Murdoch. In: *The Philosophical Writings of Descartes,* 2 vols. Cambridge University Press. (Originally published 1641.)

Desmond, Adrian, and Moore, J. R. 1991. *Darwin.* London: Joseph.

[Diderot, Denis]. 1755. Droit naturelle. In: *Encyclopédie ou dictionnaire raisonnée,* vol. 5, ed. D. Diderot and J. d'Alembert, p. 131. Paris. Reprinted (1966): New York: Readex Microprint.

——— 1765. Loi naturelle. In: *Encyclopédie ou dictionnaire raisonnée,* vol. 9, ed. D. Diderot and J. d'Alembert, pp. 665–6. Neûchatel. Reprinted (1966): New York: Readex Microprint.

Duchet, Michèle. 1995. *Anthropologie et histoire au siècles des lumières,* new edition with postscript by Claude Blanckaert. Paris: Michel.

Eddy, John. 1984. Buffon, organic alterations, and man. *Studies in History of Biology* 7: 1–45.

Fox, C., Porter, R., and Wokler, R. (eds.). 1995. *Inventing Human Science: Eighteenth Century Domains.* Berkeley: University of California Press.

Frost, Alan. 1976. The Pacific Ocean: the eighteenth century's "New World." *Studies in Voltaire and the Eighteenth Century* 152: 779–822.

Haakonssen, Knud. 1996. *Natural Law and Moral Philosophy.* Cambridge University Press.

Jardine, Nicholas, Secord, J., and Spary, E. (eds.). 1996. *Cultures of Natural History.* Cambridge University Press.

Lamarck, Jean-Baptiste. 1994. *Philosophie zoologique,* reprint edition. Paris: Flammarion. (Originally published 1809.)

Lanessan, J. L. (ed.). 1884–5. *Buffon: Oeuvres complètes,* 14 vols. Paris: Le Vasseur.

MacIntyre, Alisdair. 1981. *After Virtue.* Notre Dame University Press.

Manier, Edward. 1977. *The Young Darwin and His Cultural Circle.* Dordrecht: Reidel.

Marshall, P. J., and Williams, G. 1982. *The Great Map of Mankind: Perceptions of New Worlds in the Age of Enlightenment.* Cambridge, MA: Harvard University Press.

Maupertuis, Pierre de. 1756. Lettre sur le progrès des sciences (1752). In: *Oeuvres de Mr. de Maupertuis,* new ed., 4 vols. Lyon: Bruyset.

Moran, Francis, III. 1995. Of pongos and men: orangs-outang in Rousseau's discourse on inequality. *Review of Politics* 57: 641–64.

Morel, J. 1867. Bernardin de Saint-Pierre. In: *Nouvelle biographie générale,* vol. 43, ed. M. Le Hoeter, pp. 83–91. Paris: Firmin Didot.

Piveteau, Jean (ed.). 1954. *Buffon: Oeuvres philosophiques.* Paris: Presses Universitaires de France.

Rey, Roselyne. 1988. Le vitalisme de Julien-Joseph Virey. In: *Julien-Joseph Virey, naturaliste et anthropologue,* ed. C. Benichou and C. Blanckaert, pp. 31–59. Paris: Vrin.

Roger, Jacques, and Laurent, G. (eds.). 1991. *Lamarck: Articles d'histoire naturelle.* Paris: Belin.

Rosenfield, Leonora Cohen. 1941. *From Beast Machine to Man Machine: Animal Soul in French Letters from Descartes to la Mettrie.* Oxford University Press.

Saint-Pierre, Jacques-Henri Bernardin de. 1773. *Voyage à l'Ile de France, à l'Ile de Bourbon, au Cap de Bonne-Espérance, etc.: Avec des observations nouvelles sur la nature & sur les hommes,* 2 vols. Amsterdam: Merlin.

    1797. *Studies of Nature,* 5 vols., trans. Henry Hunter. Boston: Nancrede. (Originally published 1784–8.)

    1798. *De la nature de la morale; fragment d'un rapport sur les mémoires qui ont concouru pour le prix de l'Institut national, dans sa séance publique du 15 messidor de l'an 6, sur cette question: Quelles sont les institutions les plus propres à fonder la morale d'un peuple?* Paris: Librairie du Cercle-Social. Reprinted in Saint-Pierre (1818).

    1818. *Oeuvres complètes de Jacques-Henri Bernardin de Saint-Pierre,* 13 vols., ed. L. A. Martin. Paris: Méquignon-Marvis.

    [1900]. *Oeuvres choisis de Bernardin de Saint-Pierre.* Paris: Firmin-Didot.

Schneewind, J. B. (ed.). 1990. *Moral Philosophy from Montaigne to Kant: An Anthology,* 2 vols. Cambridge University Press.

Sloan, Phillip R. 1973. The idea of racial degeneracy in Buffon's *Histoire naturelle. Studies in Eighteenth-Century Culture* 3: 293–321.

    1981. The question of natural purpose. In: *Evolution and Creation,* ed. E. McMullin, pp. 121–50. University of Notre Dame Press.

    1995. The gaze of natural history. In: *Inventing Human Science: Eighteenth Century Domains,* ed. C. Fox, R. Porter, and R. Wokler, pp. 112–51. Berkeley: University of California Press.

Staum, Martin. 1980. *Cabanis: Enlightenment and Medical Philosophy in the French Revolution.* Princeton, NJ: Princeton University Press.

Stocking, George. 1982. French anthropology in 1800. In: *Race, Culture and Evolution,* ed. G. Stocking, pp. 15–21. University of Chicago Press.

Suarez, Francisco. 1944. Tractatus de legibus, ac Deo legislatore. In: *Francisco Suarez: Selections from Three Works,* ed. and trans. J. B. Scott. Oxford: Clarendon Press. (Originally published 1612.)

Van Kley, Dale. 1996. *The Religious Origins of the French Revolution.* New Haven, CT: Yale University Press.

Wokler, Robert. 1978. Perfectible apes in decadent cultures: Rousseau's anthropology revisited. *Daedalus* 107 (Summer): 107–34.

    1988. Apes and races in the Scottish Enlightenment: Monboddo and Kames on the nature of man. In: *Philosophy and Science in the Scottish Enlightenment,* ed. P. Jones, pp. 145–68. Edinburgh: Donald.

    1993. From *L'Homme physique* to *L'Homme morale* and back: towards a history of Enlightenment anthropology. *History of the Human Sciences* 6: 121–38.

1995. Anthropology and conjectural history. In: *Inventing Human Science: Eighteenth Century Domains,* ed. C. Fox, R. Porter, and R. Wokler, pp. 31–52. Berkeley: University of California Press.

Wood, Robert. 1996. The science of man. In: *Cultures of Natural History,* ed. N. Jardine, J. Secord, and E. Spary, pp. 197–210. Cambridge University Press.

4

French Evolutionary Ethics during the Third Republic: Jean de Lanessan

Paul Lawrence Farber

"French evolutionary ethics" might, at first glance, appear to be an oxymoron,[1] for evolutionary ethics has been a philosophical tradition primarily in English-speaking and German-speaking countries. Russian and other variants have existed, but British, American, and German writers have constituted the bulk of its serious supporters as a school of ethics. For example, Cora Williams's general survey, *A Review of the Systems of Ethics Founded on the Theory of Evolution* (1893), examined the work of a dozen authors – six of them Anglo-Americans, and six Germans (Williams 1893). That the study of evolutionary ethics developed where it did is hardly surprising, for the scientific communities in those countries were the ones in which the biological theory of evolution was most widely accepted.[2] By contrast, French philosophers who were concerned with the life sciences were notoriously hostile to Darwin and to other British evolutionary theories, as well as to German versions of evolution.[3] Later in the nineteenth century, evolutionary ideas began to be accepted in France, but even then they were more closely related to the ideas of Lamarck than to Darwinism. Many English and American academics have been surprised to see the inscription on the base of Lamarck's statue at the north entrance of the Jardin des Plantes: "Fondateur de la Doctrine de l'Evolution." The "transformationist" biology that became popular in France rejected the Darwinian concept of selection as the key creative force in producing new species, and, moreover, it favored a teleological view of nature in place of the nondirectional perspective common to Darwinians. Perhaps most important, Lamarck's emphasis on the ability of organisms to adapt directly to a changing environment was strongly supported.[4]

With evolutionary ideas kept largely at bay for several decades, and with a decidedly French interpretation given to evolutionary ideas when they finally did achieve a level of respectability in the French scientific community, one well might wonder how significant evolutionary ethics was in France.

Although never a major component of French ethical thought, it nonetheless had importance, for it was part of a larger tension between religious and secular thought. Nineteenth-century philosophy supplied a public forum for the wider battle over the place of the church in the public and private domains. As a topic of investigation, the study of evolutionary ethics in France also has historical interest for comparative purposes. Recently, I traced the Anglo-American tradition of evolutionary ethics from its beginnings in the work of Charles Darwin and Herbert Spencer to recent writings inspired by sociobiology (Farber 1994). One of the generalizations that emerged from that study was that when carefully examined, most of the authors who had claimed to have discovered a set of values in nature had instead read those values into nature, or had attempted to justify currently held beliefs by reference to their alleged biological origin. From Herbert Spencer to Julian Huxley and then to Edward O. Wilson, the story is the same. The values shifted, of course, and the theory of evolution changed, but the same process of reading one's own values into nature has been a constant in the history of evolutionary ethics. This is not to say that biological investigation does not have the potential to yield information that may be of considerable interest for moral issues, or that science doesn't have great relevance for ethical concerns. Rather, the point (disputed by supporters of evolutionary ethics) is that the claim that the theory of evolution could serve as a foundation for ethics was highly problematic, and the major historical attempts to construct an evolutionary ethics neither justified beliefs held on prior grounds nor served to reveal new values. By looking at French evolutionary ethics, we can see these criticisms in even greater relief, because of the contrast in French evolutionary ideas and French values as compared with those across the Channel or across the Rhine.

Although Darwin was never elevated to the same level of mythical importance in French science as in other countries, evolutionary thought has had a long history in France. Eighteenth-century writers like Georges-Louis Leclerc, comte de Buffon, described the changing appearances, over time, of life forms on Earth and believed that descent with modification had occurred. They were followed in the early nineteenth century by Jean-Baptiste de Monet de Lamarck and Étienne Geoffroy Saint-Hilaire, who offered evolutionary explanations for the diversity of contemporary life forms. Both Lamarck and Geoffroy were overshadowed in Paris by Georges Cuvier's strident opposition to evolutionary ideas. Evolutionary ideas were also attacked by the founders of French experimental physiology, who rejected what they considered "speculative" or "metaphysical" thinking in the life sciences. Nonetheless, the early French evolutionists left a legacy that never completely died and reemerged in the last decades of the nineteenth century to become

85

an important intellectual force (Laurent 1987). Édmond Perrier, Jean de Lanessan, Alfred Giard, and Félix Le Dantec popularized a neo-Lamarckian biology that stressed the inheritance of acquired characteristics and the influence of the environment on living organisms. Like their contemporary American neo-Lamarckians (Edward Drinker Cope, Alpheus Hyatt, Alpheus Packer), they favored teleological perspectives that were easily extended to cover human physical and cultural evolution (Bowler 1983). It was among those French neo-Lamarckians that interest in evolutionary ethics developed.

Illustrative of the naturalists and anthropologists who were attracted to a neo-Lamarckian perspective on ethics is Jean-Marie-Antoine de Lanessan (1843–1919). He, more than any of the others, wrote extensively on ethics and therefore is of special importance in understanding French evolutionary ethics (Lagarde 1979). Lanessan studied medicine, wrote on a variety of biological subjects, and was a prominent public figure whose career included major positions, such as governor of Indochina and minister of the navy. His views were widely disseminated in books, articles, and reviews, and he published works on political and social issues, as well as on scientific topics. Like other reformers of the Third Republic, Lanessan waged a vigorous campaign against the influence of the church and for a rational, scientific worldview as a replacement of church teachings. He also resisted what he took to be a Darwinian emphasis on the "struggle for existence" for understanding the living world or the human world. That individualistic perspective, he believed, was typical of British thought, but foreign to the modern French spirit, which stressed ideas of association and cooperation. In a number of works he explained how nature reflected the value of association, and such association was the natural tendency, even in the physical world (Lanessan 1881, 1883, 1903): Isolated individuals were destroyed, but in association they could survive. Lanessan did not deny the reality of the struggle for existence and a process of natural selection in nature, but he did not agree with Darwin that it could be a creative force: Where it operated, it worked as a force of destruction, rather than of creation.

Lanessan was critical of British and German writers, but the target of his strongest invective was closer to home. True to the ideals of the Third Republic, he fervently supported a rational worldview and the values of "association." The church and its philosophical allies posed a greater threat than foreign thinkers. In his writings he attempted to construct a substitute for the outdated teachings of the clergy and their supporters among contemporary French philosophers. His most extended discussion is found in *La Morale naturelle* (1908), which begins with a survey of what he took to be the misguided opinions on ethics being expressed by traditional writers on religion and metaphysics. He went on to erect a sweeping natural morality based on the evolution of living beings and human society. For Lanessan, no reconciliation between the theological conception of morality and the scientific knowledge of evolution was possible, and "it is not possible to explain the upward evolution of morality other than by the organization and the relations of humans and their animal ancestry and by the education that all human and animal individuals receive" (Lanessan 1908, pp. 1–2).

Although Lanessan believed that basic moral positions grew out of the very existence of matter and its associations, as well as the fundamental nature of human experience, he contended that no absolute ethics was possible. Rather, moral ideas varied "with place, time, social milieux, and individuals," and "there are as many moralities as men" (Lanessan 1908, p. 4). Like others of his contemporaries, such as Edward Westermarck, Lanessan believed that study of the history of ethical ideas revealed a dramatic divergence of thinking: Some societies had considered respect for elders a duty, but others had killed them when they were considered no longer useful. Lanessan sketched the history of western moral thought to show that most of the religious and philosophical foundations of specific ethical positions had incorrectly attributed to humans a divinely created soul that had the faculty of freely choosing recognizably good or evil actions. He praised Darwin as the first to recognize the relationship between animals and humans with respect to morals, but he was highly critical of Darwin's explanation of the selectionist evolution of a moral sense in man, as well as an inherent confusion in Darwin's writings on morality. He rejected the Darwinian notion that only humans were moral beings and equally took exception to Darwin's emphasis on the inheritance of instincts and sentiments. How, Lanessan asked, could humans be the only moral beings when anyone could see evidence of a social instinct in animals? Darwin had merely taken ideas from philosophers and read them into nature (Lanessan 1908, pp. 28–30).

Spencer, according to Lanessan, erred equally. Although a more consistent evolutionist than Darwin, Spencer also had read into nature the existence of a moral intuition that was strikingly like what philosophers had established on a priori grounds or on theological grounds, but which was equally fallacious. Although Lanessan appreciated Spencer's opinion that egotism and altruism coexisted in human action, he rejected Spencer's utilitarian argument that seeking pleasure and avoiding pain were the prime motivating forces in human action (Lanessan 1908, pp. 30–3).

Lanessan believed that in the place of merely reifying moral sentiments, a truly natural morality was possible. In examining his ideas, we can see that although he criticized British writers for reading their values into a picture of nature (which he also rejected), he was guilty of an analogous effort. His starting point was dismissal of the belief that moral ideas were innate. Instead, he argued for the view that moral ideas developed in individuals by education and by their interactions with their environment. All living beings, according to Lanessan, had certain basic needs. The first two, nutrition and reproduction, were fundamental, and the evolution of life was intimately linked to satisfying them. Less discussed was a third need: activity. Because life basically was the movement of matter, albeit of a special and complex sort, it followed that activity would manifest itself wherever there was life. Lanessan took a leap from the definition of life as a particular form of matter in motion to the conclusion that all living beings would naturally seek activity. Activity could be muscular or intellectual (*activité cérébrale*), and education needed to attend to the exercise of both.

The primal needs of organisms determined the formation of organs, mental faculties, and ideas, but the environment played an important role in modifying them. Lanessan traced the recognition of that environmental influence from Buffon, through Lamarck, to the writings of contemporary naturalists such as Édmond Perrier, director of the Paris Muséum (Lanessan 1980, pp. 55–8). Lanessan argued that an understanding of the physical organization of living beings in their environment would suffice to explain how they went about satisfying their needs. Consequently, there was no need for extraneous concepts like "free will" or "divine action." His materialism was strident and important for understanding his ethics. Lanessan rejected terms like "voluntary" and instead relied on an often-used reconciliation between material determinism and the psychological awareness of volition: Humans accomplished acts with "intention" in the sense that they reflected on them and were aware of conflicting influences, but in the end they obeyed the strongest and most powerful influences (Lanessan 1908, p. 83).

Lanessan, then, set out to replace what he considered Darwin's confused and contradictory notion that humans were the only true moral beings, because they could reflect on the moral acts they chose, with an equally problematic notion that humans were materially determined: "If one agrees with evolutionary materialism that all evolution, all transformation, and all acts are causally determined, one is then certain of being able to provide the means of creating in each man, through physical, intellectual, or moral education, an organic and psychic state which will necessarily result in good or bad conduct" (Lanessan 1908, p. 83).

Humans and animals were determined in their actions, but they were also conscious and had ideas. Lanessan did not accept the Cartesian view that animals were automatons, and humans were the only fully conscious animals. Drawing on contemporary writers who had described animal behaviors, Lanessan discussed the intellectual capabilities of animals, including memory, attention, and judgment. He recognized that many mental acts were instinctive, but that they were equally modified by education and intelligence. All agency, however, was material, and for Lanessan that had important implications for an understanding of moral development. Lanessan, indeed, voiced the optimistic viewpoint that "it is not more difficult to create an honest man than it is a smart dog" (Lanessan 1908, p. 113).

What about an honest dog? Lanessan contested the commonly held view of his day that morality belonged exclusively to humans. He claimed that animals had moral ideas and that the production of moral ideas in animals and humans grew out of the natural needs of nutrition, reproduction, and activity. Because humans evolved late in the history of the earth, there was even a continuity of moral ideas: The earliest humans had received a moral education from their animal ancestors.

Lanessan used the term *idée morale* in a broad sense to include mental and intellectual states, as well as notions of what constituted good and bad. He first derived ideas of fear, territory, and foresight from the need for nutrition in animals and humans. The derivations were typical of the speculative and anecdotal armchair animal-psychology writings of the late nineteenth and early twentieth centuries. For example, Lanessan "derived" the idea of fear from the natural reactions of animals that were prey to others. In support of his view, he noted that in those areas where herbivores lived without the presence of predators, they displayed little or no fear of other animals. Similarly, voyagers had reported on the docile and friendly nature of native populations who had not previously encountered Europeans (Lanessan 1908, pp. 115–17). Fear, then, was the result of animals and humans learning to take precautions for the sake of protection. Similarly, in a manner that anticipated arguments that are used today in some pop sociobiology, he derived the concept of territory from the difficulty that carnivores experienced in maintaining themselves in a state of nature. They needed large territories because of the scarcity or defiance of their prey, and consequently they had developed a jealously guarded isolation (Lanessan 1908, pp. 118–19). Lanessan was vague as to the underlying mechanism responsible for producing those ideas, as well as to how they were transmitted from generation to generation. In general, however, he downplayed the role of heredity and stressed the individual's "education" and its interaction with the environment.

Lanessan stressed that the origin of many ideas in humans was the same as in animals, and he extended his argument to encompass abstract ideas such as individual liberty and happiness. Individual liberty, for example, developed because of the necessity for frequent moves to search for food, to avoid predators, and to satisfy the need for activity. The resulting idea of individual liberty could be seen in animals that were restricted by being caged, for in consequence they would attempt to escape or would act abnormally. He also noted that the idea of liberty was subservient to the need for nutrition and that by capturing animals at a young stage, one could domesticate them. The animals would forgo liberty if provided with ample food. Lanessan stressed the continuity between humans and animals and drew a parallel between domesticated animals and slaves. Just as animals can be raised to be content in a domestic setting, so, too, slaves, according to Lanessan, could be quite content if they were well provided for and their labor was not too harsh (Lanessan 1908, p. 128).

Happiness, which figured so centrally in the Anglo-American tradition of ethics, played a much smaller role in Lanessan's thought. The idea of happiness arose primarily from the physical pleasure animals derived from proper nutrition. Satisfying the need for activity was another source of pleasure, as was reproduction, although the latter was less important because it was not operative for the entirety of the animal's existence. Lanessan did not use the idea of seeking happiness to build a moral theory, and in fact he blamed it for leading to the deleterious passions of gluttony and alcoholism. The pleasure derived from eating and drinking could easily lead animals and humans to consume more than necessary, and consequently they could develop habitual overconsumption.

The *idées morales* relating primarily to the need for nutrition were considered by Lanessan motivated by self-interest (*egoïsme*). The need for activity contributed other equally egoistic ideas (Lanessan 1908, pp. 153–5). The moral ideas associated more closely with reproduction, however, in addition to explaining other egoistic ideas, accounted for the origin of some altruistic ideas, some of which will be discussed later. The battle to attract mates gave rise to the ideas of force, beauty, and courage. Jealousy could be seen in female domestic dogs who suspected their mates of infidelity. And as the pleasure among humans associated with satisfying the need for nutrition could result in gluttony or alcoholism, so, too, the desire to satisfy the reproductive urge could result, especially among males, in a "veritable frenzy" (Lanessan 1908, p. 144). The need could be transformed into a passion involving grave physical, intellectual, and moral disorders, as Émile Zola had so devastatingly depicted in *Nana*.

Reproduction led to family life in some animals, and with the development of the family, even short-lived ones, came ideas of obedience, respect, punishment, and recompense. Lanessan digressed to point to the common mating of siblings among birds in order to comment on the false idea, in his opinion, that consanguineous unions were unhealthy "to the progress of the race" (Lanessan 1908, p. 170). Of greater importance, he believed that more complex and altruistic, positive moral ideas resulted from the need for reproduction. Among humans, and even some animals, affection developed between male and female. Marital fidelity developed from the female need for protection and from the male sense of pride. Years of living together would create a wealth of common experiences that gradually would cause "love to take a less physical character" (Lanessan 1908, p. 174). Maternal love was a natural biological sentiment that engendered a filial love from the child. In turn, that was extended to the father and other household members. The love for the father, of course, was more complex than the affection for others, because it was compounded with the child's respect for his authority. That authority, according to Lanessan, had been codified in religious beliefs and into the law. Women, however, had struggled over the years to assert their authority, and Lanessan contended that the battle between the sexes had contributed to the progress of humankind.

Family ties of affection were at the root of the concept of duty. The moment a mother neglected a child, or mistreated it, she brought upon herself the censure of everyone. When a child showed a lack of respect for the father, that likewise provoked a negative response in those who observed it. Thus the bonds of affection, in time, became a tacit convention to which all individuals felt an obligation, and when those obligations became sufficiently clear and the language attained a level of precision, the concept of family duty could be enunciated. Thus those universally recognized duties had their origin not in sacred texts nor in ancient poets nor in the writings of philosophers, but in natural facts. But they were often in conflict with the egoistic demands of the individual, and Lanessan argued that the object of moral education was to reconcile those conflicting elements in such a way as to further human progress.

What of the conflict between individual and society? What duty extended beyond the family? Was there a "struggle for existence" that drove social progress and demanded a broader moral vision? Lanessan wrote extensively about *la Lutte pour l'existence* (Lanessan 1881, 1883, 1903, 1908). He admitted that all of nature was intimately involved in a struggle for survival. The oak tree, the tiger, and man all had to resist constant threats to their existence. The consequence of that universal and unremitting struggle was that the less robust, those less able to escape the causes of destruction, and the less fecund tended to disappear. Lanessan credited Buffon as the first to clearly recognize the consequences of the struggle for existence.[5] For Darwin, that struggle for existence was the force responsible for the evolution of living forms. The differential reproduction rates of individuals, he thought, would in time lead a population to adapt to changed environmental conditions, or to alter its habit so as to exploit other portions of the environment. Lanessan was typical of the majority of French naturalists who rejected that selectionist perspective on the living world. Instead, he focused on how individuals and species had resisted the powers of destruction. In neo-Lamarckian fashion, he wrote of how animals responded to the environment by developing new defensive organs, coloration, and so forth, but what he stressed most was the power of "association" to resist destruction. That was a subject to which Lanessan had given considerable thought and on which he had published extensively. Like his Russian contemporary Petr Kropotkin, who stressed cooperation in nature, Lanessan saw association as the principal force resisting destruction in nature (La Vergata 1992). He went so far as to claim that association applied even in the inorganic world: Large isolated rocks standing apart from the cliffs on the coast were worn away in a manner that the larger "association" of rocks could withstand. Groups of individuals of the same species, or of different species, plants and animals, formed associations that provided greater safety in the fight for survival. As in the case of the family, social existence gave rise to new moral ideas.

Although the family was the most rudimentary form of association, it was not the basis for society, but rather was basically in conflict with society. Societies consisted of groups of families that lived together in a permanent fashion. In some primitive clans, such as those in Australia, families scarcely existed. Because the environment there was so harsh, it worked against males making any long-standing attachments, but it nonetheless served to bring together individuals who without group hunting would not have been able to procure sustenance, or who would have been unable to protect themselves from assault by other clans (Lanessan 1908, p. 209). Such societies were extreme examples, but Lanessan's point was that societies consisted of individuals, most of whom also had family ties, and therefore a complex tension existed, resulting from individual, family, and social demands.

Lanessan sketched a global history of mankind and concluded that to ensure their existence human societies had produced social morals that stressed patriotic duty. Individuals and families were inspired to make sacrifices for the larger good. That altruism, unfortunately, had been exploited by religious, military, and government leaders to their own profit. In contrast, Lanessan lauded the goals of the Third Republic, which he claimed would work for the general good of all its citizens. He contrasted it with those countries where the idea of the "struggle for existence" dominated social thought. In them, unbridled competition allowed the physically or mentally strongest to dominate at the expense of the weaker and gave rise to irresponsible ideas like Nietzsche's supermen, who "would leave behind them only ruins inhabited by slaves and idiots" (Lanessan 1908, p. 226).

Lanessan argued that a society that evolved in a manner that conformed to human nature would promote the maximum liberty and the greatest contentment. The Third Republic was far from that ideal, but was a step in the correct direction. What was necessary next was that the French learn to raise moral individuals who would develop an altruism that would balance individual, family, and social demands and who would strive toward creating a government in which individual needs would be satisfied in harmony with the general needs of France and of all other countries. To accomplish that end, Lanessan synthesized his ideas on heredity, social development, psychology, and ethics to map a blueprint for the future.

The cornerstone in Lanessan's ambitious plan to create a generation of truly moral individuals was his belief that education was more powerful than heredity. He dismissed the notion of innate ideas and hereditary dispositions. Although Lanessan recognized that mental organization was inherited, and as a consequence temperaments could be transmitted, he stated that how one's temperament would be expressed would depend on education and environment. Lanessan claimed, for example, that it was impossible to cite a single example of a child whose parents had been burglars or assassins and who in turn became a burglar or an assassin after that child had been separated from his parents and raised by honest folk, instructed in a profession, and permitted to live in an honest milieu (Lanessan 1908, p. 248).

Education, therefore, was central, and it was from that insight that moral education should be guided. Animal and human infants were profoundly egoistic. They cared only about satisfying their needs for nutrition and activity. But they also imitated what they saw and heard, and parents, therefore, could teach by example. A happy and smiling mother would stimulate the infant to smile and to experience a sense of joy. A cheerful and happy surrounding would promote a proper moral education. All the existing written manuals on morals and catechisms should be ignored, because they were based on false and misleading assumptions. Instead, a mother needed to attend to developing a positive sensibility in the child. Children needed to be given a sense of being loved and of self-worth. Notions of the inherent sinfulness of man, by contrast, would become self-fulfilling prophesies. In like manner, attempts to suppress the passions would only backfire. Rather, it was essential that young people learn to moderate their emotions. Lanessan believed that they could do so by dampening them through wide experience (a form of habituation) and by developing the power of rational reflection. Together, those two forces should allow young people to exercise control over their emotions and, combined with a positive and gay environment, should prevent the development of antisocial feelings. A balanced upbringing would avoid the excesses that might lead to gluttony or alcoholism.

What sanctions were necessary to assure the moral development of the child? For the young child, the pleasure or displeasure of the parent was sufficient. The first teachers of the child should establish their authority, but not in a brutal fashion. Three qualities should reign in the upbringing of children: kindness, fairness, and firmness. A child raised in a supportive environment and given adequate nutrition and the opportunity for physical and mental exercise would develop in a natural manner into a mature individual. Ideally such a child would be tolerant, prudent, and understanding. Such a child would have a sense of loyalty to its family and its country based on sincere love, not as a result of having been cynically manipulated. Children should be educated in coeducational settings so that boys and girls would understand one another better and would be more likely to choose better matches. Ideally, unions should be based on love, rather than interest, for such unions would result in a deeper commitment and more successful marriage (Lanessan 1908, pp. 306–12).

Lanessan believed that even the best marriages were often severely tested by the arrival of children, and for that reason he felt that it was important that young women be instructed by their mothers in how to balance their natural maternal feelings with their conjugal responsibilities. He elaborated on numerous other "natural" principles, such as the inadvisability of wet nursing, the role of the father as a source of justice tempered by indulgence and affection (even though he was assumed to be away from the house during most of the day), and the responsibility of the upper social classes to cease limiting the number of their children and to have more offspring.[6]

What did Lanessan accomplish in his attempt to derive a natural morality? His goal had been to demonstrate how moral ideas had developed from the evolution of living beings and to use those natural morals as a basis for outlining an ethical vision for his contemporaries in the Third Republic. In doing so, he believed he could liberate ethics from the stranglehold of religion and philosophy. The picture of the living world that Lanessan held was one typical of the neo-Lamarckians of the late nineteenth and early twentieth centuries. It rejected the selectionist perspective of Darwinists, and in its place posited a progressive evolution, stressing the influence of the environment and the ability of organisms to adapt to shifts in it. Lanessan further believed that higher animals possessed an intelligence that allowed them to formulate all the basic moral ideas usually ascribed only to humans. The origin of those morals was in the primordial needs of living beings: nutrition, reproduction, and activity. The most basic moral ideas were egoistic. Altruistic ideas resulted from reproductive urges that united couples and families. The family served as a starting point for social development and subsequent moral ideas, but the family was always in tension with larger social units.

In examining the *idées morales* that Lanessan claimed to have derived from nature, what we discover is that his values were taken from the reform politics of the Third Republic. He used science to ward off the power of the church and its philosophical supporters, and he attempted to erect a rational ethics to guide responsible citizens to raise orderly, responsible, loyal off-spring. But it is not clear that Lanessan succeeded in demonstrating the extent to which nature dictated his specific ideas (e.g., that men should work all day and supply justice in the home, or that for domestic peace women should be educated to balance their conjugal duties with maternal ones). Nor did he provide any ultimate sanction to justify his moral dictates. Lanessan informed his readers how children should be raised so as to instill certain virtues and avoid certain vices in them. But he did not demonstrate why society *ought* to adopt those values. They were his values and were of obvious merit, perhaps, to many of his readers, but his prescriptions did not constitute philosophically compelling arguments. Given that Lanessan stated that there were no absolute ethics, merely ethics relative to time and place, on what grounds could he claim that his values were superior to the values of the church or to foreign mores? Moreover, his materialistic determinism made it difficult to understand how any individual could be responsible for his actions. If human behavior was determined in the way that Lanessan claimed, in what sense were people responsible for it?

Lanessan's writings are typical of the arguments found in French evolutionary ethics. Like his counterparts in Britain and America, he attempted to use biology as an explanation for the origin of morals and as a justification for his beliefs. Also like them, he read his own values into nature only to discover them there. What makes Lanessan interesting, however, is that unlike the writers across the Channel and across the Atlantic, he was using a different theory of evolution and a different set of values. The Anglo-American tradition favored a more selectionist approach to nature and a more utilitarian approach to ethics.[7] The associationist philosophy of Lanessan rejected that selectionist approach. He equally rejected the individualistic emphasis that characterized the utilitarian philosophy that underlay most British and American evolutionary ethics of the period. Instead, Lanessan stressed the value of association, which for him was an evolutionary principle as well as a moral guide.

Comparison of Lanessan to Anglo-American writers is instructive because it underscores the confusion that has surrounded discussions of evolutionary ethics as an adequate basis for morality. Given a set of values and a picture of nature, a clever and imaginative writer can always construct a consistent match. For Anglo-American writers, evolutionary ethics provided a secular foundation on which to graft a set of Christian values. Lanessan attempted to use his *morale naturelle* polemically to legitimate the values of the Third Republic and to undercut some of his political opponents who favored a return to a monarchy and a society following the dictates of the church. As polemic, evolutionary ethics has worked; however, as philosophy, it has not been satisfactory. Contemporary authors who look to evolution for knowledge applicable to ethics need to be mindful of past attempts to use evolution as a foundation for ethical systems, and thereby they may avoid duplicating the mistakes of the past.

1. "Evolutionary ethics" generally refers to attempts to use the theory of evolution as a foundation for ethics. The theory of evolution can be narrowly construed to mean those ideas stemming from Darwin and Wallace's theory, or more broadly interpreted to include general evolutionary philosophies, such as that of Herbert Spencer. In either case, appeals to scientific foundations for justification are central.
2. See Glick (1972) and Kohn (1985) for articles dealing with Darwin's reception in different countries.
3. Yvette Conry (1974) has carefully documented the dismal story of Darwinism in France. She has explained the origins of the poor reception by the distinguished French scientific community (which stands out as an anomaly in the history of evolution theory) by exploring the complexity of the French scientific enterprise and by examining the factors that determined the rejection. Also see Farley (1974), Buican (1984), and Stebbins (1965). Linda Clark (1984) has attempted to show some social significance of Darwin's ideas in France.
4. For a discussion of French neo-Lamarckianism, see the articles in *Revue de synthèse,* 1979, 95–96: 279–468. Its later development has been discussed by Buican (1984).
5. Lanessan had edited an important edition of Buffon, and he often quoted Buffon.
6. That anti-Malthusian view reflected a common preoccupation of many French medical and political figures with the decline of the birth rate in France. For a discussion of the problem, see Bardet and LeBras (1988). For an interesting discussion of the cultural expression of that concern, see Nye (1993).
7. During the second half of the nineteenth century there were some Americans who favored a neo-Lamarckian approach. A selectionist perspective, however, dominated much of the evolutionary writing in Britain and America.

REFERENCES

Bardet, P., and LeBras, H. 1988. La Chute de la fécondité. In: *Histoire de la population française,* vol. 3, pp. 351–402. Paris: Presses Universitaires de France.
Bowler, P. 1983. *The Eclipse of Darwinism. Anti-Darwinian Evolution Theories in the Decades around 1900.* Baltimore: Johns Hopkins University Press.
Buican, D. 1984. *Histoire de la génétique et de l'évolutionnisme en France.* Paris: Presses Universitaires de France.

Clark, L. 1984. *Social Darwinism in France*. Tuscaloosa: University of Alabama Press.

Conry, Y. 1974. *L'introduction du Darwinisme en France au XIXe siècle*. Paris: Vrin.

Farber, P. L. 1994. *The Temptations of Evolutionary Ethics*. Berkeley: University of California Press.

Farley, J. 1974. The initial reaction of French biologists to Darwin's *Origin of Species*. *Journal of the History of Science* 7 (2): 275–300.

Glick, T. F. 1972. *The Comparative Reception of Darwinism*. Austin: University of Texas Press.

Kohn, D. 1985. *The Darwinian Heritage*. Princeton, NJ: Princeton University Press.

Lagarde, A. 1979. Jean de Lanessan (1843–1919). *Revue de synthèse* 95–96: 337–51.

Lanessan, J. de. 1881. *La Lutte pour l'existence et l'Association pour la lutte*. Paris: Dion.

    1883. *Le Transformisme évolution de la matière et des êtres vivants*. Paris: Dion.

    1903. *La Lutte pour l'existence et l'évolution des sociétés*. Paris: Félix Alcan.

    1908. *La Morale naturelle*. Paris: Félix Alcan.

Laurent, G. 1987. *Paléontologie et évolution en France de 1800 à 1860. Une Histoire des idées de Cuvier et Lamarck à Darwin*. Paris: Editions du Comité des Travaux historiques et scientifiques.

La Vergata, A. 1992. Les bases biologiques de la solidarité. In: *Darwinisme et société*, ed. P. Tort, pp. 56–87. Paris: Presses Universitaires de France.

Nye, R. 1993. *Masculinity and Male Codes of Honor in Modern France*. Oxford University Press.

Stebbins, R. E. 1965. French reactions to Darwin, 1859–1882. Ph.D. dissertation, University of Minnesota.

Williams, C. M. 1893. *A Review of the Systems of Ethics Founded on the Theory of Evolution*. London: Macmillan.

5

The State and Nature of Unity and Freedom
German Romantic Biology and Ethics

Myles W. Jackson

Two of the most fundamental issues grappled with by German romantics, *Naturphilosophen,* and investigators of nature during the late eighteenth and early nineteenth centuries were the nature of freedom and the freedom of nature. Fichte, Schelling, Goethe, Schiller, the Schlegel brothers, Schleiermacher, and Novalis all sought to restructure civic order based upon natural order and freedom. Each of those scholars turned to the philosophy of Kant in order to reformulate the relationship between nature and culture. Fichte and Schelling, in particular, set out to establish systematic studies of human freedom, transcending the boundaries set by their intellectual hero from Königsberg. Filled with the Jacobin spirit kindled by the French Revolution, Fichte's *Wissenschaftslehre,* as he acknowledged, owed much to the ideology of the French Republic. It was "the first system of freedom," and "as that nation [revolutionary France] tore man loose from his outer chains, Fichte's system of knowledge removed the Kantian fetters of the things-in-themselves and declared man to be ontologically free as well" (Fichte 1970, p. 298; Morgan 1991, p. 27).

Schelling echoed Fichte's emphasis on political freedom and support for the French Revolution. Indeed, Schelling implicitly linked the two revolutions, Kantian and French. In his obituary for Kant in 1804, Schelling wrote that "it was one and the same spirit, which was long in forming, that created – according to the differences of nation and circumstance – the atmosphere there [in France] for a real revolution and here [in the German territories] for an ideal revolution" (Schelling 1992, vol. 3, pp. 585–94). According to Schelling, it was a consequence of Kantian philosophy that German intellectuals (save Goethe) all rallied around the events of 1789, just as it was a consequence of the new belief in rights and constitutions that a knowledge

I would like to thank Michael Ruse, Robert Richards, and an anonymous referee for their helpful comments on earlier versions of this essay.

of Kantian philosophy became a necessity for statesmen of the world (Morgan 1991, p. 27). Schelling's *Naturphilosophie,* much along the lines of Fichte's and Friedrich Schiller's beliefs, celebrated an inner freedom and claimed that nature was the symbol of the path to such freedom (Morgan 1991, p. 28).

Those German romantics (very broadly defined) certainly generated a "romantic morality" that differed, quite crucially, from the morality of the French Enlightenment philosophes and the ethics of Bentham, for example. However, upon closer inspection, a fine-grained analysis reveals that there existed interesting and informative differences between the moral views of individual romantics. Although connections between nature and ethics were fundamental to all of their weltanschauungen, the moral lessons that they drew certainly were not the same. A good example of the differences in the moral outlook among the romantics is seen in the contrast between Lorenz Oken and Johann Wolfgang von Goethe. For both men, solutions to questions of moral order were also solutions to questions of natural order. Neither Goethe nor Oken believed that moral knowledge was distinct from natural knowledge; indeed, they both asserted the opposite. Because nature was, for German romantics, the source of morality, in researching nature one was simultaneously probing into human ethics. Although the laws that formed the basis of Oken's embryology were very similar to the laws governing Goethe's morphology, their ethical stances could not have been further apart. Whereas Oken called for a political revolution throughout the German territories in the hope of forming a unified German republic, Goethe recalcitrantly clung to a reformed version of enlightened despotism and court culture. More importantly, their political and ethical differences resulted, in part, in Goethe's dismissal of Oken from the University of Jena. This essay traces the relationship between history and ethics in the works of Oken and Goethe in order to argue that similar views in biology did not necessarily imply similar ethical and political orientations.

LORENZ OKEN

The German romantics asserted that there was a fundamental relationship between nature research and the freedom of the human spirit. Schelling, for example, stated that "the outer world has opened out for us, in order to find within it the history of our spirit" (Morgan 1991, p. 31). Nature was a history of the path to freedom. Precisely because of links among nature, history, and freedom, Oken's nature research must also be viewed as a historical activity. Indeed, the goal of Oken's *Naturphilosophie* was primarily historical.

99

Nature was needed to depict the origins of the universe and its subsequent development from God's archetypal plan. Oken believed that the study of embryology enabled such a reconstruction. In the drawing of parallels between human gestation and the history of the universe (the typically romantic practice of establishing relationships between macrocosm and microcosm), knowledge of human gestation was simultaneously knowledge of the history of the cosmos and earth (Richards 1991, pp. 130–43). For Oken, the same natural laws that governed the development of humans also governed the evolution of the universe. Hence, *Naturphilosophen* probing the embryo could also elucidate the sequential development of the animistic heavens. In that respect, Oken's views mirrored those of Schelling, who argued that "it is certain that whoever could write the history of one's own life from its very ground, would have thereby grasped in a brief conspectus, the history of the universe" (Schelling 1942, p. 94).

In his *Elements of Physiophilosophy,* Oken wrote as follows:

> Physio-philosophy is, therefore, the generative history of the world, or, in general terms, the History of Creation. . . . Man is the summit, the crown of nature's development, and must comprehend everything that has preceded him, even as the fruit includes within itself all the earlier developed parts of the plant. In a word, Man must represent the whole world in miniature. Now since in Man are manifested self-consciousness or spirit, Physio-philosophy has to show that the laws of spirit are not different from the laws of nature, but that both are transcripts or likenesses of each other. . . . The whole of physio-philosophy depends, consequently, upon the demonstration of the parallelism that exists between the activities of nature and spirit. [Oken 1847, p. 2]

Because nature and spirit were "transcripts of each other," the freedom with which the human spirit separated itself from the world's objects must also exist in nature. The separation of the human spirit from the world's objects allowed the generation of natural knowledge. Oken demanded that "Physio-philosophy be called the Science of the Conversion of Spirit into Nature" (Oken 1847, p. 656). Because the laws that governed the human spirit were the same as those governing nature, the *Naturforscher* (investigator of nature) was morally obligated to illustrate the laws of freedom in both.

A second obligation of the *Naturforscher,* according to Oken, was to depict the unity of nature. As a student in 1802, Oken penned his *Compendium of Nature Philosophy, the Theory of the Senses and a System of Animal Classification Based on the Senses,* which argued that unity was the chief criterion of the new "universal science" invented by Kant (Mullen 1977, p. 383). Indeed, much of his scholarship was dedicated to offering an elaborate classification of the multifarious forms of life. A typical example of

Oken's emphasis on the unity of nature was his theory of the formation of the vertebral skeleton, the theme of which he presented in his doctoral dissertation, which maintained that the succession of individual bones represented a recapitulation in nature. That is to say, all bones were modifications of an archetypal, fundamental plan – a primal vertebral segment. Hence the cranial bones were merely extensions of that simple, unifying vertebral segment. Analogous and homologous organisms were pivotal to Oken's insistence on the unity of nature. He reiterated that theme in his inaugural address as professor of zoology at the University of Jena, delivered in 1809, entitled "On the Worth of Natural History, Particularly for the Formation/Education (*Bildung*) of the Germans." Restating Schelling's earlier attempts, Oken asserted that philosophy needed to return to its roots in *Naturphilosophie*. The organic whole was the basis for all scientific endeavor, which was not limited to the plant and animal kingdoms, but also necessarily included the relationship between humankind and the state.

Oken did not restrict his use of the themes of unity and freedom to his investigations of nature. Those themes formed the basis of his rather active political philosophy. For Oken, freedom and unity were part and parcel of human ethics, which in turn was inextricably linked to biology. As he proclaimed, "a[n] Ethicks apart from Physio-philosophy is a nonentity, a bare contradiction, just as a flower without a stem is a non-existent thing" (Oken 1847, p. 656).

The first two decades of the nineteenth century were difficult times for Germans. The destruction of the German territories (particularly damaging was the Prussian defeat at the battle of Jena-Auerstedt in October of 1806) at the hands of Napoleon's armies and the subsequent occupation of those territories by French troops had profound ramifications for the intellectual community in the Duchy of Saxe-Weimar-Eisenach. The French Revolution, which Fichte, Schelling, Schiller, the Schlegel brothers, and Hegel had all originally heralded as the beginning of a period of freedom and liberation from tyranny, ushered in an unprecedented era of absolutism. Oken, too, was distressed. Previously he had praised the unified, centralized, and cohesive characteristics of French science and education. He had contrasted France's eminence with the German territories' backwardness: "I find everything in Saxony hideous, without taking into my estimate the poverty that stares at you from every window. As to Jena, it is dead; there it lies, minus its head and riddled like a sieve." Of Göttingen, Oken lamented that

> on all occasions the most lively Protestant feeling prevails here – which all amounts to observing economy in the kitchen: no one will meddle with anything which he cannot put into his mouth – such is the case both in upper and lower grades of society – and no student has any higher notion of knowledge (*Wissenschaft*). [Ecker 1883, pp. 91–2, 101]

But after Napoleon's conquest, Oken's francophilic rhetoric ceased rather abruptly. Like many of the early romantics, such as Friedrich Schlegel, Oken became obnoxiously chauvinistic. German unification, sparked by an intellectual revolution and based on Germanic traditions of the High Middle Ages, was the only recourse for Germans seeking to thwart French domination. In 1814 Oken wrote a series of politically charged pamphlets outlining and analyzing German unification. He called the series *New Armaments, New France, New Germany.* It listed new political tactics, patriotic uprisings, and scientific and technological inventions that all could assist in unification of the German territories. Hence, two themes that united Oken's zoology with his political and moral enterprise were unity and freedom. In Oken's view, the German investigator of nature was *morally* obligated not only to depict and emphasize those themes in his research but also to bring about those cherished goals within a new political framework of post-Napoleonic "Deutschland."

GOETHE: ADMINISTRATING NATURAL KNOWLEDGE

Goethe's science represented a managerial moral (Jackson 1992, 1994). The natural world of Saxe-Weimar-Eisenach was managed in the same manner as the civil administration of that region, and Goethe was the *Hofmeister.* Goethe, using Kant's definition of genius in *Kritik der Urteilskraft* as a resource, fashioned himself as the genius who could interpret natural laws and manage their application to civil cases (Goethe 1887–1919, pt. 1, vol. 29, p. 146).

The organizing principles of nature were the same, for Goethe, as the methods for investigating it. Goethe's administrative rhetoric, featuring words such as "budget," "balance," "economy," "law," and "order," was also applicable to the relationships inside both the organization of the state and the structure of natural philosophy. The role of a late-eighteenth- and early-nineteenth-century *Naturforscher* tied those relationships together. A *Naturforscher*'s job as court official was necessarily political, and laws that governed human knowledge in eighteenth-century Weimar were, according to Goethe, natural and not artificial.

As is widely known, Goethe's work on morphology in particular, and biology in general, was vitalistic in nature. Although he readily admitted that physics and chemistry could help elucidate the properties of living organisms, he vehemently opposed any attempt to reduce biological phenomena to physico-chemical principles. In his essay *Bildung und Verwandlung* ("Formation and Transformation"), Goethe argued that "analytical efforts, if continued indefinitely, have their disadvantages. To be sure, the living thing is separated into its elements, but one cannot put these elements together again and give them life" (Goethe 1887–1919, pt. 2, vol. 6, p. 8; 1952, p. 23). In his *Vorarbeiten zu einer Physiologie der Pflanzen* ("Preliminary Notes for a Physiology of Plants"), written in the mid-1790s, Goethe declared that

> from the physicist in the strict sense of the term, the science of organic life has been able to take only the general relationships of forces, their position and disposition in the given geographical location. The application of mechanical principles to organic creatures has only made us more aware of their perfection; one might almost say that the more perfect living creatures become, the less can mechanical principles be applied to them. [Goethe 1887–1919, pt. 2, vol. 6, p. 295; 1952, p. 89]

Goethe was attacking the mechanical philosophy of French rationalism that had dominated European thought for more than a century and a half. He despised the French depiction of organisms as complex machines. In *Dichtung und Wahrheit,* Goethe recalled his reaction to the mechanistic nature presented throughout Baron d'Holbach's *Système de la Nature* published in 1770:

> . . . we found ourselves deceived in the expectation with which he had opened the book. A system of nature was announced, and therefore we truly hoped to learn something of nature, our idol. . . . But how hollow and empty we did feel in this melancholy, atheistic half-night (*atheistische Halbnacht*), in which the earth vanished with all its images, the heaven with all its stars. [Goethe 1887–1919, pt. 1, vol. 28, pp. 69–80]

It was that mechanical ideology of the French that had driven the younger Goethe to Shakespeare, where he had discovered the notion of genius, the creative spirit of man, and the creative spirit of nature (Goethe 1887–1919, pt. 1, vol. 28, pp. 71–7).

Goethe was, of course, not alone in his condemnation of the French weltanschauung. His *Sturm und Drang* cohorts, and later the early romantics, echoed Goethe's disdain and indignation. All of them fundamentally opposed the notion of mechanical laws of nature. Nature was to the late-eighteenth-century German-speaking investigators of nature and *Naturphilosophen* a dynamic force, not a static entity. Although they were united in their belief in a living, nonmechanical nature, not all of them agreed on the governing, primary principles of nature's system. In short, there were at least two responses to the proposal that nature was ruled by mechanical laws. One was to reject the notion of natural law altogether. The other was to base nature on organic laws.[1] Goethe, the privy councillor, believed that organic laws were the principles governing nature. Goethe's organic laws accounted for natural order. His linkage between the state and the organism, as discussed later, illustrates the claim that laws served as an organizational principle both for his views of nature and for his administration of Saxe-Weimar-Eisenach.

Like all of his views on nature, Goethe's organicism was based on the law of polarity. Nature was a unity of opposing forces and entities. For example, in his *Zur Farbenlehre,* Goethe classified all of the colors under two categories, plus and minus, and also ascribed corresponding attributes to each side (Goethe 1887–1919, pt. 2, vol. 1, p. 277).

> Considered in a general point of view, color is determined towards one of two sides. It thus presents a contrast which we call a polarity, and we may designate by the expression plus and minus.

plus:	*minus:*
yellow	blue
action	negation
light	shadow
force	weakness
warmth	coldness
proximity	distance
repulsion	attraction
affinity with acids	affinity with bases

Polarity, which Goethe claimed to be a natural property, was of fundamental importance to his botanical and morphological studies. For example, in his 1790 treatise *Die Metamorphose der Pflanzen,* Goethe contrasted the two major forces he saw in plants: expansion and contraction. The cotyledon and corolla of a plant were worked on by an expansive force, whereas the calyx and corona were affected by a contractive force:

> By repeating here a remark made earlier, that styles and stamens represent the same stage of development, we can further clarify the cause of this alternate expansion and contraction. From seed to fullest development of stem leaves we noted first an expansion; thereupon we saw the calyx developing through contraction, the petals through expansion, and the sexual organs again through contraction; and soon we shall become aware of the maximum expansion in the fruit and the maximum contraction in the seed. In these six steps nature ceaselessly carries on her eternal work of reproducing the plants by means of the two sexes. [Goethe 1887–1919, pt. 2, vol. 6, pp. 62–3; 1952, pp. 60–1]

Returning to his essay "Preliminary Notes for a Physiology of Plants," Goethe, in a section entitled *Organische Entzweiung* ("Organic Duality"), contrasted the root and leaf of a plant:

> Origin of root and leaf. They are united by origin; indeed, the one cannot be imagined without the other. They are also by their origin opposed to each other ... root embryo develops downward and the leaf embryo upward, by saying

that they are opposed, in keeping with the general dualism of Nature, which here becomes specific. . . . We find that the roots require moisture and darkness to develop; the leaf requires light and acridity. Thus, from the beginning to end these needs are opposed. . . . Chief differences between the root and leaf embryo: the former always remains simple. . . . The leaf embryo, on the other hand, develops most diversely, and step by step approaches perfection. Light and dryness foster elaboration. Moisture and darkness retard it. . . . We reach the climax of organic duality in the division into the two sexes. [Goethe 1887–1919, pt. 2, vol. 6, pp. 307–8; 1952, pp. 95–6]

Those passages are noteworthy for two reasons. First, polarity illustrated nature's balancing abilities. Equal and opposite forces existed in nature. Nature, therefore, was balanced.

Second, those passages reveal that Goethe believed in the existence of external laws (i.e., laws that were external to the organism). French mechanistic accounts of nature held that mechanical laws governed living organisms and that those laws governed only the internal organization of an organism. Goethe countered such a depiction of nature first by claiming that natural laws governing life were organic, not mechanical, and second by asserting that natural laws also governed the relationship between the organism and its environment. Goethe's belief in internal, organic laws, such as the *Bildungstrieb,* or formative force, was to ensure that nature was not simply a projection of man's ego, which Schelling would later insist upon until the first years of the nineteenth century.

The concept of the formative force did not originate with Goethe. In fact, Goethe's vitalist view belonged to the Kant-Blumenbach tradition (Lenoir 1989, pp. 17–35; Jardine 1991, pp. 22–8). Kant, rather famously, argued in the second part of his *Critique of Judgement* that

an organized being is, therefore, not a mere machine. For a machine has solely motive power, whereas an organized being possesses inherent formative power, and such, moreover, as it can impart to material devoid of it – material which it organizes. This, therefore, is a self-propagating formative power, which cannot be explained by the capacity of movement alone, that is to say, by mechanism. [Kant 1988, p. 2, p. 23]

An organism, in Kant's view, was a being in which the parts were reciprocally means and ends for each other. Indeed, one of the major features of vital materialism and morphology was a central commitment to those vital forces, such as Blumenbach's *Bildungstrieb.* Much of Goethe's work on morphology resulted from his anatomical investigations with Blumenbach (Bräuning-Oktavio 1956, pp. 79–86).

In his work of 1781, entitled *Über den Bildungstrieb und das Zeugungsgeschäfte,* Blumenbach described the various directions that the *Bildungstrieb* could take in the multifarious life forms, both internally and externally (Jardine 1991, pp. 27–8). Within certain limits, external stimuli could produce variations in the formative force, thereby creating slight modifications in the structure of the organism (Lenoir 1989, p. 22). Goethe lauded Blumenbach for basing development on an anthropomorphic entity, the *nisus formativus,* or *Trieb,* rather than reverting to the notion of *Kraft,* which Goethe condemned as being mechanical and "a dark incomprehensible point" (*ein dunkler unbegreiflicher Punkt*) in the organization of living matter (Goethe 1887–1919, pt. 2, vol. 7, p. 72).

As in Blumenbach's account, every organism in Goethe's scheme had an allotted sum of formative force that remained constant throughout the organism's lifetime. The formative force was allocated to different organs, depending on the needs of that organism in response to its environment. In his essay *Erster Entwurf einer allgemeinen Einleitung in die vergleichende Anatomie, ausgehend von der Osteologie* ("The First Outline of a General Introduction in Comparative Anatomy, Starting from Osteology") of 1795, Goethe spoke of an organism's inherent ability to budget its *Bildungstrieb:*

> The rubric of its (the *Bildungstrieb*'s) budget (*Etat*) in which it divides up the cost, are prescribed to the organism. What it will direct to each part is, to a point, up to the organism itself. If it wants to direct more [force] to one part, it is not completely hindered as long as it simultaneously takes away the same amount from another part. Thus, nature can never be in debt nor can it become bankrupt (*bankrutt*). [Goethe 1887–1919, pt. 2, vol. 8, p. 16]

He continued by elaborating on nature's budgets:

> To illustrate this notion of budgetary give-and-take (*haushältische Gebens und Nehmens*), let us cite a few examples. The snake is a highly organized creature. It has a strongly marked head, with a perfectly developed auxiliary organ – an extensible lower jaw. Its body, on the other hand, is, as it were, infinite; and this is possible because it needs to waste neither substance nor strength on auxiliary organs. No sooner do such auxiliary organs appear in some form or others – the short arms and legs of the lizard, for example – when the infinite length shrinks into a much shorter body. The long legs of the frog greatly foreshorten into body, and by the same law (*Gesetz*) the shapeless toad is extended in breadth. [Goethe 1887–1919, pt. 2, vol. 8, pp. 18–19]

Goethe spoke of the "preponderance" (*Übergewicht*) of certain organs with respect to others. For example, "the neck and extremities of a giraffe are encouraged [to develop] at the cost of the giraffe's body, whereas just the opposite occurs in the mole" (Goethe 1887–1919, pt. 2, vol. 8, p. 16). Goethe called that budgeting ability of nature the organic law of compensation, or the correlation of the parts to the whole. That law (*Gesetz*) declared that "no part can gain something without another part losing something, and *vice versa*" (Goethe 1887–1919, pt. 2, vol. 8, p. 16). At the end of that section in the essay, Goethe reiterated the importance of external laws that kept the internal organic law of compensation in check: "The *Typus* must comply to a certain degree with the general external laws (*allgemeine äußere Gesetzen*)" (Goethe 1887–1919, pt. 2, vol. 8, p. 19). He continually spoke throughout his morphological writings of the natural principle of balance (*Gleichgewicht*) and preponderance of parts of an organism and the principle of giving and taking (*der Prinzip des Gebens und Nehmens*) (Goethe 1887–1919, pt. 2, vol. 8, pp. 309, 312). According to Goethe, no internal part of an organism could be superfluous when an external part of that organism was useless (Goethe 1887–1919, pt. 2, vol. 8, p. 309). In his preparatory notes for the aforementioned essay, Goethe sketched out the following notion:

1. Law (*Gesetz*)

The *Typus* has a certain amount of forces, which is independent of its size. This mass of forces must be used by nature. Nature cannot exceed the limit, nor can it fall below that limit. . . . The sum of the forces of one animal is the same as the sum of the force of another [of the same *Typus*]. [Goethe 1887–1919, pt. 2, vol. 8, p. 316]

The best-known example of Goethe's law of compensation was the inverse relationship between the development of horns and front teeth in the upper jaw in certain animals. Goethe argued that because the lion, like all members of the cat family, had both upper incisors and large canine teeth, it could not possess any horns (Goethe 1887–1919, pt. 2, vol. 8, p. 60). Hence, the *Bildungstrieb* of the lion was spent on the development of teeth. The ox, on the other hand, had neither upper incisors nor canine teeth, but did have horns. Likewise, sheep had neither upper incisors nor canines, but did have horns (Goethe 1887–1919, pt. 2, vol. 8, pp. 17–18). The budgeted allotment of *Bildungstrieb* for oxen and sheep was expended on horns (Wells 1978, pp. 20–1).

The law of compensation became, for Goethe, a 'fundamental truth' (*Hauptwahrheit*) of nature. In 1830, while critiquing the debate between Baron Cuvier and Geoffroy Saint-Hilaire, Goethe wrote that "economical nature (*haushältische Natur*) has prescribed a budget (*einen Etat, ein Budget*) in which . . . the main sum (*Hauptsumme*) remains the same, for if too much has been given on one side, it subtracts it from the other side and balances out in no uncertain manner" (Goethe 1887–1919, pt. 2, vol. 7, pp. 205–6; Nisbet 1972, p. 21).

In short, morphology provided investigators of nature an investigative method that could depict nature's perfect economy. By recording the allotments of *Bildungstrieb* to the differing organelles, nature researchers could provide a *Bauplan.* That *Bauplan* enabled Goethe to illustrate nature's diversity while simultaneously allowing him to proclaim the unity of nature. All organisms possessed *Bildungstrieb* in varying amounts, and the variance among the organisms was due to the distribution of *Bildungstrieb* as directed by the external laws, the climate. Budgets were the methodological tools used to illustrate nature's complex relationships and inherent order. They also permitted investigators of nature to trace a natural, morphological network by depicting the different ways in which the *Bildungstrieb* manifested itself. Similarly, administrators utilized an accounting budget to link various administrative enterprises that formed a financial network. Goethe hoped that his artificial budgets would approximate the accuracy of the natural budgets.

GOETHE VERSUS OKEN

Goethe's work on morphology was very similar to Oken's. Both were vitalistic and fought to subvert the mechanical depiction of nature. Both also sought to find unity in nature's multifarious appearances. Indeed, their researches were so similar that Goethe and Oken were embroiled in priority disputes in morphology. Goethe was enraged that Oken did not credit Goethe's demonstration that all bones were serial representations of an archetypal segmentation (i.e., Goethe's work on the intermaxillary bone in humans, which, although performed in the early 1790s, was not published until after Oken's announcement of the discovery) (Bräuning-Oktavio 1959). Although their investigations of nature overlapped considerably, their political and ethical ideologies could not have been further apart.

As I have argued, Goethe saw in nature the regime that he served: enlightened despotism. Although he wished to reform court culture, he certainly did not wish to abolish enlightened despotism. As he quite famously remarked to Schiller, he wished to find law and order in nature, whereas Schiller emphasized nature's freedom. It followed that preservation of law and order was the moral obligation of the German citizen.

Oken, on the other hand, was a revolutionary. In 1871, looking back at the origins of the German scientific community, the pathologist Rudolf Virchow proclaimed that

> the old Oken – as you know – can be considered to be a great revolutionary of his time, because he dared to dream of the vastness of a future German Empire. He was driven out beyond the borders of the Fatherland, and as a result, needed to search for protection from Switzerland, much like Ulrich von Hutten had to be buried in a foreign land. But then his [Oken's] friends from the various German territories collectively called out for him to unite [all German

108

Naturforscher] in a gathering of investigators of nature, not only to negotiate with him the affairs of science as science, but also with a view to awaken the inner-connected thoughts of the scattered sons of the larger Fatherland so that they actually could build together a future, united Empire. [Virchow 1871, pp. 73–4]

Six years earlier, Virchow had proclaimed the importance of Oken in the history of German science at the Nationalversammlung deutscher Naturforscher und Ärzte. He credited Oken with using the Nationalversammlung as a model for unifying the German spirit (Virchow 1866, p. 57). Virchow credited Oken as being the first to realize "that there must be a German science, and that science must be placed in the closest relationship to the life of the nation" (Virchow 1866, p. 59). Virchow labeled Oken "the great revolutionary, because he could dare risk to search for constitutional models for the life of our state and for free national models for our science" (Virchow 1866, p. 59). Virchow's emphasis on a unified effort in science was no accident, for 1871 had witnessed the birth of a unified German *Reich,* powered in part by German scientists. His depiction of Oken was quite accurate. Oken had organized the first meeting of the Versammlung deutscher Naturforscher und Ärzte in 1823. It was hoped that political delegates from the various territories would meet annually, like their scientific compatriots, in order to discuss German unification. Oken insisted that political freedom could be achieved only by a unified Germany. Such hopes, however, were not to be realized during Oken's lifetime.

The year after Napoleon's defeat seemed the perfect time for Oken to launch his campaign to advance German science, political and scientific unity, and political freedom. He focused his energies on the freedom-of-the-press laws (Mullen 1977, p. 381). In July of 1816 he announced the birth of his new encyclopedic journal, *Isis.* It would publish on science, history, art, and industry and would "remedy a want that exists throughout Germany, that of becoming acquainted in good times with the many sided diffusions of all discoveries and the varied criticism that may be passed upon intellectual works" (Oken 1817, vol. 1, pref.).

Oken's nationalism, freely broadcast throughout the issues of *Isis,* became rampant and politically dangerous. In 1817 he argued that

you do not learn at the universities to ape the habits and knowledge of the French, English, Spanish, Russians, or Turks; but what you really would want and may become, like the rest of the German people and its princes, is nothing else than educated Germans, who are equal to one another and whose calling is everywhere free. [Oken 1817, vol. 1, p. 195]

He was one of the favorite professors of the *Burschenschaften,* student organizations similar to modern fraternities that were ferociously nationalistic.

Also, Prussia and Russia were often satirized in Oken's political articles in
Isis. Journal woodcuts depicting asses were often intended to represent university faculties, which Oken loathed.

Weimar officials, Goethe in particular, admonished Oken, as his journal
was published in nearby Jena. The Duchy of Saxe-Weimar-Eisenach was
closely affiliated with Prussia, as Grand Duke Carl August was a relative of
the king of Prussia. Goethe warned Oken that continued insults to Prussia
could result in suspension of the journal. Oken, of course, viewed the admonition as encouragement to continue. He reported a *Burschenschaft* meeting at Wartburg where youths had burned what they considered to be objects
of repression: A Prussian military uniform, the Prussian police code, and
books written by Jews. The Prussian government was outraged and officially
protested to the duchy's officials for permitting the publication of such an
incendiary journal. Goethe discussed the matter with a very concerned Met-
ternich in Carlsbad. Then, two years later, things came to a head when Oken
publically insulted a Russian official, and Carl Sand, a former Jena student
and follower of Oken, chose that moment to murder the poet, courtier, and
Russian police spy Kotzebue in Mannheim. Goethe could then take advantage of paragraph 78 of the Carlsbad decrees, which dealt with the "misuse of
position in influencing the minds of the young," to allow Oken's dismissal
to become inevitable (Bräuning-Oktavio 1959, pp. 75–95; Paulen 1991, p. 17).

To the German romantics of the late eighteenth and early nineteenth centuries,
then, knowledge of nature was simultaneously moral and political knowledge.
Naturphilosophen such as Schelling and Fichte sought to emphasize the intellectual freedom in investigating nature and the freedom of the individual
vis-à-vis the state brought about by the Kantian revolution and the French
Revolution, respectively. Oken fit squarely into that context. The freedom
and unity inherent in nature, according to Oken, needed to be applied to the
moral and political circumstances of the post-Napoleonic German territories.
Goethe also believed in the fundamental relationships among nature, politics, and ethics. But, as we have seen, his support for reformed enlightened
despotism and his attack against revolutionary ideologies, both in politics
and in depictions of nature, represented a very different ethic.

NOTE

1. This would not seem to have been an option at all for German romantics, and indeed for nearly all of them it was not. However, it should be noted that Friedrich
   Schlegel toyed with the idea. See his *Kölner Vorlesungen 1804–5* (Schlegel 1958ff,
   vol. 12, p. 417). And, of course, Kant argued in his *Critique of Judgement* that na-

110

ture was ruled by mechanical laws, although many relationships were rendered intelligible only through the supposition of teleological principles.

REFERENCES

Bräuning-Oktavio, H. 1956. Vom Zwischenkieferknochen zur Idee des Typus: Goethe als Naturforscher in den Jahren 1780–1786. *Nova Acta Leopoldina* 18: 79–86.

1959. *Oken und Goethe im Lichte neuer Quellen.* Weimar: Hermann Böhlaus Verlag.

Ecker, A. 1883. *Lorenz Oken: A Biographical Sketch.* London: Kegan Paul.

Fichte, J. G. 1970. *Johann Gottlieb Fichte. Briefwechsel,* ed. Reinhard Lauth and Hans Jacob. Stuttgart: Bad Cannstadt.

Goethe, J. W. von. 1887–1919. *Goethes Werke. Herausgegeben im Auftrage der Grossherzogin Sophie von Sachsen,* 143 vols., Weimar edition. Weimar: Hermann Böhlaus Verlag.

1952. *Goethe's Botanical Writings,* ed. Bertha Mueller. Honolulu: University of Hawaii Press.

Jackson, M. W. 1992. The economy of Goethe's nature and the nature of his economy. *Accounting, Organizations and Society* 17: 459–69.

1994. Natural and artificial budgets: accounting for Goethe's economy of nature. *Science in Context* 7: 409–31.

Jardine, N. 1991. *The Scenes of Inquiry: On the Reality of Questions in the Sciences.* Oxford: Clarendon Press.

Kant, I. 1988. *The Critique of Judgement,* ed. James Creed Meredith. Oxford: Clarendon Press.

Lenoir, T. 1989. *The Strategy of Life: Teleology and Mechanics in Nineteenth-Century German Biology.* University of Chicago Press.

Morgan, S. R. 1991. Schelling and the origins of his *Naturphilosophie.* In: *Romanticism and Science,* ed. Andrew Cunningham and Nicholas Jardine, pp. 25–37. Cambridge University Press.

Mullen, P. C. 1977. The romantic as scientist: Lorenz Oken. *Studies in Romanticism* 16: 381–99.

Nisbet, H. B. 1972. *Goethe and the Scientific Tradition,* publication 14. London: Institute of Germanic Studies.

Oken, L. (ed.). 1817. *Isis oder encyclopädische Zeitung,* vol. 1. Jena.

1847. *Elements of Physiophilosophy,* ed. Alfred Tulk. London: Ray Society.

Paulen, R. 1991. *Goethe, the Brothers Grimm and Academic Freedom.* Cambridge University Press.

Richards, E. 1991. "Metaphorical mystifications": the romantic gestation of nature in British biology. In: *Romanticism and Science,* ed. Andrew Cunningham and Nicholas Jardine, pp. 130–43. Cambridge University Press.

Schelling, F. W. J. 1942. *The Ages of the World,* ed. F. de Wolfe Bolman, Jr. New York: AMS Press.

1992. *Schellings Werke,* 3rd ed., 12 vols., ed. Manfred Schröter. Munich: Beck-'sche Verlagsbuchhandlung.

Schlegel, F. 1958ff. *Kritische-Friedrich-Schlegel-Ausgabe,* vol. 12, ed. E. Behler, J.-J. Anstedt and H. Eichner. Zürich: Thomas.

Virchow, R. 1866. Über die nationale Entwicklung und Bedeutung der Naturwissenschaften. In: *Amtlicher Bericht über die vierzigste Versammlung Deutscher Naturforscher und Ärzte zu Hannover im September 1865,* ed. C. Krause and K. Karmarsch. Hannover: Hahn'sche Hofbuchhandlung.

1871. Über die Aufgaben der Naturwissenschaften in dem neuen nationalen Leben Deutschlands. *Tageblatt der 44. Versammlung Deutscher Naturforscher und Ärzte in Rostock 1871.* 5: 73–82.

Wells, G. A. 1978. *Goethe and the Development of Science, 1750–1900.* Alphen aan den Rijn: Sijthoff & Noordhoff.

6

Darwin's Romantic Biology

The Foundation of His Evolutionary Ethics

Robert J. Richards

> I am at present fit only to read Humboldt; he like another Sun illumines everything I behold.
>
> Charles Darwin, Beagle Diary

On learning that Alexander von Humboldt's health had worsened, Darwin wrote to his friend Joseph Hooker, then in Paris:

> I grieve to hear Humboldt is failing; one cannot help feeling, though unrightly, that such an end is humiliating: even when I saw him he talked beyond all reason. – If you see him again, pray give him my most respectful & kind compliments, & say that I never forget that my whole course of life is due to having read & reread as a Youth his Personal Narrative. [Darwin 1985–, vol. 3, p. 140]

Alexander von Humboldt – romantic adventurer, friend of Goethe, and the very doyen of German science in the first half of the nineteenth century – had preceded Darwin to the Americas. He wrote of his explorations in *Personal Narrative of Travels to the Equinoctial Regions of the New Continent, during the Years 1799–1804,* and elaborated their findings in his greatly celebrated *Cosmos.* These two multivolume books had a tremendous impact on Darwin and helped forge the romantic conception of nature that underlay his theory of evolution. That romantic conception reveals itself most perspicuously in the moral evaluation of nature and man that Darwin plaited through his two principal works: the *Origin of Species* and the *Descent of Man.*

Darwin's conceptions of nature and man have, of course, typically been regarded as the very antithesis of the romantic.[1] The moral theory most often ascribed to him harks back to Hobbes: ethical propositions are presumed to be merely flimflam for efforts at selfish aggrandizement. Michael Ghiselin offered a self-revelatory example of this assumption when he asserted that, in the Darwinian view, "an 'altruistic' act is really a form of ultimate self-interest" (Ghiselin 1973, p. 967). This perception of Darwin's moral theory flowed from a further presumption, namely, that he eliminated from nature the kinds of values that naturalists of an earlier generation had thought secreted therein by a beneficent God. Human character, having a perfectly mundane origin, would thus reflect the features of deracinated nature. As one recent analyst of the *Origin* put it: "Darwin does not moralize when describing nature. . . . Nature is harsh and strictly amoral. Notions of good and evil do not exist in the natural world" (Bulhof 1992, p. 95). And why was Darwinian nature indifferent to human moral aspirations? Well, because "nature brings forth her products without ever being distracted from her task, untiring, without regard for persons, blind and cruel – in one word [*sic*]: machine-like, mechanical" (Bulhof 1992, p. 85).[2]

That conception of Darwinism and its implications is vintage, though more like strong vinegar than fine wine. At the beginning of this century, George Bernard Shaw sounded the once and future theme in the preface to his play *Back to Methuselah,* wherein he disclosed what he thought to be the real meaning of Darwinian nature: "When its whole significance dawns upon you, your heart sinks into a heap of sand within you. There is a hideous fatalism about it, a ghastly and damnable reduction of beauty and intelligence, of strength and purpose, of honor and aspiration." "If it could be proven that the whole universe had been produced by [Natural] Selection," he thought "only fools and rascals could bear to live" (Shaw 1961, pp. 33–4, 44). The singular difference between Shaw's and Ghiselin's assessments amounts to this, that the dramatist abhors what he takes to be the Darwinian construction of nature and morals, whereas the scientist revels in it.

In this essay I shall try to demonstrate that the usual interpretation of Darwinian nature is quite mistaken, that Darwin's conception of nature was derived, via various channels, in large measure from the romantic movement and that consequently his theory functioned not to suck values out of nature but to recover them for a detheologized nature. I shall attempt to show that his romantic conception led him to attribute to human beings a moral conscience that sought not selfish advantage but rather sought to respond altruistically to the needs of others. Darwin's evolutionary ethics, I shall finally suggest (just short of arguing), meets the usual challenges of metaethical analysis and recommends itself as a justified moral theory.

THE ROMANTIC MOVEMENT

During the 1790s in Germany, particularly at Jena and Berlin, like-minded thinkers, by design and by chance, fell in together, and through an indefinable magic their various conceptions and personalities mutually exerted strongly attractive romantic forces – and later, repulsive, but always decidedly romantic forces. Friedrich von Schlegel (1772–1829) and his brother August Wilhelm von Schlegel (1767–1845), literary historians and critics, formed the galvanizing pole for the activity of that circle of friends, which at various times included the following: their quite remarkable wives, respectively Dorothea Mendelssohn Veit Schlegel (1763–1839), daughter of the Kantian philosopher Moses Mendelssohn, and Caroline Michaelis Böhmer Schlegel (1763–1809), who later added the surname Schelling; the poets Friedrich Leopold von Hardenberg (1772–1801), writing under the name "Novalis," and Ludwig Tieck (1773–1853), a man of infinite jest and linguistic celerity; the great idealist philosopher Friedrich Wilhelm Joseph von Schelling (1775–1854), the very model of the romantic thinker; and the theologian Friedrich Schleiermacher (1768–1834), whose views would gently sway the metaphysical notions of Ernst Haeckel (1834–1919). They often gathered at Wilhelm and Caroline's home in Jena to "symphilosophize" and "sympoetize," as Friedrich Schlegel liked to put it. Their circle also embraced, or was just tangent to, the ideas and personalities of several other major thinkers of the period. Johann Gottlieb Fichte (1762–1814) was a friend of Friedrich Schlegel and initial mentor to Schelling; the transcendental idealism developed by all three provided the metaphysical and epistemological foundation for the romantic movement. The critical work of the great poet Friedrich von Schiller (1759–1805), who also taught at Jena, helped form, out of Kantian resources, the central aesthetic doctrines of the group. And that guiding star from which all concerned took their bearings, Johann Wolfgang von Goethe (1749–1832), supplied the morphological proposals that gave fundamental direction to nineteenth-century evolutionary theory.

Limitations of both space and readers' patience prevent a detailing of the specific ideas that constituted the intellectual heritage of the romantic movement.[3] Let me here simply provide a very general overview of the conceptions that sprang from the heads and hearts of those figures, so as to provide some notion of the river of resources from which Darwin drew.

The kind of scientific perspective that emerged in Germany, particularly in Jena, Weimar, Halle, and Berlin, during the period 1790–1810 usually bears the names *Naturphilosophie* and "romantic science." Initially, though, it will be helpful to distinguish the somewhat different phenomena that careful usage of these terms would demand, for these designations are not the historian's confections but were introduced by German authors at the end of the eighteenth century for specific semantic purposes. Based on their respective histories and the constituent ideas with which they were associated, I wish to suggest that romantic biology be taken as a species of the wider genus of German nature-philosophy. Thus all romantic biologists were *Naturphilosophen,* but not all *Naturphilosophen* were romantics, at least in the way I propose more singularly to use the terms. Romantic thinkers added to the content of ideas traveling under the rubric of *Naturphilosophie* aesthetic and moral elements. Friedrich Schlegel, who coined the term *romantisch,* used it to indicate a specific kind of poetic and morally valued literature. I shall use the term to distinguish a type of science that retained this aesthetic and moral heritage.

Naturphilosophie

Many of the main figures of early-nineteenth-century German biology – Carl Gustav Carus (1789–1869), Lorenz Oken (1779–1851), Karl Burdach (1776–1847), Ignaz Döllinger (1770–1841), and Karl Ernst von Baer (1792–1876), for instance – adopted a conception of the unity of natural types that had its foundations in the philosophy of nature developed by Kant, and especially by Schelling, and in the anatomical and aesthetic studies of Goethe.[4] "Archetype" theory, as it became known, regarded living nature as exhibiting fundamental organic types (*archetypi, Urtypen, Haupttypen, Urbilden,* etc.). In France, Étienne Geoffroy Saint-Hilaire (1772–1844) and Georges Cuvier (1769–1832), both of whom knew the German literature, also assumed this doctrine of unity of type, though without connections to the idealist metaphysics that underlay the German version of the theory. Cuvier distinguished what he took to be the four most basic animal structures – radiata (e.g., starfish and medusae), articulata (e.g., insects and crabs), mollusca (e.g., clams and octopuses), and vertebrata (e.g., fish and human beings) – and by the 1830s these types had become canonical. Both Darwin and Haeckel, for instance, accepted the idea that nature exhibited such structural patterns; their task, of course, was to account for them. The archetype of, say, the vertebrata formed the *Bauplan* for backboned animals. It had ideal reality, according to Schelling, and formed the metaphysical ground for the infinite varieties it potentially contained. Through application of the ideal type as an investigative standard, a researcher would be able to understand the intimate physical logic of organic development. As Goethe put it, "the creative force produced and developed a more complete organic nature according to a common scheme; and in light of this archetype (*Urbild*), which is perceived only by the mind and not the senses, we might work out, as by a norm, our descriptions" (Goethe 1954, p. 189). These archetypes, as realized in particular organisms, were thus understood to exfoliate extraordinary subtype variations. For instance, the articulata bulged with numerous classes of insects – and species of beetles beyond reckoning. These types also exhibited a progressive hierarchy – so, for instance, within the vertebrata, horses were thought more progressively developed organisms than fish. The *Naturphilosophen* usually invoked special causal forces to explain the instantiation of archetypes and their progressive variations. These scientists and philosophers, however, did not consider such forces incompatible with more mundane physical powers; rather, the forces often were conceived as special transformations of physical powers (e.g., Schelling's polar forces) or emergent out of them (e.g., *Lebenskraft, Bildungstrieb,* natural selection).[5]

Since the *Naturphilosophen* adopted the metaphysical position of monism, in which matter and *Geist* (understood indifferently as mind or spirit) were regarded as two features of the same underlying *Urstoff,* the causal activities of either had ultimately to express a unified force. The natural world – with its various organic types and their vertiginous varieties, its underlying substantial unity, and its particular forces – displayed, it was thought, higher-ordered patterns. Thus, particular types of vegetation might be found at comparable latitudes and altitudes on the different continents, and similar animal forms would be associated with similar plant forms in the New World and Old World. Nature, to use Alexander von Humboldt's term, was a *cosmos* – a harmoniously unified network of integrally functioning parts.

The *Naturphilosophen* commonly thought individual organisms and nature as a whole to be teleologically structured. In the German context this did not mean what it was taken to mean in the British. Kant argued that we had to understand biological organisms *as if* they had been designed so that disparate parts functioned reciprocally as means and ends and together contributed to the well-being of the entire individual. The biological researcher had to understand, for example, the functioning of a mammal *as if* its heart contracted for the purpose of circulating the blood, and *as if* its blood circulated for the purpose of supplying vital elements to the parts, including the heart, so as to maintain the whole organism. In this way the researcher would treat organisms teleologically – something necessitated, Kant argued, by reason of the epistemological situation, not by nature. After Kant, and especially because of the influence of Goethe and Schelling, biologists came to hold the teleological structure of nature not simply *as if* but as intrinsic: nature, whether in the individual or at large, really was purposively organized. But the *Naturphilosophen,* unlike the British natural theologians, did not appeal to a separate creator who imposed final order on recalcitrant matter. Rather, they conceived nature in Spinozistic fashion – it was *Deus sive Natura:* God and nature were one. This meant that the teleological structuring of biological organisms modeled the conceptual structuring of the ideas in terms of which nature was understood. Evolutionary theory would happily accommodate this kind of teleology.

Robert J. Richards

Romantic Biology

Those German scientists to whom I shall refer as romantic biologists – for instance, Schelling, Goethe, Humboldt, Burdach, Oken, Carus, and Haeckel – generally accepted the metaphysical and epistemological propositions of *Naturphilosophie*. They took more to heart, however, Kant's analysis of the logical similarity between teleological judgment and aesthetic judgment, which he developed in the *Kritik der Urteilskraft*. Romantic biologists came to regard these kinds of judgment as complementary approaches to nature, approaches that penetrated to the same underlying structures. This meant artistic experience and expression might operate in harmony with scientific experience and expression: the laws of nature might also be apprehended and represented by the artist's painting or the poet's metaphor. Further, romantic biologists maintained, sometimes explicitly, often implicitly, that the aesthetic comprehension of the entire organism or of the whole interacting natural environment would be a necessary preliminary stage in the scientific analysis of respective parts: both in art and in science, comprehension of the whole had to precede that of the parts. Initially for the biologist, then, ineffable aesthetic experience had to open the way to articulate scientific understanding. And for the reader, this meant that the sketches, drawings, figures, and metaphors that graced biological monographs could not be relegated to the status of dispensable pedagogical aids: images carried a scientific content often impossible to render precisely in words. Art became thus employed in the logic of scientific demonstration. And with Darwin, as I shall attempt to show, the artful use of metaphor gave structure to a conception of nature far different from that which modern neo-Darwinists employ.

Romantic thinkers considered the activities of the scientist comparable to those of the artist: both employed creative imagination. And when addressing nature, both found in their object a source of similar creativity. Nature's forms – various, unexpected, delightful, but exhibiting a deep unity – had to be regarded as creative expressions as well. Romantic biologists thus concluded that in nature herself "from so simple a beginning endless forms most beautiful and most wonderful have been, and are being, evolved."

Kant also uncovered a deep logical similarity between the structure of aesthetic judgment and that of moral judgment: in each we render a judgment on the basis of mere form (i.e., in respect to the structure of a beautiful object or the structure of a moral maxim); and each such judgment produces a certain feeling, which is presumed valid for everyone. So, according to Kant, when evaluating a painting as beautiful or an act as moral, we simultaneously demand that others reach the same conclusion and experience the same feelings toward the object or act. These isomorphisms of judgment implied that the scientific and aesthetic comprehension of nature also involved a moral component. After Kant, the precise connections among scientific, aesthetic, and moral judgments received various constructions. But for all those whom I am calling romantic biologists, the scientific understanding of nature led to moral evaluation. Romantic biologists thus understood nature to be the repository not only of lawful regularities and aesthetic delights but also of moral structures.

These general definitions of *Naturphilosophie* and "romantic biology" lack, of course, specificity and a context that would make them more intelligible and precise. They need to be rewoven within the mesh of the actual history that gave them substance. I am attempting that in another publication, but perhaps this general scheme will have sufficient integrity to pin down Darwinian theory so that its surface can be cut away to reveal its romantic heart. I shall, however, devote some space to a piece of the history of early romanticism, because it formed a direct link between the ideas of the romantic movement and Darwin's conception of nature.

Alexander von Humboldt

In the late 1790s, Alexander von Humboldt (1769–1859), along with his brother Wilhelm (1767–1835), spent considerable time in Jena involved with a circle intersecting that of the Jena romantics. His more intimate configuration included Goethe and Schiller, but he knew the Schlegels and Schelling quite well. Later he would write Schelling of his admiration for the philosopher's conception of nature. Humboldt imbibed a particularly heady atmosphere of philosophy, science, and poetry while at Jena, and these helped form his scientific essays and books into both resources for the Jena romantics and, indeed, representations of the particular kind of science emanating from that small German redoubt.

In 1799 Humboldt began the first leg of a journey for which he had been preparing for several years. He and his friend Aimé Bonpland set off for the New World to engage in research that would range over botany, zoology, anthropology, geology, and meteorology.[6] Like Darwin's own adventure, Humboldt's trip lasted five years, though his itinerary differed a bit. His travels took him through the northern parts of South America, the Central Americas, and finally through to North America and a visit to Thomas Jefferson in Washington. And like Darwin's, his voyage yielded several volumes of specialized zoological and botanical studies, a geological monograph (and others on astronomy and anthropology), as well as a large book describing his travels and researches, his *Personal Narrative of Travels to the Equinoctial Regions of the New Continent, during the Years 1799–1804*. The *Personal*

Narrative, like the travel book of his friend Georg Forster (1754–1794) –
who with his father accompanied Captain Cook on his second voyage across
the Pacific – takes the reader chronologically on a journey of adventure and
science, but Humboldt brought to fruition what were only suggestions in the
work of Forster, for Humboldt's volumes also incorporated meteorological
data, geological measurements, zoological observations, anthropological ex-
aminations, and comprehensive studies of the biogeography of plants.

Humboldt, like Goethe, thought that plants exhibited different fundamental forms, of which he identified some sixteen. He regarded the different
species and subspecies of plants as playing variations on these constant
themes. The same types, he thought, would be found in similar environmental conditions (of temperature, pressure, moisture, etc.) across the globe. For
just as common geological and mineral formations displayed comparable
patterns in different, far-flung locations, so too did plants:

> All formations are, therefore, common to every quarter of the globe and as-
> sume the like forms. Everywhere basalt rises in twin mountains and truncated
> cones; everywhere trap-porphyry presents itself to the eye under the form of
> grotesquely shaped masses of rock, while granite terminates in gently rounded
> summits. Thus, too, similar vegetable forms, as pines and oaks, alike crown
> the mountain declivities of Sweden and those of the most southern portion of
> Mexico. [Humboldt 1850, p. 218][7]

The discrimination of form, according to Humboldt, was an aesthetic task,
not a classification done according to the usual criteria of botanical systems,
such as that of Linnaeus. Rather, one should be guided by the painterly eye,
which would highlight the distribution of leaves, the forms of stems and
branches, the height and breadth of the entire plant. Moreover, the geographical distribution of plants depended on surrounding organisms; certain
types usually occurred together in assemblages – palm and banana trees, for
example, were always found in the same places. The assemblages would have
distinctive characters that would give a geographical location its individual
feel. "Swiss scenery" or "Italian sky" were produced by distinctive combi-
nations of elements: "The azure of the sky, the effects of light and shade, the
haze floating on the distant horizon, the forms of animals, the succulence of
plants, the bright glossy surface of the leaves, the outlines of mountains, all
combine to produce the elements on which depends the impression of any
one region" (Humboldt 1850, pp. 217–18).

Humboldt understood all of nature, and the laws governing her, to unite
in a vast complex of interrelationships – "a Cosmos, or harmoniously ordered
whole, which, dimly shadowed forth to the human mind in the primitive ages
of the world, is now fully revealed to the maturer intellect of mankind as the

result of long and laborious observations" (Humboldt n.d., vol. 1, p. 24).[8] Not only did this whole of nature reveal itself through the scientific instruments that Humboldt carried with him into the jungles along the Orinoco River in the Amazon and to the biological, geological, and meteorological studies he compiled in vast statistical tables – but nature also revealed her most intimate self, her deepest lawful relations, to his aesthetic perceptions. "Everywhere," he exclaimed, "the mind is penetrated by the same sense of the grandeur and vast expanse of nature, revealing to the soul, by a mysterious inspiration, the existence of laws that regulate the forces of the universe" (Humboldt n.d., vol. 1, p. 25). It was this sort of aesthetic penetration into nature that led Darwin to exclaim about *Cosmos* that it provided "a grand coup d'oeil of the whole universe" (Darwin 1985–, vol. 4, p. 135).

Humboldt had intended that a reader like Darwin should have exactly this kind of complete engagement with his text, that the reader should respond not only to the science conceptually articulated therein but also to the aesthetic display of nature's other face. Humboldt made this intention explicit in *Cosmos,* where he argued that the natural historian had the duty not only to present the scientifically comprehensible aspects of nature but also to re-create in the reader aesthetic experiences – through artful language – of the sort the naturalist had himself undergone in his immediate encounters with nature. A worthy naturalist, Humboldt thought, ought not leave any means "unemployed by which an animated picture of a distant zone, untraversed by ourselves, may be presented to the mind with all the vividness of truth, enabling us even to enjoy some portion of the pleasure derived from the immediate contact with nature." A poetic natural historian, in this romantic mold, ought to utilize appropriate language to make palpable to an audience what had emanated "from the intuitive perception of the connection [in the naturalist's mind] between the sensuous and the intellectual, and of the universality and reciprocal limitation and unity of all the vital forces in nature" (Humboldt n.d., vol. 2, p. 81).[9] As we shall see, Darwin's own prose, which vibrated with poetic appreciation of nature's inner core, had the comparable end: to deliver to the reader an aesthetic assessment that lay beyond the scientifically articulable.[10]

DARWIN'S ROMANTIC CONCEPTION OF NATURE

Few would deny, I think, that Darwin's perception of nature, his feeling of its lived reality, his understanding of the connections between organic and geological phenomena achieved critical form during the five years of his *Beagle* voyage. But it would be a mistake to believe that Darwin, like a cheerful

Kurtz, went into the jungle and viewed his surroundings with naked and be-wildered eyes, simply registering the beauty and mystery of the flora and fauna and the exotic behavior of its human inhabitants. His letter to Humboldt expressed a deep truth, which an examination of his writing from this period confirms: Darwin experienced the South American environment, the interconnectedness of its various aspects, the sublimity of its scenes, and the moral behavior of its peoples – all filtered through a Humboldtian discourse on these very subjects.

While at Cambridge in the years 1828–31, aside from dinner parties, fox hunting, beetle collecting, and the occasional study of Euclid and Paley – certainly preparation suitable enough for the life of the country parson that he planned – Darwin did spend time on more serious, scientific pursuits. In the lectures on botany he took from John Henslow (1796–1861), as well as over tea at the professor's house, the young student was introduced to questions stimulated by German biologists concerning the vital forces that distinguished the organic from the inorganic. As Phillip Sloan has meticulously demonstrated, Darwin became intrigued by the controversy over whether vital force, as Henslow thought, came extrinsically to potentially living matter or whether such force, as the Germans contended, was intrinsic to animate organization (Sloan 1986, pp. 369–445). The role of vital force became poignant for the young Darwin as he attempted to puzzle out the nature of pollen granules and their relation to similar substances found in simple invertebrates. These questions, which whetted the sensibilities of the yet green undergraduate, would be pursued while on the *Beagle* voyage, and most assiduously thereafter, as Darwin began working out his hypotheses about species change.

But these more minute research topics for the novice played only circumspectly in the background of his concerns. During Darwin's first year at Cambridge, he began reading Humboldt's *Personal Narrative*.[11] That young German adventurer knew how to compose a tale that would entice a medley of readers (Haeckel also among them), but particularly the young and those whose bones ached, like Humboldt's own, for escape from suffocating surroundings. And if a reader, say one cloistered in the decaying traditions of an ancient university, had yet discovered a taste for nature and huddled against the warming desire to do something in a scientific mode – a desire stoked, perhaps, by the low expectations of a famous father – then could he fail to escape in his imagination with this polymathic romantic? Even the twentieth-century reader, of requisite disposition, might feel a stir on reading what Darwin read:

> If America occupies no important place in the history of mankind, and of the ancient revolutions which have agitated the human race, it offers an ample

field to the labours of the naturalist. On no other part of the Globe is he called upon more powerfully by nature, to raise himself to general ideas on the cause of the phenomena, and their natural connection. I shall not speak of that luxuriance of vegetation, that eternal spring of organic life, those climates varying by stages as we climb the flanks of the Cordilleras, and those majestic rivers which a celebrated writer [Chateaubriand] has described with so much graceful precision. The means which the new world affords for the study of geology and natural philosophy in general are long since acknowledged. Happy the traveler who is conscious, that he has availed himself of the advantages of his position, and that he has added some new facts to the mass of those which were already acquired! [Humboldt 1818–29, vol. 1, pp. xlv–xlvi][12]

So taken with Humboldt's tale was Darwin that he copied out long passages to read aloud to Henslow (Darwin 1969, pp. 67–8). Like Humboldt, Darwin in his last year at Cambridge was of an age, twenty-one, "when life appears an unlimited horizon," an age "when we find an irresistible attraction in the impetuous agitations of the mind, and the image of positive danger" (Humboldt 1818–29, vol. 1, p. 3). Quickly the young Englishman became thrall to the idea of undertaking a similar trip of adventure, if not all the way to the New World, then at least to the Canary Islands, with an active volcano and ancient and magnificent dragon tree, all of which Humboldt had so richly described.

Darwin, of course, got his chance. He shipped out on the *Beagle* on 27 December 1831, carrying among his supplies a small library, including Humboldt's *Personal Narrative,* which had been given him as a parting gift by Henslow.[13] Ironically the ship had to bypass the Canaries because of quarantine restrictions. The *Beagle* finally reached the Brazilian coast on 28 February, much to Darwin's great relief, for his seasickness in transit had been monumental. During the voyage, when not retching, Darwin read from his Humboldt. And just after landing, he quickly wrote his father of the exciting new land and of the proper way to appreciate it: "If you really want to have a notion of tropical countries, study Humboldt. – Skip the scientific parts & commence after leaving Teneriffe. – My feelings amount to admiration the more I read him" (Darwin 1985–, p. 204).

Though Darwin would pay close attention to the kinds of scientific studies Humboldt had undertaken – and keep the sort of notes that would enable him later to produce a travel book precisely modeled after Humboldt's narrative – it was undoubtedly the sensuous and imaginative descriptions to which Darwin most immediately responded. Shortly after landing in Bahia, Darwin took to the jungle. When he first entered under the emerald canopy, he underwent that experience bathed in a soft Humboldtian light. As he wrote in his diary on the day of disembarking,

I believe from what I have seen Humboldts glorious descriptions are & will for ever be unparalleled: but even he with his dark blue skies & the rare union of poetry with science which he so strongly displays when writing on tropical scenery, with all this falls far short of the truth. The delight on experiences in such times bewilders the mind. . . . The mind is a chaos of delight, out of which a world of future & more quiet pleasure will arise. – I am at present fit only to read Humboldt; he like another Sun illumines everything I behold. [Darwin 1988, p. 42]

Throughout the diary that Darwin kept on his voyage, entry after entry invokes the name of Humboldt on such varied subjects as the passing of time in the tropics, the profusion and forms of vegetation, the constellations of the southern sky, volcanic formations, the color of landscape through the softening effects of the atmosphere, sickness in the tropics, the ingressions of Christianity, mountaineering, biogeographical relationships, and much more.[14] All of these subjects, braced by the authority of his German predecessor, Darwin lightly transformed into the more public record of his voyage, his *Journal of Researches into the Geology and Natural History of the Various Countries Visited by H.M.S. Beagle* (1839). But it wasn't simply Humboldt's individual descriptions or scientific calculations that arrested the young naturalist's attention. His very experience passed through the lens provided by Humboldt.

In September of 1836, during the long voyage home, Darwin had opportunity to leaf through his diary and reflect on his experiences. He judged that Humboldt had provided the mold that had shaped his observations, his perceptions, and his emotional reactions to the many tableaux that had passed before him. He recorded in his diary this judgment and then transferred it virtually unchanged into the *Journal of Researches:* "As the force of impression frequently depends on preconceived ideas, I may add that all mine were taken from the vivid descriptions in the Personal Narrative which far exceed in merit anything I have ever read on the subject" (Darwin 1988, p. 443). The nature that Darwin experienced with the aid of Humboldt was not a machine, a contrivance of fixed parts grinding out its products with dispassionate consequence. The nature that Darwin experienced was a cosmos, in which organic patterns of land, climate, vegetation, animals, and man were woven into a vast web pulsating with life. Darwin's aesthesized experience, rendered in Humboldtian terms, delivered to him a vision of nature and man that would subtly form his later theory, a vision that can be glimpsed in a concluding passage of his diary:

Among the scenes which are deeply impressed on my mind, none exceed in sublimity the primeval forests, undefaced by the hand of man, whether those of Brazil, where the powers of life are predominant, or those of Tierra del Fuego,

where death & decay prevail. Both are temples filled with the varied productions of the God of Nature: – No one can stand unmoved in these solitudes,
without feeling that there is more in man than the mere breath of his body.
[Darwin 1988, p. 444]

These sentiments indicate that by the end of Darwin's voyage, the God of
Paley had begun to give way to the God of nature, soon to be completely
transmogrified into the romantics' *Deus sive Natura.*

After Darwin had published his *Journal of Researches,* he then, with extreme hesitation and trepidation, sent a copy to Humboldt himself. The great
man responded with a generosity and perceptiveness that leaped beyond any
of Darwin's most secret hopes. Obviously Humboldt had no reason to flat-
ter Darwin, for by that time Humboldt had achieved an eminence in Europe
exceeded by no other naturalist – but he found in Darwin a kindred spirit
(Darwin 1985–, vol. 2, pp. 218–22). Indeed, Darwin's aesthetic sensitivity
and expressiveness made an indelible impression on that Nestor of science.
In *Kosmos,* published almost a decade after Darwin's *Journal,* Humboldt's
admiration had not abated. He likened the Englishman's "aesthetic feelings"
and "vividly fresh images" to those of his own great friend Georg Forster,
who first opened the way to "a new era of scientific travel and who aimed at
a comparative study of peoples and lands" (Humboldt 1845–58, vol. 2, p. 72;
n.d., vol. 2, p. 80).[15]

It might yet be thought that the influence of Humboldtian romanticism
on Darwin was of only slight significance, that his fundamental conception
of nature, at least as presented in the *Origin of Species* and in the *Descent of
Man,* had shuffled off metaphysical and preciously aesthetic decorations, so
that only the cold machinery of the living world might lie exposed. Needless
to say, I do not think this conception of Darwin's accomplishment to be
sound. Before considering directly his evolutionary ethics, as completed in
the *Descent of Man,* I would like briefly to examine the character of the nature that Darwin portrayed in the *Origin of Species,* because his own tacit
evaluations of that larger nature support his analyses of the ethical dimensions of that smaller nature, man. I shall touch on three aspects of his representation: the archetypal unity displayed by nature, her organic and non-
machine-like aspect, and finally her moral features.

ROMANTIC NATURE IN THE *ORIGIN OF SPECIES*

Theory of the Archetype

Archetype theory became a central doctrine for German romantic writers.
Schelling, Goethe, Oken, and Carus, along with such *Naturphilosophen* as

von Baer, developed the idea that nature exhibited fundamental unities, of which species and individuals played out the variations. In England, the theory was embraced by Joseph Henry Green (1791–1863), Hunterian lecturer at the Royal College of Surgeons, and most famously by Green's protégé, Richard Owen (1804–1892). Green had studied in Germany, and like Schelling and Goethe, both of whom he assiduously read, he argued that nature exhibited basic patterns or rational ideals. These ideals, as existing in the mind of God, served as "archetypes and preexisting models," but as "acts of the Divine Will manifested in nature," they functioned as "laws" (Green 1840, pp. xxv–xxvi).[16] Quite in conformity to the "objective idealism of Schelling," Green construed an archetype as

> a causative principle, combining both power and intelligence, containing, pre-determining and producing its actual result in all its manifold relations, in reference to a final purpose; and realized in a whole of parts, in which the Idea, as the constitutive energy, is evolved and set forth in its unity, totality, finality, and permanent efficiency. [Green 1840, p. xxv]

As in the theories of Schelling, Kielmeyer, and other German biologists at the beginning of the nineteenth century, Green's archetype theory included the notion that individual organisms higher in the scale of development recapitulated the same forms achieved during the evolution of those lower in the hierarchy of animals (Green 1840, pp. 39–40).[17]

Richard Owen, perhaps the most influential biologist in England during the first half of the century, likewise followed the German lead in constructing his own, quite celebrated, theory of the archetype. In his *Report on the Archetype* (1847) and *On the Nature of Limbs* (1849), Owen turned his analytical attention to the osteological patterns that lay at the foundation of the vertebrate skeleton. The archetype of the vertebrata, in Owen's construction, was simply a string of vertebrae. According to his theory, different vertebrate skeletons manifested modifications of this basic plan. So, for instance, the bones of the head would be regarded as a development of the several anterior vertebrae, and the ribs, pelvis, and limbs as developments of different processes of more caudal vertebrae. This conception allowed the anatomist to compare the "same" bones in different species or in different higher taxa. Thus the bones in the human hand, in the wing of a bat, and in the paddle of a porpoise would be considered homologous, that is, bones having, as Owen put it in his Germanophilic way, the same *Bedeutung* or meaning (Owen 1849, pp. 2–3). Such bones might be adapted to particular purposes (e.g., tool use, flying, or swimming), but their structure would express their common nature.

Owen drew particularly on the work of Oken and Carus in constructing his theory of the archetype. Like these romantic anatomists and his mentor

Green, Owen thought of the archetype as more than merely a standard by which to conduct comparative zoology. For him, the archetype reared up as a vital force in nature. In the metaphysical considerations occurring at the end of his *Report on the Archetype,* Owen made a distinction between two opposing vital forces: "besides the ιδεα, organizing principle, vital property, or force, which produces the diversity of form belonging to living bodies of the same materials . . . there appears also to be in counter-operation during the building up of such bodies the polarizing force pervading all space." The Platonic 'idea' operated to establish the species-specific form impressed on organic matter, whereas the polarizing force, which he (a bit confusingly) identified with the archetype, constrained activity to produce similarity of form among species (general homology) and repetition of parts within species (serial homology) (Owen 1847, pp. 339–40). Perhaps because of the pantheistic implications of two oppositional forces creating living beings, three years later, in his little book *On Limbs,* Owen collapsed the vital forces into one, "answering to the 'idea' of the Archetypal World in the Platonic cosmogony, which archetype or primal pattern is the basis supporting all the modifications of such part [as the limb] for specific powers and actions in all animals possessing it" (Owen 1849, pp. 2–3). Construing the archetype as a Platonic 'idea' presumably made easier its identification with an idea in the mind of the creator God, and thus also easier for Owen to mute any suspicions that he allowed nature herself to fashion creatures from her own resources (Rupke 1994, pp. 196–204). Such suspicions, though, could not be quelled by a subtle alteration of arcane metaphysics, especially when Owen himself invited the idea that he had, in Germanic fashion, attributed to nature herself powers of development. At the conclusion of *On the Nature of Limbs,* he perorated as follows:

> To what natural laws or secondary causes the orderly succession and progression of such organic phaenomena may have been committed we as yet are ignorant. But if, without derogation of the Divine power, we may conceive the existence of such ministers, and personify them by the term 'Nature,' we learn from the past history of our globe that she has advanced with slow and stately steps, guided by the archetypal light, amidst the wreck of worlds, from the first embodiment of the Vertebrate idea under its old Ichthyic vestment, until it became arrayed in the glorious garb of the Human form. [Owen 1849, p. 86]

Orthodox readers of this passage perceived that Owen trod gingerly along a dangerous pantheistic path laid by German romantics who had promulgated the doctrines of naturalistic development and transformation. Those guardians of theological rectitude became quite concerned for Owen's welfare and that of naive readers.[18] The less orthodox discovered in Owen's theory of the archetype a comparable message, but felt considerably more sanguine about it.

Owen himself reacted to charges of pantheism with stiffened indignation. And though he apparently did harbor a notion of the gradual appearance of new species through a kind of progressive transformation, his oblique theory seems to have been more in harmony with Schelling's conception, that is, a transformational theory, but not a genealogical one. Natural powers (directed by God, of course) might introduce new species according to a plan that fossils – those archetypal vestiges – disclosed, but such species would not evolve genealogically out of one another. The genealogical view, Owen thought, was characteristic of the German writers of the "transcendental school." In his Hunterian Lecture of 1837, he expressly denied the two genealogical doctrines he saw linked in painful excess: a theory of embryonic recapitulation and a theory of species transformation. In regard to anyone who held to the notion that the embryos of more developed organisms went through the morphological stages of those lower in the line of development, Owen believed that such a foolish naturalist would succumb immediately to the even more pernicious idea that during the past history of the globe, species likewise would have developed out of one another: "The doctrine of Transmutation of forms during the Embryonal phases," he cautioned, "is closely allied to that still more objectionable one, the transmutation of Species." He thought both propositions would be "crushed in an instant when disrobed of the figurative expressions in which they are often enveloped; and examined by the light of a severe logic" (Owen 1992, p. 92). Well, one of Owen's friends did examine these ideas, naked and unadorned by the language of transcendentalism, and did not blush to apply a creative logic that confirmed both of the notions.

Darwin had been working on a theory of development since shortly after his return from the *Beagle* voyage. In the initial entry of his first transformation notebook, this under-laborer in the natural history field made the precise linkage that Owen had execrated. On the very first page of his "Notebook B," begun in July 1837, a short time after Owen's lecture series, Darwin considered two kinds of generation. One sort produced identical individuals asexually through budding or division. The other sort was the "ordinary kind," as he termed it, "the new individual passing through several stages (typical, [of the] or shortened repetition of what the original molecule has done)."[19] Darwin here proposed that during ontogeny the individual would pass through the same morphological stages, by way of a shortened repetition, that the first living speck of life went through in its evolutionary trajectory. He continued to work out the details of this very German idea of evolutionary recapitulation through the period leading up to the composition of the *Origin of Species.*[20]

Darwin's friend and future antagonist Owen had initially brought several objections to recapitulation theory. In conformity to von Baer's similar strictures, Owen contended that the fetus of a more developed animal never displayed, during gestation, the individual adult form of lower animals. Von Baer had epitomized this objection by postulating that fetal development passed from a more general morphological state (e.g., the basic mammal form) to the more specialized structures of the species and particular individual. Hence the fetus at early stages of development could represent only generalized animal patterns, not the adult forms of definite, primitive species. Darwin worked through objections of this sort in the many scattered manuscript notes that he left on the theory of recapitulation prior to the publication of the *Origin.* Perhaps his most delicious rejoinder came in a note he penciled on the back flyleaf of his copy of Owen's *On the Nature of Limbs.* There Darwin wrote: "I look at Owen's Archetypes as more than idea, as a real representation as far as the most consummate skill & loftiest generalization can represent the parent form of the Vertebrata."[21] Darwin thus suggested that the generalized archetype of the vertebrates did not lie hidden away as an idea in the mind of God; rather, it was the form of a creature that walked the earth many generations ago. Embryos of descendants would consequently pass through the stage of a generalized vertebrate, but that generalized vertebrate would have been a real, individual creature. In the *Origin of Species,* especially in the later editions, Darwin made perfectly clear the central role archetype and recapitulation theory played in his conception of evolution.

As an instance of the molding force exercised by archetype theory on the deep structure of the *Origin,* one might point to Darwin's chronic presumption that transformation of species occurred without common descent. In his essay of 1844, for example, he assumed that evolution would occur within the archetypes of, say, the articulata, radiata, mollusca, and vertebrata, but allowed that no common ancestor would be found for these branches of the animal kingdom. He thus concluded that for the animal and plant kingdoms, "all the organisms yet discovered are descendants of probably less than ten parent forms" (Darwin 1909, p. 252). In the *Origin,* Darwin advanced the same conviction that "animals have descended from at most only four or five progenitors, and plants from an equal or lesser number." This hypothesis, he maintained, was sufficient for his general theory. He did venture, however, that analogy suggested that "probably all the organic beings which have ever lived on this earth have descended from some one primordial form, into which life was first breathed" (Darwin 1859, p. 484).

The principle of recapitulation sank equally deep into the foundations of Darwin's theory. Of the many expressions of the principle in the *Origin,* perhaps the most straightforward comes toward the end of the penultimate chapter of the final edition: "As the embryo often shows us more or less plainly the structure of the less modified and ancient progenitor of the group, we can see why ancient and extinct forms so often resemble in their adult state the embryos of existing species of the same class" (Darwin 1959, p. 704).

Thus central and captivating blooms of Darwin's theory of evolution opened from ideas initially cultivated in romantic *Naturphilosophie*. But these intricate parts of the composition could not have taken their form if the very root of his theory, his idea of nature, had not also drawn vitality from that ever-fertile soil.

Nature as Organic and Value-Laden

No phrase comes so trippingly to the lips of contemporary biologists as "the *mechanism* of natural selection." Almost reflexively we think of natural selection, paradoxically, in nonorganic terms. And the nature that selection creates, that too appears to the contemporary scientist as something best fixed, at least in principle, in the hard, mechanistic language of chemistry and physics. Yet the belief that nature was nothing but a vast machine and that natural selection operated according to mechanical principles – all of that remained quite distant from the mind that originally composed the *Origin of Species*. Darwin never referred to or conceived of natural selection as a mechanistic principle, and the nature to which selection gave rise was perceived in its parts and in the whole as a teleologically self-organizing structure.[22]

The most colorful instance of Darwin considering the integral operations of the web of life – the mutual ends–means relationships existing among different organisms – occurred when he pondered, almost as a whimsical question, how cats in a district might control the growth of red clover (i.e., cats prey on mice, who destroy the nests of the humble bees that pollinate the clover) (Darwin 1859, p. 73). Life, to use his very rich metaphor, could be compared to an entangled bank in which bushes, plants, snails, and birds all exerted finely balanced forces upon each other to produce "the proportional numbers and kinds" of life-forms exhibited in the complex. These were the sorts of entangling relationships that Humboldt insisted upon and that Darwin, under the genial guidance of his predecessor, had described in great detail in his *Journal of Researches*. The deeply nonmechanistic character of Darwin's theory is, however, most conspicuously displayed in the principle of natural selection itself.

Consider how Darwin compared in the *Origin* the selection practiced by human beings and the selection practiced by nature, natural selection:

Man can act only on external and visible characters: nature cares nothing for appearances, except in so far as they may be useful to any being. She can act on every internal organ, on every shade of constitutional difference, on the whole machinery of life. Man selects only for his own good; Nature only for that of the being which she tends. . . . It may be said that natural selection is daily and hourly scrutinizing, throughout the world, every variation, even the slightest; rejecting that which is bad, preserving and adding up all that is good; silently and insensibly working, whenever and wherever opportunity offers, at the improvement of each organic being in relation to its organic and inorganic conditions of life. [Darwin 1859, pp. 83–4][23]

The productions of nature, Darwin resonately observed, were "far 'truer' in character than man's productions." They plainly bore, he averred, "the stamp of far higher workmanship" (Darwin 1859, p. 84). But did that biblical echo refer to the stamp of a machine? Most assuredly not, as the archaeology of these passages indicates.

Darwin's description here of the activities of natural selection can be easily traced back to the large manuscript – to be called *Natural Selection* – that he abandoned to undertake an abridged version – the *Origin of Species* – after he had received Alfred Russel Wallace's letter outlining a theory close to his own. In that somewhat older version, Darwin described the operations of natural selection this way:

[Man] selects any peculiarity or quality which pleases or is useful to him, regardless whether it profits the being. . . . See how differently Nature acts! She cares not for mere external appearance; she may be said to scrutinize with a severe eye, every nerve, vessel & muscle. . . . Can we wonder then, that nature's productions bear the stamp of a far higher perfection than man's product by artificial selection. With nature the most gradual, steady, unerring, deepsighted selection – perfect adaptation to the conditions of existence. [Darwin 1975, pp. 224–5][24]

These passages, which describe natural selection as peering into the very fabric of a creature, selecting altruistically that which is good and casting out what is bad, a natural selection that operates perfectly (as the earlier manuscript has it) – these passages hardly describe the operations of Locke's spinning jenny or even the clatter and wheeze of a Manchester mechanical loom. When we probe still further back into Darwin's earliest speculations about transformational forces, then we see even more clearly that his conception of natural selection sprang from the head of a divinized nature. His 1844 essay and his preceding 1842 essay, which sketched out his theory in detail for the first time, reveal his model for understanding the "truer" work of natural selection. It is crucially important to understand that in these essays

Darwin came *to explain to himself* the operations of selection using aesthetic devices and that these devices helped him intuitively explore the possibilities of his embryonic idea.

In these early essays, Darwin, struggling through inchoate notions, grasped in metaphor and images the slowly forming structure of natural selection. These images, I believe, distinctly shaped the deep logic of his formulation. In his 1844 essay we can see more clearly the concept in gestation:

> Let us now suppose a Being with penetration sufficient to perceive differences in the outer and innermost organization quite imperceptible to man, and with forethought extending over future centuries to watch with unerring care and select for any object the offspring of an organism produced under the foregoing circumstances; I can see no conceivable reason why he should not form a new race (or several were he to separate the stock of the original organism and work on several islands) adapted to new ends. As we assume his discrimination, and his forethought, and his steadiness of object, to be incomparably greater than those qualities in man, so we may suppose the beauty and complications of the adaptations of the new races and their differences from the original stock to be greater than in the domestic races produced by man's agency. [Darwin 1909, p. 85]

Here, as well as in the 1842 essay, Darwin worked out for himself the character of natural selection, and that character was cast in the image of a divine being, whose "forethought" might teleologically produce creatures of great "beauty" and with progressively intricate "adaptations." Natural selection, in its original, metaphorical conception, was hardly machinelike, but rather godlike.

When Darwin's earlier understanding of natural selection is revealed, we can see more clearly the import of those descendant ideas that animate the *Origin.* In the passages quoted earlier, natural selection is depicted as operating on the least shade of difference in organic life, penetrating to the very core of that life, working unceasingly, intricately, aesthetically, and teleologically. Moreover, natural selection picks out the bad and preserves the good, not selfishly, but altruistically ("Man selects only for his own good; Nature only for that of the being which she tends"). The good that nature fosters, in Darwin's conception, contributes to that ever-growing, morally progressive state that the history of evolution exemplifies and, in the long run, produces, for "as natural selection works solely by and for the good of each being, all corporeal and mental endowments will tend to progress toward perfection" (Darwin 1859, p. 489).

The moral character of nature's actions in regard to her own creations is whispered, as in those passages just mentioned, throughout the *Origin of Species,* the muted notions slipping out of the deeper recesses of Darwin's basic conception of nature. That moral perception, however, led also to Darwin's more explicit recognition of a great paradox, a paradox that had bedeviled generations of theologians – How could a morally good, infinite being produce evil? – for the destruction wrought by selection certainly appeared evil. Darwin, I think, found the key to the solution of this metaphysical conundrum in a quite poetical source.

During his *Beagle* voyage, both on ship and as he traveled into the interior of South America, Darwin always had his copy of Milton's *Paradise Lost* with him. Milton, that favorite of both the German and English romantics, had captured in sublime poetry the solution to the problem of evil. In one passage, as Satan approaches the Garden of Eden, he is stopped by an entangled bank of bushes and undergrowth, not unlike the barriers Darwin constantly met in his own jungle-garden adventures. Milton wrote (*Paradise Lost*, IV, 172–201):

> Now to the ascent of that steep savage hill
> Satan had journeyed on, pensive and slow,
> But further way found none, so thick entwined,
> As one continued brake, the undergrowth
> Of shrubs and tangling bushes had perplexed
> All path of man or beast that passed that way . . .
> Thence up he flew, and on the Tree of Life,
> The middle tree and highest there that grew,
> Sat like a cormorant, yet not true life
> Thereby regained, but sat devising death
> To them who lived, nor on the virtue thought
> Of that life-giving plant, but only used
> For prospect what, well used, had been the pledge
> Of immortality.

With the fall of Adam and Eve, the progenitors of us all, came, however, the happy possibility of our salvation. So through the pain and desolation visited upon this earth by what seems an imperfect law, the Redeemer comes, and with his own death the world is transformed (*Paradise Lost*, XII, 548–51):

> From the conflagrant mass, purged and refined,
> New Heavens, new Earth, Ages of endless date
> Founded in righteousness and peace and love,
> To bring forth fruits, joy and eternal bliss.

This orthodox explanation of evil – that it is only apparent and really is the guise of the good – echoes through Milton's lyrics: out of death and destruction comes life more abundantly, life transformed. And this is exactly the resolution that nature, in Darwin's divinized reconstruction, offers: out of struggle and death comes the greatest perfection, the higher creatures. We should thus be assured, comforts Darwin, that the struggle for existence is yet gentle and for that higher purpose: "When we reflect on this struggle, we may console ourselves with the full belief, that the war of nature is not incessant, that no fear is felt, that death is generally prompt, and that the vigorous, the healthy, and the happy survive and multiply" (Darwin 1859, p. 79). And in the end, the purpose of nature will be fulfilled, the transformation of the lowly and debased into higher beings. In the last paragraph of the *Origin,* the entanglements of nature continue to protect against evil; and in Miltonic cadences, Darwin draws the explicit lesson:

> It is interesting to contemplate an entangled bank, clothed with many plants of many kinds, with birds singing on the bushes, with various insects flitting about, and with worms crawling through the damp earth, and to reflect that these elaborately constructed forms, so different from each other, and dependent on each other in so complex a manner, have all been produced by laws acting around us. These laws, taken in the largest sense, being Growth with Reproduction . . . a Ratio of Increase so high as to lead to a Struggle for Life, and as a consequence to Natural Selection, entailing Divergence of Character and the Extinction of less-improved forms. Thus, from the war of nature, from famine and death, the most exalted object which we are capable of conceiving, namely, the production of the higher animals directly follows. There is grandeur in this view of life, with its several powers, having been originally breathed into a few forms or into one; and that, whilst this planet has gone cycling on according to the fixed law of gravity, from so simple a beginning endless forms most beautiful and most wonderful have been, and are being, evolved. [Darwin 1859, pp. 489–90]

O felix culpa! Or rather, *O felix natura!* – nature, that thoroughly organic being which expresses aesthetic and moral values.[25]

By the time Darwin began to compose the *Origin of Species,* his images and metaphors had shed much of the rich fabric with which they had originally been adorned. Yet his simpler expressions nonetheless disclosed, via a deeper aesthetic logic, a morally saturated nature. That logic thus belies the usual view of Darwin's accomplishment, which was supposed to show, according to Susan Cannon, nature to be "morally meaningless" (Cannon 1978, p. 276).

Darwin's conception was of an intelligent and moral nature, very much like the nature formulated by Schelling and purveyed by Humboldt and Owen. His conception makes sense of certain aspects of his general evolutionary theory that seem inexplicable on the assumption of a nature clanking along in the manner of a nineteenth-century steam engine. Were natural processes really machinelike, ought not the products be identical – same mold, same cookie? But the products of nature, characterized by an underlying theme, to be sure, were yet infinitely varied, exuding the great abundance of life. Moreover, machines, at least those of Darwin's acquaintance, could hardly produce traits of near perfection. From the time he read Paley's *Natural Theology,* Darwin never doubted that organs like the eye (Paley's favorite example) were adaptations of extreme perfection, hardly the sort of thing a machine could produce. Darwin later confessed that whenever he thought of the eye, his blood ran cold. But if the agency producing an eye were virtually a lesser god, then such production might well be intelligible. Nature's agency displayed a refined sense of possibility in Darwin's construction. She very gradually and over long periods of time refined and shaped her creatures. Darwin's friend Huxley (like some present-day Darwinians) urged that nature must be conceived as hopping along in fits and starts, producing stuttering advances in species forms instead of the slow gradual progress Darwin's theory supposed. Huxley's insistence would have been appropriate if Darwin had indeed construed nature as a machine, nineteenth-century variety. But his nature conceptually grew according to the model of the departed divinity; hence it could move gradually, slowly, and majestically toward the perfection it teleologically sought. There are many other features of Darwin's conception of nature that hardly mesh with the gears of a machine; but perhaps I've suggested enough to make the case that the usual conception of Darwin's accomplishment simply fails to treat seriously the canonic expression of his theory in the *Origin of Species.*

Given Darwin's conception of nature, which in large part, I have argued, was expressive of the kind of romanticism cultivated originally in Germany and imported to England under various guises – given that conception, would it be surprising that Darwin did not eviscerate human nature of moral capacity?

DARWIN'S ETHICAL THEORY

Theories of Ethics prior to the Origin of Species

Darwin not only reinfused nature with value but also was especially keen not to leave man morally naked to the world. This, of course, is not the general view of Darwin's construction of morality. Most commonly he is taken to have advocated something like Ghiselin's brand of sociobiology, a kind of selfish utilitarianism disguised in decorous Victorian language. Yet such an evaluation completely misses what Darwin himself thought to be distinctive of his biology of morality, namely, that it overturned utilitarianism.

During his five-year voyage on H.M.S. *Beagle,* Darwin experienced the extremes of moral behavior, from the brutality he frequently observed among the South American gauchos to the nobility of the Indians whom they slaughtered. And he reacted with particular moral revulsion at the institution of slavery. Humboldt had readied him for his encounter with that trade in which both "copper-colored Indians" and "African Negroes" suffered the kind of injury that resulted "in rendering both the conquerors and the conquered more ferocious" (Humboldt and Bonpland 1818–29, vol. 3, p. 3). But his response to the reality was hardly less genuine for being filtered through the somber lens of his German predecessor. He felt simultaneously depressed and furious as he witnessed African families being separated at slave auction, and slaves being beaten and degraded (Darwin 1839, pp. 22, 27–8). Though it was only in the tranquility of his study, as he composed his *Journal of Researches,* and then recomposed it in a second edition ([1845] 1962), that the Humboldtian framework focused his reflections on the peculiar institution. In that second edition, Darwin added the following in the conclusion to his book:

> On the 19th of August we finally left the shores of Brazil. I thank God, I shall never again visit a slave-country. To this day, if I hear a distant scream, it recalls with painful vividness my feelings, when passing a house near Pernambuco, I heard the most pitiable moans, and could not but suspect that some poor slave was being tortured, yet knew that I was as powerless as a child even to remonstrate. . . . It is argued that self-interest will prevent excessive cruelty; as if self-interest protected our domestic animals, which are far less likely than degraded slaves, to stir up the rage of their savage masters. It is an argument long since protested against with noble feeling, and strikingly exemplified, by the ever illustrious Humboldt. [Darwin 1962, pp. 496–7]

As he reflected on the nature of moral behavior, which formed a topic of his considerations from the very beginning of his work on the species question, Darwin's patience for utilitarian arguments became exhausted. Initially, though, prior to his reflective consideration of Humboldt and prior to the formulation of natural selection, utilitarianism did seem the right way to think about morals.

While an undergraduate at Cambridge, Darwin had to get up for his exams William Paley's *Moral and Political Philosophy.* The utilitarian position worked out by that Anglican divine provided Darwin a preliminary framework through which to weave his emerging biological theory of behavior.

On 8 September 1838, about three weeks before he read Malthus, who ignited the spark for the idea of natural selection in his imagination, Darwin reread Paley, who also kindled an idea. In his "Notebook M" he considered "Paley's rule." In *Moral and Political Philosophy* Paley offered this rule of "expediency" as the central axiom of his ethics:

> Whatever is expedient is right. But then it must be expedient on the whole, at the long run, in all its effects collateral and remote, as well as in those which are immediate and direct. [Paley n.d., p. 40]

Darwin, as was his habit, gave this rule a biological interpretation:

> Sept 8th. I am tempted to say that those actions which have been necessary for long generation, (as friendship to fellow animals in social animals) are those which are good & consequently give pleasure, & not as Paley's rule is then that on long run *will* do good. – alter *will* in all such cases to *have* & *origin* as well as rule will be given. [Darwin 1986, p. 552]

Darwin constructed that interpretation of Paley's rule prior to having formulated his principle of natural selection. Up to that time he had been persuaded that habits, practiced over several generations, could become instinctive, that is, innate. Continued exercise of instincts – such as that of ancient birds' swimming out on a pond – might alter anatomy (e.g., the birds, by stretching their toes in swimming, might produce webbing, and so ducks would be born). This habit-instinct device became, for Darwin, prior to reading Malthus, the principal explanation for the alteration of species over time.[26] In this light, his interpretation of Paley's rule becomes a bit more clear. He suggested that those useful and expedient habits that had been necessary to preserve animals' groups, allowing them over long periods to propagate and protect their young (such habits as friendship, parental nurture, etc.), were what we had come to call morally good. The continued practice of such useful social behaviors would produce moral instincts that would conform to a temporally readjusted rule of expediency: what has been good will become interred in an animal's bones and thus will continue to be what animals and their offspring, including man, regard as good.

From the beginning of his speculations about the changes in species, Darwin fully understood that his theory would require a reconceptualization of human nature, one that would naturalize man's moral behavior. For had he allowed an exception to a naturalistic explanation of that very defining trait of man – moral behavior – then he would have left an opening for the return of the Creator. Hence, in late September 1838, when, as a result of reading Thomas Malthus's *Essay on Population,* he hit upon natural selection as his chief device to explain the evolution of species, he had also to adjust his account of the origins of moral behavior. He was aided in that task by his conversations with James Mackintosh (the brother-in-law of Darwin's uncle, Josiah Wedgwood) and by his reading of Mackintosh's *Dissertation on the Progress of Ethical Philosophy* (1836).

Darwin began reading Mackintosh's *Dissertation* on and off from summer

1838 to spring 1839. In this historical and critical treatise, Mackintosh provided a survey of ethical theories up to his own time. He objected particularly to Paley's utilitarianism, because it allotted human beings a nature that responded only to the urgings of pleasure and pain for self. Mackintosh believed human beings to be endowed with a more generous nature, one that might act altruistically for the good of others. In this proposal he aligned himself with the views of Shaftesbury, Butler, and Hutchinson. He argued that human nature came equipped with a moral sense for right conduct. That sense might be educated through experience, but its roots were innately embedded in the human constitution. This meant that men could be and were motivated by urges other than those of selfish pleasure. They rather spontaneously sought to improve the welfare of others and approved of such behavior when exhibited by acquaintances. Mackintosh did not completely disavow Paley's or Bentham's utilitarianism. He granted that in a cool hour, when we considered the *criterion* for right behavior, we would recognize the utility of acts that would promote the common welfare and the disutility of behavior that would compromise it. Mackintosh insisted, however, that the *moral sense* for right conduct, that is, our immediate perception of what we ought to do in a situation, would not depend on any rational calculation of pleasures and pains, utilities and disutilities. We would simply recognize innately what behaviors would be morally required in a situation.

Mackintosh's theory of moral behavior did have one glaring residual difficulty: he had no good explanation (at least to Darwin's mind) of the conjunction of the moral sense and the moral criterion. Just why was it that what a person might do spontaneously and without reflection would conform to what appeared, after sufficient reflection, to be the most useful act that could have been chosen in the situation? This was a difficulty for which Darwin would suggest an ingenious answer.

Mackintosh's attack against notions that human beings could be motivated only by selfishness, even in matters of morals, made a considerable impression on the young naturalist. This is evidenced in three different sets of documents: various passages in Darwin's "Notebook N" (kept between fall 1838 and spring 1840); remarks left in a bundle of notes he labeled "Old and Useless Notes"; and a nineteen-page manuscript on Mackintosh's views that he composed in spring 1839.[27] In these various jottings, Darwin gradually sketched out a theory of moral behavior and conscience that marked another stage in his developing considerations about these subjects. The manuscripts show him amalgamating his earlier Paleyesque ideas to Mackintosh's scheme, and thus laying the foundation for the ethical theory presented in the *Descent of Man*.

Fundamental to Mackintosh's moral-sense theory was the proposal that

a certain kind of knowledge lay buried in the human soul, a proposal anti-
thetic to the main empiricist stream of British thought. This conception of
innate knowledge received a theoretical boost, at least for Darwin, by an essay that appeared in the *Westminster Review* in 1840. The essay summarized
and evaluated the general philosophy of Samuel Taylor Coleridge, who had
borrowed many of his ideas from Schelling. The appreciation by the anonymous author (John Stuart Mill), though critical, recognized that the romantic philosophy of Coleridge pervaded the minds and hearts of a significant
portion of British intellectuals. That philosophy maintained that the fundamental principles of morals (along with those of religion, mathematics, and
the basic laws of physics) came to reason not through experience but through
deeper channels of the soul.[28] Darwin's own response to the essay showed
him, typically, finding a biological interpretation for a theory with which he
was in sympathy. When Mill objected that only experience could be the object of our knowledge, Darwin reflected "is this not almost a question whether
we have any instincts, or rather the amount of our instincts – surely in animals according to usual definition, there is much knowledge without experience, so there *may* be in men – which the reviewer seems to doubt" (Darwin
1986, p. 610). Instinct, Darwin would insist, formed the basis of our moral
sense and anything we might call innate knowledge.

The theory of moral behavior that Darwin developed in his early notebooks – and, with a few modifications, presented much later in the *Descent
of Man* – proposed that our judgments about appropriate behavior (the sort
of behavior we call moral) stemmed from a kind of innate knowledge, from
an instinct for right action. In his theory, Darwin distinguished two general
kinds of instincts: (1) impulsive instincts, really emotional reactions, such as
a stab of lust or a flash of anger that impels an immediate response, and (2) the
more calm and persistent social instincts, which, though powerful, in the
longer term do not have that same immediate force. But it was instincts of
this latter sort, Darwin reflected, that kept a mother bird patiently tending
her brood. Sometimes, however, the good mother might, while out foraging
for her nestlings, catch sight of the migrating flock and spontaneously fly away
to better climes, without adverting to her young. In a sunnier environment,
if this reprobate mother had mind enough to recall her starving chicks, her
social instincts would again take hold. She would feel that pull to aid her offspring, though the instinct would remain unsatisfied. A rational animal in such
circumstances, Darwin supposed, would confess a troubled conscience. All
that was required to turn an animal into a moral creature, according to this
scheme, was an intellect approaching that of man. On 3 October 1838, just
a few days after he had read Malthus, Darwin reformulated his theory of moral
conscience and recognized with relish its importance:

Dog obeying instinct of running hare is stopped by fleas, also by greater temptation as bitch. . . . Now if dogs mind were so framed that he constantly compared his impressions, & wished he had done so & so for his interest, & found he disobeyed a wish which was part of his system, & constant, for a wish which was only short & might otherwise have been relieved, he would be sorry or have troubled conscience. – Therefore I say grant reason to any animal with social & sexual instincts & yet with passion he *must* have conscience – this is capital view. Dogs conscience would not have been same with mans because original instincts different. [Darwin 1986, p. 536]

In his manuscript on Mackintosh, Darwin further developed this account of conscience. He considered that the useful habits that members of a species would develop in social circumstances – cooperative foraging, group defense, parental nurturing, and the like – would continue to be practiced over generations. Those other behaviors directed to particular and individual desires would, however, not be constantly practiced from one generation to the next and therefore would not become deeply ingrained in the heritable substance of the animal. Those remaining practices directed to the common good, on the other hand, would seep into the heritable core of the individual and be passed on as instincts. In the human species, characterized by sufficient intellect to provide a memorial foundation for instincts to remain active, even in the absence of their original eliciting situations, these social instincts would become moral motives. The moral sense of individuals would thus evolve along with their other species-characteristic traits. When human beings reflected on their behaviors and moral impulses, they would, quite naturally feel a particular kind of quiet pleasure in the satisfaction of such instincts and simultaneously perceive the utility in their exercise. Rational creatures would thus take as criterion the utility subsequently recognized in such reflections. In this fashion Darwin had solved Mackintosh's problem: he found a biological explanation for the agreement between spontaneous moral behavior and reflective judgment about the utility of that behavior for the community at large.

Darwin's account of moral behavior in his early notebooks relied entirely on the principle of the inheritance of acquired characteristics. But during the roughly two decades prior to publication of the *Origin of Species,* he slowly brought his device of natural selection to explain all sorts of species-characteristic traits, including instinct. However, a stubborn problem surfaced when he applied his device to such instincts as those promoting moral behavior. Moral behavior, unlike other traits that might be selected for, conferred benefit not on their agent but on their recipient. The social and moral instincts were fundamentally altruistic. Yet it appeared that natural selection operated only on traits that gave their agent an advantage, not their recipient. How to explain it?

140

In the mid-1840s the problem of the social instincts and their natural-selection explanation became quite poignant. It captured Darwin's attention when he sought to explain the distinctive instincts and anatomical traits of the social insects, especially those of the workers in beehives and ant nests. Soldier bees, for example, would guard the hive and even give up their lives in its defense. But these casts of insects displayed another feature that seemed to make natural selection inapplicable: they were neuters, and left no offspring to inherit any advantageous traits. As I have tried to show elsewhere, Darwin worried about this difficulty for some time, fearing that it would undo his general theory of evolution altogether (Richards 1987, pp. 127–56). He attempted various resolutions, but could not shake the chilling thought that he himself might have exposed the fatal flaw in his theory. It was only during the first months of 1858, while laboring over the manuscript that he intended to publish under the title of *Natural Selection,* that he fixed upon a solution that would be highlighted in the *Origin of Species,* which became the more compact successor of that earlier manuscript. In those first months of 1858, Darwin sketched out his new theory:

> In the eighth chapter, I have stated that the fact of a neuter insect often having a widely different structure & instinct from both parents, & yet never breeding & so never transmitting its slowly acquired modifications to its offspring, seemed at first to me an actually fatal objection to my whole theory. But after considering what can be done by artificial selection, I concluded that natural selection might act on the parents & continually preserve those which produced more & more aberrant offspring, having any structure or instincts advantageous to the community. [Darwin 1975, p. 510]

In the *Origin of Species,* Darwin reiterated this solution, namely, that natural selection operated not on the individual workers to provide their unusual traits but on the whole hive or community, which would contain relatives of the workers. (And in the fifth edition of the *Origin* he extended the idea of group selection to any assemblage of social animals, including man.)[29] Thus the altruistic behavior of a soldier bee in sacrificing its life for the nest could be explained as the result of community selection. Natural selection could indeed, then, be brought to account for moral behavior. And in the *Descent of Man,* this is precisely what Darwin did, and then some.

Ethical Theory in the Descent of Man

Darwin had originally intended to consider human evolution in the *Origin of Species.* The haste with which he composed the book, however, and, more importantly, his desire to avoid unnecessary provocation, counseled postponement. In the wake of the success of the *Origin,* when he decided to write on matters of domestication, variation, and heredity, he mischievously mentioned to Alfred Russel Wallace that he would include in his contemplated volume an essay on man, because that creature seemed "an eminently *domesticated* animal" (Wallace 1916, vol. 1, p. 181). Darwin's *On the Variation of Animals and Plants under Domestication* (1867) did not, however, carry the intended essay. Instead, Darwin had decided, as a result of his dispute with Wallace over sexual selection, to treat human evolution in a volume dedicated to that subject. As late as 1869 he seems to have had no serious intention of discussing human morality in the planned book; he wished only to consider those traits of males and females that could be attributed to sexual selection (e.g., the male's greater musculature, secondary sexual characteristics, and other minor traits). However, several of his friends, notably Charles Lyell (1797–1875) and Asa Gray (1810–88), had during the mid-1860s publicly suggested that natural selection could not explain the distinctive features of human beings, especially their moral nature.[30]

The greatest challenge in this regard, however, came from the co-founder of the theory of evolution by natural selection, Alfred Russel Wallace (1823–1913). Darwin's friend had undergone a kind of spiritualistic conversion, and in the *Quarterly Review* (1869) Wallace had pressed the objection that whereas the animal kingdom had arisen through the power of natural selection, human mental and moral traits could not have:

> Neither natural selection or the more general theory of evolution can give any account whatever of the origin of sensational or conscious life. . . . But the moral and higher intellectual nature of man is as unique a phenomenon as was conscious life on its first appearance in the world, and the one is almost as difficult to conceive as originating by any law of evolution as the other. [Wallace 1869, p. 391]

Darwin felt something akin to despair over his friend's abandonment of natural selection in the case of man. But Wallace's new attitude, along with the stated reservations of Lyell and Gray, provided just the stimulus to alter Darwin's intentions for the new book. He then decided hastily to resurrect his early notes on human moral evolution, but to reformulate those youthful ideas in light of his theory of community selection. The result was an articulated theory of moral conscience and a firm hypothesis as to the origins of that faculty. Darwin's labors yielded two volumes, which appeared in 1871 under the title *The Descent of Man and Selection in Relation to Sex*.

The theory of conscience that Darwin presented in the *Descent* bore the distinctive marks of his earlier considerations. He maintained that human moral judgment lay anchored in social instinct, the kind of instinct that would have urged the mother bird to forage for her nestlings. Human conscience, of course, had to be more than simple social instinct. Darwin argued that that faculty would arise through four overlapping stages: First, protohumans had to develop a set of social instincts strong enough to bind them together into a society. Second, members of this society had to have acquired sufficient intellect to recall a social instinct that might momentarily have been swamped by a more insistent urge. Third, language would be required to codify and communicate the needs of other society members. Finally, members of this community would have to develop habits of attending to the needs of others, and with this stage they would have attained the truly human state, that of a moral creature (Darwin 1871, vol. 1, pp. 72–3).

Darwin had essentially planted Mackintosh's ethical theory in biological ground. Moral sense consisted in the variety of social instincts, that is, altruistic instincts, that had been evolutionarily acquired by a social group, instincts that its members could reflect upon, codify through language, and stabilize in habit. As was characteristic of Darwin's inclination to overdetermine his explanations of empirical events, he constructed three complementary accounts of the origins of the altruistic instincts. First, he proposed that as the reasoning powers of members of a tribe improved, individuals would learn that if they helped others in need they might receive help in return, a process we call reciprocal altruism. He also suggested that the effects of praise and blame would stamp into individuals, as a kind of second nature, habits of social conduct. But the mode of acquisition he thought the most effective, and the one he took pride in advancing, was based on community selection. He put it this way:

> It must not be forgotten that although a high standard of morality gives but a slight or no advantage to each individual man and his children over the other men of the same tribe, yet that an advancement in the standard of morality and an increase in the number of well-endowed men will certainly give an immense advantage to one tribe over another. There can be no doubt that a tribe including many members who, from possessing in a high degree the spirit of patriotism, fidelity, obedience, courage, and sympathy, were always ready to give aid to each other and to sacrifice themselves for the common good, would be victorious over most other tribes, and this would be natural selection. [Darwin 1871, vol. 1, p. 166]

Darwin conjectured that our ancestors had lived in small tribal communities that had been in competition with one another, not unlike groups of social-insect hives. The communities that had reaped the propagative advantage had been those whose members had exhibited, by chance, altruistic impulses that had directed their behavior to the welfare of the whole. Darwin supposed that intellectual acquisitions, learned customs, and advances in knowledge that a group might have enjoyed would have focused the altruistic instincts of members on actions that would have worked for the benefit of others. Over many generations, as a community's fund of knowledge had increased and its understanding of effective instrumental procedures had improved (e.g., the discovery of the value of inoculation), the altruistic impulses would have been directed toward conferring ever more useful benefits. He thus wedded notions of cultural progress in learning with his theory of community selection to produce a conception that made human beings intrinsically moral animals, but animals whose ethical prescriptions were informed by increasing knowledge of what really was in the best interests of their community.

The utilitarians, Darwin observed, had claimed that "the foundation of morality lay in a form of Selfishness; but more recently in the 'Greatest Happiness principle'" (Darwin 1871, vol. 1, p. 97). His own theory, by contrast, did not suppose that moral action was motivated by self-interest or was executed to achieve the greatest happiness. Rather, human beings, he maintained, acted spontaneously, impelled by their altruistic instincts, to advance the welfare of others without counting the cost to self. They did not seek the greatest happiness of the greatest number, or even the greatest happiness of number one; rather, they directed their actions to achieve the "greatest good," which Darwin interpreted to be the "vigor and health" of the greatest possible number of community members. He thus concluded that "the reproach of laying the foundation of the most noble part of our nature in the base principle of selfishness is removed" (Darwin 1871, vol. 1, p. 98).

Darwin, that humble (and, at times, seemingly bumbling) biologist, had constructed an ethical theory of elegance, power, and nobility. He held that human behavior, in some of its forms, could and did achieve those ideals that had been enshrined in the western ethical tradition. He preserved a conception of man as an intrinsically moral being, a being whose morality tinctured the very core of his substance. Certainly human beings acted selfishly on occasion. Darwin nonetheless believed that men could recognize the needs of others and could respond unselfishly to satisfy those needs. His ethical theory, therefore, stood apart from the typical Benthamite systems of his contemporaries. That was because, I believe, from the very beginning Darwin had recognized in nature a source of moral and aesthetic value. Alexander von Humboldt had inculcated into him the kind of moral evaluation of nature that simply could not be reduced to the low utilitarianism infecting most British moral philosophy at the time.

Darwin's theory of the rise of moral behavior had an added benefit which seems to have struck him only during the composition of the *Descent*. Wallace had also maintained that man's high intellect was quite superfluous for the simple needs of survival, and hence that human faculty had to have had a source other than natural selection (Wallace 1869, p. 392). This kind of difficulty must have been emphasized for Darwin when he read a passage in Humboldt's *Personal Narrative* that suggested that under such favorable climatic and agricultural conditions as existed in the torrid zone, "the intellectual faculties unfold themselves less rapidly than under a rigorous sky, in the region of corn, where our race is in a perpetual struggle with the elements" (Humboldt and Bonpland 1818–29, vol. 3, p. 14). Next to this passage, Darwin scribbled in his copy of Humboldt: "Allude to this, when saying, the causes of the progress of intellect from Monkeys to Man is inexplicable."[31] With his theory of community selection, however, Darwin had found an explanation: if in a tribal group a genius by chance had appeared, that primitive Newton would have benefited his whole community; they would have learned his tricks, thus giving them an advantage in competition with other tribes; and because his tribe would have included his relatives, improved mind would have arisen among our ancestors (Darwin 1871, vol. 1, p. 161).[32]

Darwin's account of morals and mind had a common root in his theory of community selection. This sort of explanation had resonance with theories found among the German romantics, particularly in Schelling's thesis that absolute mind produced individual mind and its moral structures. To argue for a direct descent of ideas here would certainly press my case for the influence of German romantics on Darwin beyond the endurance of the most tolerant readers. Yet there seems to have been, at least in this instance, something like convergent evolution: the Humboldtian legacy created an environment that favored Darwin's construction of individual morals and mind as the result of the selection of group mind.

CONCLUSION

The venerable Darwin who peers out from John Collier's posthumous portrait, done in 1884, has the visage of a terrible Old Testament prophet. Photographs taken during his last years confirm that the artifice embraced the man, not merely the painting. These are the images of Darwin we remember most vividly – hardly the kind of figure one would think of as a romantic revolutionary. Yet, that he was a revolutionary there can be no doubt. Nor, I believe, can we deny, at least when the written evidence is carefully considered, the deep romantic strain in his thought.

Darwin came by his attitudes much as the earlier German romantics had, through prolonged contact with exotic nature – but nature as filtered through a certain literature. In Darwin's case, the literature was singularly provided by the conceptually and aesthetically lush works of Alexander von Humboldt, who taught him how to experience the sublime and how morally to evaluate the nature he met in the jungles, mountains, and plains of South America. That early experience, formed and shaped under the guiding images provided by Humboldt, settled deeply into the conceptual structure of the *Origin of Species* and the *Descent of Man*. The sensitive reader of Darwin's works, a reader not already completely bent to late-twentieth-century evolutionary constructions, will feel the difference between the nature that Darwin described and the morally effete nature of modern theory. Darwin's early attitudes obviously became subject to numerous other conceptual influences – he was not simply young Werther in a blue frock coat and yellow vest, reading his Homer and suffering unrequited love, albeit in a jungle clearing. But neither was he that unflinching mechanist who would deprive nature of her soul of loveliness.

Darwin's nature, like that of other German romantics, exemplified archetypal patterns beneath the wild frenzy of their variations. These patterns gradually changed under the aegis, not of Paley's God and certainly not of a Victorian stamping-press, but of a productive nature (*Natura naturans*) – ever fruitful and rich in possibilities, realizing those possibilities in the best interests of her creatures. Darwin's nature, like that of the other romantics, progressively produced organisms of ever greater value, "the higher creatures," as he labeled them. His nature acted altruistically: unlike human beings, she tended her creatures for their own sake, improving their lot and that of the whole interconnected assembly constituting her frame. The intrinsic moral aspects of nature also imbued her most developed creatures, human beings.

Darwin's ethical theory stood in contrast to those of most of his British contemporaries, who advanced various versions of utilitarianism. He contended that human beings could indeed act altruistically, that moral conscience did not veil a pleasure-seeking machine. In making his argument, Darwin erected a theory that might be judged a singular response to the philosophical and cultural environment of his time, but one that must fail under the stronger metaethical analysis typical of our time. By way of concluding this account of Darwin's accomplishment, let me briefly assess the fortunes of Darwin's ethical theory in light of two kinds of criticism.

The first general sort of criticism to which his ideas have been subjected, by both friend and foe, is exemplified by the construction of Ghiselin, who maintained that Darwin's conception really was, at root, a theory of utilitarian selfishness (despite Darwin's own explicit judgment to the contrary). This position supposes either that Darwin had an ancillary theory of selfish genes (or perhaps "gemmules," his version of our genes) or that the device of community selection – which Darwin employed to explain human altruism – was just a higher type of utilitarianism. The former supposition has no textual justification and, of course, simply reads contemporary ideas back into Darwin's work. The latter supposition has more weight, but is mitigated or obviated by two further observations concerning Darwin's ethical view. First, Darwin contended that though altruistic impulses might have been selected originally in the competition of protohuman clans, such impulses were no longer confined to kin. As human beings advanced in intelligence and knowledge, they would come to judge all men as brothers and so would invest their care-giving behavior beyond the narrower pale of immediate relatives. Second, Darwin recognized that what counted in judging an action moral was the intention that guided the act: he denied that any utilitarian calculation of benefit to self or one's self-interest (e.g., in terms of relatedness of kin) would be manifested to consciousness in such an act.

The second kind of criticism that has been leveled at ethical theories like Darwin's is that they commit the naturalistic fallacy: this criticism supposes that human beings may have evolved to regard certain acts as enjoined, but that does not mean that their decisions in particular cases could be justified in terms of what they "ought" to have done. Michael Ruse has contended that although humans can and do act decently toward one another, and though they may even murmur to themselves that they are acting for conscience's sake or for duty, they nonetheless make those decisions based on an empirical mechanism that has operated in a utilitarian mode (Ruse 1986). I have attempted to answer this sort of criticism of Darwin's theory – actually my own revised version of Darwin's theory – in several places and at great and tedious length (Richards 1986a,b; 1987, 2nd app.; 1989; 1993).[33] Let me here briefly add another defensive counterobjection. It is simply this: If we have evolved to regard certain behaviors (altruistic ones) as especially valuable, and if within the deepest recesses of our consciousness we find that principle of altruism beckoning, can we do anything other than regard those behaviors as obliged, as what we *ought* to do? And if we have evolved as suggested, will we not be in the delicate Kantian position of having our mental apparatus so constructed as to judge any such behaviors as required, indeed, as virtually constituting us as human beings? In a cool hour of reflection, when I might contemplate insistent urges to act in an altruistic way, would I not apply the only criterion available to me (i.e., that of altruism) and judge them morally sanctioned? Simply having altruistic instincts will not, perhaps, make us moral, but subjecting those impulses to a standard of moral appropriateness and then judging that indeed they ought to be followed – well, what else could be asked to make a system morally authentic? In this case, we will not be hoodwinked into a selfish act. We will be enlightened as to our duty.

George Bernard Shaw wrote splendid plays, but simply had no head for intricate scientific and philosophic argument. And as was true of many, his historical sense undoubtedly became distorted by his contemporary circumstances. Perhaps only "fools and rascals" could bear to live in the nature created by late Victorian science, the science of his acquaintance. But Darwin's nature did not have that provenance. His was a nature born of a lineage ripe in early German romanticism. When his conception of nature is replaced into that generating context, then all of us, I think, can bear to live within its grasp.

NOTES

1. David Kohn (in press) has provided one of the very few exceptions to this generalization. Kohn, with great penetration, has analyzed two salient metaphors in the *Origin of Species,* the "entangled bank" trope, which occurs on the penultimate page of the book, and the "wedging" simile, which characterizes the way in which selection forces organisms into the economy of nature.
2. Even historians engaged in the study of romanticism's impact on science usually do not strain against these conventional assessments of Darwin's accomplishment. In an essay that briefly touches (but hardly wrestles with) this subject, the author affirms that "Darwin's great synthesis in the *Origin of Species* was not rooted in Romanticism but in the very different tradition of Paley and Thomas Malthus" (Knight 1990, p. 22).
3. I am currently attempting this in a book manuscript bearing the tentative title *Romantic Biology: from Goethe to the Last Romantic, Ernst Haeckel.*
4. See especially Immanuel Kant's *Kritik der Urteilskraft* (1957, vol. 5, pp. 526 [A346–47, B350–51] and 538 [A363–64, B368]). The most accessible account of Schelling's archetype theory can be found in his dialogue *Bruno oder über das göttliche und natürliche Prinzip der Dinge* (Schelling 1992). Goethe discussed archetypes in his "Vorträge, über die drei ersten Kapitel des Entwurfs einer allgemeinen Einleitung in die vergleichende Anatomie, ausgehend von der Osteologie" (Goethe 1954, pp. 193–209).
5. Elsewhere I have discussed the origins and development of archetype theory (Richards 1992, pp. 17–61).
6. Susan Faye Cannon (1978, pp. 73–110) has provided a still-instructive sketch of the character of Humboldt's science.
7. Darwin read Humboldt's *Views of Nature* in February 1852. See Darwin's "Reading Notebooks" (1985–, vol. 4, p. 487).
8. Humboldt's *Kosmos* (1845–58) was translated into English three times in the nineteenth century. Darwin disliked the initial, pirated translation of the first two volumes, but enjoyed that of Elizabeth Sabine. See Darwin's comments on Humboldt's *Cosmos* (Darwin 1985–, vol. 3, p. 342).
9. Humboldt's aesthetic philosophy is based on Kant's "Third Critique" and on Schelling's transcendental philosophy. At the beginning of the second volume of *Cosmos* (p. 19), Humboldt contrasted his "objective" physical account of nature with his observations about the way nature impressed the sensations and fancy of the naturalist. In reading this one might assume that Humboldt's aesthetics of nature was entirely "subjective" (a term, incidentally, that he did not use). If we make a

148

simple distinction between epistemic and ontological claims, on the one hand, and the objective or subjective bases of those claims, on the other (e.g., "The wine is tart" would be ontologically subjective but epistemically objective), then we shall have to interpret Humboldt's aesthetics of nature as epistemically objective, though it deals with the *universal* reactions of the subject.

10. Darwin's own prose became so structured by that of Humboldt that his sister Caroline began to complain. She wrote him in 1833: ". . . as to your style. I thought in the first part (of this last journal) that you had, probably from reading so much of Humboldt, got his phraseology, & occasionally made use of the kind of flowery french expressions which he uses, instead of your own simple straight forward & far more agreeable style. I have no doubt you have without perceiving it got to embody your ideas in his poetical language & from his being a foreigner it does not sound unnatural in him" (Darwin 1985–, vol. 1, p. 345).

11. This early date for Darwin's reading of Humboldt is suggested by a friend's discussion of the *Personal Narrative* with him (Darwin 1985–, vol. 1, p. 51).

12. Humboldt's companion Bonpland supplied many observations, but the tale unfolded in the first-person of Humboldt.

13. Darwin took at least the first two volumes of the narrative with him, a gift from Henslow. He likely added, on his own, the remaining five volumes. For the trip, he also packed up several other books by Humboldt, chiefly geological. See the list of books taken on the voyage, published in the *Correspondence* (Darwin 1985–, vol. 1, pp. 553–66).

14. For these subjects, see Darwin (1988, pp. 34, 48, 54, 67, 70, 72, 267, 288, 308).

15. Humboldt's enthusiasm for Darwin's aesthetically sensitive descriptions was shared by many of the reviewers of the *Journal of Researches*. Typical was Broderip's judgment that Darwin's account "fill[ed] the mind's eye with brighter pictures than a painter can present" (Broderip, 1839, pp. 194–234).

16. The dynamical aspect of Green's theory followed the path into "objective Idealism" laid by Schelling (Green 1840, pp. xxix–xxx). He specifically cited Goethe, Oken, Spix, and Carus as developing the archetype theory that he himself was further elaborating (Green 1840, pp. 57–8).

17. Green did believe in the historical evolution of species, as I have argued (Richards 1992, pp. 72–9). But his theory was similar to Schelling's "dynamical evolution (*dynamische Evolution*)." Schelling's theory of dynamical evolution was not a genealogical theory, but it did maintain that new species appeared over time, each advancing toward the perfect realization of organism in general, which Green interpreted as the appearance of the human form. See Schelling, *Erster Entwurf eines Systems der Naturphilosophie* (1992, vol. 1, p. 64).

18. The *Manchester Spectator* (8 December 1849) sounded a cautionary note about the quite obvious dangers to the scientifically ignorant in Owen's apparent adoption of Germanic *Naturphilosophie*. Owen responded (22 December 1849) to the *Spectator* article saying that his language in his conclusion was "figurative," something allowed at the end of a lecture. He contended that the unity of plan displayed in organisms "testifies to the oneness of their Creator."

19. Charles Darwin, "Notebook B" (Darwin 1986, p. 170). The words in square brackets were deleted in the manuscript. Darwin was reflecting on his grandfather's book *Zoonomia*, and likely the passage in which Erasmus Darwin, after reflecting on the morphological similarity of creatures, asked "Would it be too bold to imagine, that in the great length of time, since the earth began to exist,

... that all warm-blooded animals have arisen from one living filament, which THE GREAT FIRST CAUSE endued with animality, with the power of acquiring new parts, directed by irritations, sensations, volitions, and associations; and thus possessing the faculty of continuing to improve by its own inherent activity, and of delivering down those improvements by generation to its posterity, world without end!" See E. Darwin (1794–6, vol. 1, p. 505).

20. I have traced this history elsewhere (Richards 1992, pp. 91–166).

21. This inscription is on the back flyleaf of Richard Owen's *On the Nature of Limbs,* held in the Manuscript Room of Cambridge University Library.

22. The term "machine" in any of its forms appeared only once in the *Origin of Species,* and even then in a context that vitalized its significance (Darwin 1859, p. 83).

23. The term "machinery" in this quotation is the only instance of any form of that word appearing in the *Origin.* The context makes clear that it has no semantically significant role in his description.

24. After the sentence "See how differently Nature acts!" Darwin inserted a new and perhaps ameliorative emendation into the manuscript: "By nature, I mean the laws ordained by God to govern the Universe."

25. Kohn (in press) pursues at some length Darwin's metaphor of the entangled bank; he shows its connection with Darwin's own experience in the jungles of South America. Kohn does this with great facility, though he does not suggest the moral-theological paradox or the resolution that I here urge.

26. For a more extensive account of Darwin's pre-Malthusian theories of species change, see Richards (1987, pp. 83–98).

27. These notebook and manuscript pages have been transcribed and published (Darwin 1986, pp. 563–96, 599–629). The Mackintosh manuscript is included among the "Old and Useless Notes" (Darwin 1986, pp. 618–29).

28. Mill's essays on Bentham and Coleridge have been conveniently reprinted with a splendid introduction (Mill 1950).

29. In the fifth edition of the *Origin* (1869), Darwin quietly extended the idea of selection to the entire community of individuals. What is noteworthy about this extension is that he did not stipulate that the members of the community should be closely related. He wrote: "In social animals it [natural selection] will adapt the structure of each individual for the benefit of the community; if this in consequence profits by the selected change." In the sixth edition he directly and simply advanced the proposition: "Natural selection will modify the structure of the young in relation to the parent, and of the parent in relation to the young. In social animals it will adapt the structure of each individual for the benefit of the whole community; if the community profits by the selected change" (Darwin 1959, p. 172).

30. On the theological backsliding of Darwin's friends, see Richards (1987, pp. 157–84).

31. Darwin's copy of Humboldt is held in the Manuscript Room of Cambridge University Library.

32. Darwin also argued that the development of language would have a rebounding effect on brain, so as to improve its structure, which improvement would, via inherited biological characteristics, be passed to subsequent generations (Darwin 1871, vol. 1, p. 58).

33. Many critics have taken exception to my analysis; several rejoinders can be

found in the same number of *Biology and Philosophy* as my original effort (Richards 1986a,b) and in subsequent numbers of that journal. Most recently, Paul Farber (1994) has surveyed and strongly objected to efforts at an evolutionary ethics, my own included.

REFERENCES

Broderip, W. 1839. Review of Darwin's journal of researches. *Quarterly Review* 65: 194–234.

Bulhof, Ilse. 1992. *The Language of Science, with a Case Study of Darwin's "The Origin of Species."* Leiden: Brill.

Cannon, S. 1978. *Science in Culture: The Early Victorian Period.* New York: Science History Publications.

Darwin, C. 1839. *Journal of Researches into the Geology and Natural History of the Various Countries Visited by H.M.S. Beagle.* London: Henry Colburn.

    1859. *On the Origin of Species.* London: Murray.

    1871. *The Descent of Man and Selection in Relation to Sex,* 2 vols. London: Murray.

    1909. *The Foundations of the Origin of Species: Two Essays Written in 1842 and 1844 by Charles Darwin,* ed. F. Darwin. Cambridge University Press.

    1959. *On the Origin of Species by Charles Darwin: A Variorum Text,* ed. M. Peckham. Philadelphia: University of Pennsylvania Press.

    1962. *The Voyage of the Beagle,* ed. L. Engel. New York: Doubleday.

    1969. *The Autobiography of Charles Darwin, 1809–1882,* ed. N. Barlow. New York: Norton.

    1975. *Charles Darwin's Natural Selection: Being the Second Part of His Big Species Book written from 1856 to 1858,* ed. R. C. Stauffer. Cambridge University Press.

    1985–. *The Correspondence of Charles Darwin,* 10 vols. to date. Cambridge University Press.

    1986. *Charles Darwin's Notebooks, 1830–1844,* ed. P. Barrett et al. Ithaca, NY: Cornell University Press.

    1988. *Charles Darwin's Beagle Diary,* ed. R. D. Keynes. Cambridge University Press.

Darwin, E. 1794–6. *Zoonomia, or The Laws of Organic Life,* 2 vols. London: Johnson.

Farber, P. 1994. *The Temptations of Evolutionary Ethics.* Berkeley: University of California Press.

Ghiselin, Michael. 1973. Darwin and evolutionary psychology. *Science* 179: 964–8.

Goethe, W. 1954. *Die Schriften zur Naturwissenschaft,* 1st div. Vol. 9: *Morphologische Hefte.* Weimar: Hermann Böhlaus Nachfolger.

Green, J. 1840. *Vital Dynamics: the Hunterian Oration before the Royal College of Surgeons in London, 14th February 1840.* London: William Pickering.

Humboldt, A. 1845–58. *Kosmos. Entwurf einer physischen Weltbeschreibung,* 5 vols. Stuttgart: J. G. Cotta'scher Verlag.

1850. *Views of Nature, or Contemplation on the Sublime Phenomena of Creation,* trans. E. Otté and H. Bohn. London: Henry Bohn.

n.d. *Cosmos: A Sketch of a Physical Description of the Universe,* 5 vols., trans. E. Otté. New York: Harper.

Humboldt, A., and Bonpland, A. 1818–29. *Personal Narrative of Travels to the Equinoctial Regions of the New Continent, during the Years 1799–1804,* 7 vols. London: Longman, Hurst, Rees, Orme, and Brown.

Kant, I. 1957. *Immanuel Kant Werke,* 6 vols., ed. Wilhelm Weischedel. Wiesbaden: Insel Verlag.

Knight, D. 1990. Romanticism and the sciences. In: *Romanticism and the Sciences,* ed. A. Cunningham and N. Jardine. Cambridge University Press.

Kohn, D. In press. The aesthetic construction of Darwin's theory. In: *Aesthetics and Science: The Elusive Synthesis,* ed. A. Tauber. Dordrecht: Kluwer.

Mackintosh, J. 1836. *Dissertation on the Progress of Ethical Philosophy.* Edinburgh: Adam & Charles Black.

Mill, J. 1950. *Mill on Bentham and Coleridge,* ed. F. Leavis. London: Chatto & Windus.

Owen, R. 1847. *Report on the Archetype and Homologies of the Vertebrate Skeleton.* London: Murray.

1849. *On the Nature of Limbs.* London: John Van Voorst.

1992. *The Hunterian Lectures in Comparative Anatomy, May and June 1837,* ed. P. Sloan. University of Chicago Press.

Paley, W. n.d. *The Works of William Paley.* Philadelphia: Woodward.

Richards, R. 1986a. A defense of evolutionary ethics. *Biology and Philosophy* 1: 265–93.

1986b. Justification through biolog x ical faith: a rejoinder. *Biology and Philosophy* 1: 337–54.

1987. *Darwin and the Emergence of Evolutionary Theories of Mind and Behavior.* University of Chicago Press.

1989. Dutch objections to evolutionary ethics. *Biology and Philosophy* 4: 331–43.

1992. *The Meaning of Evolution: The Morphological Construction and Ideological Reconstruction of Darwin's Theory.* University of Chicago Press.

1993. Birth, death, and resurrection of evolutionary ethics. In: *Evolutionary Ethics,* ed. M. Nitecki and D. Nitecki. Albany: State University of New York Press.

Rupke, N. 1994. *Richard Owen, Victorian Naturalist.* New Haven, CT: Yale University Press.

Ruse, M. 1986. *Taking Darwin Seriously: A Naturalistic Approach to Philosophy.* Oxford: Blackwell.

Schelling, F. 1992. *Schellings Werke,* 14 vols. Munich: Beck'sche Verlagsbuchhandlung.

Shaw, G. 1961. *Back to Methuselah.* London: Penguin Books.

Sloan, P. 1986. Darwin, vital matter, and the transformism of species. *Journal of the History of Biology* 19: 369–445.

Wallace, A. 1869. Review of *Principles of Geology and Elements of Geology* by Charles Lyell. *Quarterly Review* 120: 359–94.

1916. *Alfred Russel Wallace: Letters and Reminiscences,* 2 vols., ed. J. Marchant. London: Castile.

7

Nietzsche and Darwin

Jean Gayon

INTRODUCTION

First, a few dates. When Charles Darwin (1809–82) published *The Origin of Species,* Friedrich Nietzsche (1844–1900) was age fifteen. *The Birth of Tragedy,* Nietzsche's first book, appeared in 1872, a year after Darwin's *The Descent of Man.* It revealed no particular interest in biological theories, although it already contained the theme of vital values. Nietzsche's serious commentaries on Darwin and the Darwinians began in *Human All-Too-Human* (1878) and developed uninterruptedly from then on. The high point of that somewhat one-sided "dialogue" was the *Genealogy of Morals* (1887), a book explicitly written against the Darwinian view of the origin of morals and published shortly before Nietzsche's definitive descent into madness (early 1889). On the other side of the relationship, Darwin never alluded to Nietzsche in his published work, nor, as far as I know, in his correspondence. It is likely that Darwin did not know of Nietzsche's existence. Therefore, there was nothing like a reciprocal relationship between Nietzsche and Darwin. But there is no doubt that Nietzsche, the most famous philosopher of the second half of the nineteenth century, was concerned with Darwin. This essay aims to provide a systematic evaluation of Nietzsche's work in those areas in which he felt the necessity to position himself with regard to Darwin, or "Darwinism," as he knew it.

Strangely enough, although most philosophers may be vaguely aware of *some* relationship between Nietzsche and Darwin, analyses of that relationship have been rare in the modern philosophical literature (after 1945), and

Elisabeth Valsecchi, Jean-Philippe Gayon, and Richard M. Burian are thanked for their useful comments on the manuscript. Matthew Cobb and Marjorie Grene made linguistic corrections in the manuscript. Jean Lachapelle provided numerous quotations from English versions of Nietzsche's works. Richard Burian generously translated from German the remaining quotations that I had been unable to find in English translations.

they generally have been poorly documented. That has been true for both historians of philosophy and philosophers of biology. Contemporary historians of philosophy have tended to ignore the problem, and most generally have not even alluded to it (cf. Kaufmann 1950; Hollingdale 1965). That attitude has been characteristic of the past fifty years, but before World War II, Nietzsche's biographers frequently considered Nietzsche's relationships to biology and evolutionism (Richter 1911; Andler 1958).

The silence among modern historians of philosophy probably has been related to the appropriation of Nietzsche's philosophy by the Nazis. There exists an important literature, both philosophical and historical, on that subject. Most of it was published in German, shortly after World War II; for reviews, see Ascheim (1993) and Münster (1995). However, most recent commentators on Nietzsche have avoided public declarations on this issue. Three motives are commonly suggested for that tendency to ignore the connection between Nietzsche and the Third Reich. First, historians of philosophy have observed that Nietzsche's doctrines included elements that were incompatible with Nazi ideology (in particular, his rejection of nationalism, his criticism of national wars, and his philo-semitism). Second, they have pointed out that Nietzsche's thinking not only was appropriated by the Nazis but also has been adopted by contemporary philosophers who cannot be suspected of sympathies for national socialism (e.g., the Christian existentialist Karl Jaspers, or, in more recent times, the "libertarian" and "deconstructionist" philosophers Foucault, Deleuze, and Derrida). Third, even in the most obvious cases of convergence between Nietzsche's thinking and Nazi ideology (euthanasia and eugenics), they argue that it would be "unfair to judge a human being in the light of the uncontrolled effects his thinking may have produced on his posterity" (Münster 1995).

Philosophers of biology have been even more reticent about Nietzsche. This also is strange. Probably no major philosopher has used the word "life" as much as Nietzsche did, or given it such importance. Furthermore, with the exception of Spencer, Nietzsche was the first major philosopher who felt the need for a dialogue with Darwin, contesting his principle of the struggle for existence and his theory of the origin of morals, emphasizing the necessity for philosophers to think about "heredity," and finally adopting an extreme version of eugenics at the very time when that ideology began to emerge. The silence of modern philosophers of biology probably has two causes. One is that the philosophy of biology as it exists today (i.e., as an institutionalized field of inquiry) has been elaborated mainly by a group of English-speaking scholars whose training has been more on the side of science and analytical philosophy than has been the case for continental philosophers. Nietzsche, together with, say, Hegel, the "philosopher of nature," and Bergson, provided a perfect example of the old-fashioned "biological philosophy" that the new "philosophy of biology" has wanted to avoid at all costs.

However, there probably is another cause for the silence of modern philosophers of biology in regard to Nietzsche. As Daniel Dennett ironically has said in one of the rare books on the philosophy of biology that have alluded to Nietzsche, that philosopher definitely is not a "politically correct" author (Dennett 1995). André Comte-Sponville has recently provided a crude list of Nietzsche's unpleasant, not to say repugnant, doctrines:

> Nietzsche is one of the rare philosophers, and perhaps the only one . . . who virtually systematically sided with force against law, with violence or cruelty against gentleness, with war against peace, who defended egoism, who placed instincts above reason, . . . who claimed that there were neither moral nor immoral actions, . . . who justified castes, eugenics and slavery, who openly celebrated barbarity, disdain for the mass, the oppression of the weak and the extermination of the sick – all this one hundred years after the French Revolution, while the same philosopher spoke of women and democracy in a way that was extremely unpleasant, although not unique. [Comte-Sponville 1991, pp. 51–2]

To be sure, those assertions, or some of them, would be contested by specialists on the ground that Nietzsche's bold declarations should be understood in their proper context. However, Nietzsche did use a kind of language that literally supported the declarations in the foregoing quotation. *Literally,* at least, Comte-Sponville was right. That might well explain why Nietzsche could not be a priority for the new discipline called the "philosophy of biology," a discipline that in the 1970s and 1980s was manifestly looking for respectability. Thus Nietzsche certainly was not a priority – epistemological or ethical – for a discipline that was trying hard to elaborate a rigorous philosophical status for concepts like "natural selection" and "heredity."

Modern specialists on Nietzsche and modern philosophers of biology thus have had similar motives for not being too curious about the Darwin–Nietzsche relationship. In both cases that relationship has been largely ignored because of the horrifying developments that could be considered to have emerged in history between "social Darwinism" and "Nietzscheism." When Hitler wrote that "the State has the obligation to favor the victory of the best and of the strongest, and to impose the submission of the evil and of the weak" (Hitler 1971), he thought that he was using language that was both scientifically "Darwinian" and philosophically "Nietzschean." In reaction to that disgusting ideology, modern specialists on Nietzsche consider it absurd to evaluate him in the light of Hitler, and of course modern philosophers of biology do not evaluate Darwin in light of Hitler. As a result, the relationship between a major nineteenth century philosopher and a major nineteenth century scientist has been largely ignored. This essay is intended to shed light on that relationship. I shall not deal here with the relationship between Darwinism and Nietzscheism as an autonomous historical phenomenon developing after Darwin and Nietzsche. I shall focus on the various ways in which Darwin became a major concern for Nietzsche.

I have tried to do more than merely highlight a few interesting interactions. Rather, I have tried to provide a general description of the various dimensions of Nietzsche's writings that can be put into a Darwinian context. For this, I have used all the works published in Nietzsche's lifetime, together with the posthumous fragments, especially those published as *The Will to Power,* but not only those. (On the conventions used for references, see the introductory notes immediately preceding the References section.)

I believe that Nietzsche's concern with Darwin can be considered in terms of four dimensions. The first dimension is constituted by a remarkable series of explicit discussions of Darwin's principles of the struggle for existence and selection, most often entitled "On Darwinism" or "Anti-Darwin," and written between 1875 and 1888. In all of those cases Nietzsche criticized Darwin's principle of the "struggle for existence." Basically, "Darwin's error" was to believe that evolution favored the "strong."

The second dimension is constituted by a group of late texts explicitly devoted to "selection." Most often, Nietzsche made use of that term with the signification of voluntary (or artificial) selection in man. Although he knew of the expression "natural selection," he almost never used it. Thus, in most cases, "selection" referred to eugenics. Taken together, these first two dimensions of the Nietzsche–Darwin relationship make sense: Nietzsche doubted Darwin's theory of evolution, especially in the case of man, *and* he was an advocate of eugenics. Those two orientations pervaded the totality of Nietzsche's work, from the beginning to the end, and both of them became strengthened over time.

The other two dimensions of the Nietzsche–Darwin relationship are more diffuse and less explicit, but are intimately related to Nietzsche's most general philosophical positions. The third dimension is constituted by his ideas about the cultural evolution of mankind and the origin of morals. Nietzsche refused to reduce the problem of the origin of morals to the problem of the origin of "altruism"; nor did he accept the interpretation of the moral (and cultural) progress of mankind as a march toward universal altruism. On those two issues (origin and evaluation) Nietzsche opposed Darwin, but also Spencer and, more widely, the English utilitarian philosophers.

The fourth dimension of the Nietzsche–Darwin relationship consists in a series of reflections on the ultimate meaning of the concept of "usefulness,"

157

a concept that he used abundantly despite his opposition to the utilitarian philosophers. In his last productive period, Nietzsche adopted an integral "perspectivism." That was his term for the thesis that there is no room for absolute values in any area of culture, that knowledge and beliefs must ultimately be evaluated in terms of their survival value. It is unclear whether or not Nietzsche was aware of that obvious convergence with the Darwinian view of life.

The further sections of this essay provide evidence regarding the four dimensions of the Nietzsche–Darwin relationship that I have just outlined.

NIETZSCHE AND DARWIN:
SOME BIOGRAPHICAL LANDMARKS

Although this essay does not aim at providing a general interpretation of Nietzsche's philosophy, it is important to list a minimum of chronological landmarks. Born in 1844, Nietzsche, in 1867, received a professorship at the University of Basel (Switzerland). He was then twenty-three and had not yet defended his doctoral dissertation (in fact, he never did). His considerable publications were confined to a relatively short period from 1872 (date of publication of *The Birth of Tragedy*) to January 1889 (beginning of his definitive madness). The commonest presentation of Nietzsche's philosophy distinguishes three phases (Andler 1958, vol. 2, p. 13): First there was the phase of "romantic pessimism" (1869–76), a relatively idealistic period in which the philosopher was mostly interested in Greek tragedy, Schopenhauer, and Wagner. The main publications of that period were *The Birth of Tragedy* (1872) and *Untimely Meditations* (1873–6). The second phase was that of "skeptical positivism" (1876–81), in which Nietzsche became interested in French and English moralists, abandoned his former ideals, and vaunted the heroic freedom to be found in recognizing truth, even if it revealed an appalling reality. In that period of relative enthusiasm for science, his two main writings were *Human All-Too-Human* (1878) and *The Gay Science* (1881). In the last phase (1882–8), Nietzsche gave up any belief in truth and developed a general criticism of all ideal values in the name of "life," viewed as the unique basis for any evaluation. *Thus Spoke Zarathustra* (1882–3), *Daybreak* (1886), *Beyond Good and Evil* (1886), *On the Genealogy of Morals* (1887), *Twilight of the Idols* (1888), *The Antichrist* (1888), and *Ecce Homo* (1888) came from that phase, as did most of the posthumous fragments published as *The Will to Power*. All the writings of the last period included harsh criticisms of science (interpreted as the ultimate refuge of the "ascetic ideal").

Where do the biological sciences, and especially Darwin, come into play in this story? There is something puzzling here. Nietzsche's initial training was exclusively philological and did not include any science. Indeed, he never acquired serious scientific knowledge. But, quite early on, he became deeply aware of his ignorance of the natural sciences. In his autobiography, *Ecce Homo,* he had an interesting comment on what had happened to him at the moment of his rupture with Wagner (1875–6): "All at once it became clear to me in a terrifying way how much time I had already wasted – how useless and arbitrary my whole existence as a philologist appeared in relation to my task. . . . Ten years lay behind me in which the nourishment of my spirit had really come to a stop; I had not learned anything new that was useful. . . . My knowledge simply failed to include *realities,* and my 'idealities' were not worth a damn. A truly burning thirst took hold of me: henceforth I really pursued nothing *more* than physiology, medicine, and natural sciences" (*Ecce Homo,* "Human All-Too-Human," § 3). In fact, Nietzsche's readings in the natural sciences had begun earlier. He had enthusiastically discovered Darwin through Lange's *History of Materialism* in 1866, while still a student. A year later, when he was appointed at the University of Basel, he had formed a friendship with Rütimeyer, a major paleontologist who had a friendly relationship with Darwin and who played an important role in the introduction of Darwin's ideas into Germany. Although Rütimeyer thought that Darwin was the nineteenth century's greatest naturalist, his own evolutionism was explicitly impregnated by Lamarckism. He also manifested an extreme hostility toward Haeckel. Those features can be found intact in most of Nietzsche's reflections on biological evolution. Throughout his life, Nietzsche preferred to read neo-Lamarckian authors, and he adopted their ideas. Especially important in that respect were Oscar Schmidt and Karl von Nägeli (in the 1860s and 1870s), and above all Wilhelm Roux's *Struggle of Parts in the Organism* (1881).

Besides that interest in the properly scientific aspects of evolutionist literature, Nietzsche discovered the impact of Darwinism on moral theory when he struck up a close friendship with Paul Rée in 1876. Rée was a young Prussian scholar who was familiar with the English utilitarian philosophers, Darwin's *Descent of Man,* and Spencer's writings. In 1877, Rée published a book entitled *The Origins of Moral Feelings* in which he proposed a natural history of morals, beginning with "egoist" and "altruist" instincts and ending with moral consciousness, remorse, and free will. Nietzsche was initially impressed by that book and began to read the English utilitarian philosophers and Spencer – "a mixture of silliness and Darwinism" was his rapid verdict. That was the beginning of what became Nietzsche's major concern in his last period – the origins and value of morals. As explicitly stated in the foreword to the *Origin of Morals,* that enterprise was directed against the Darwinian-

Spencerian-utilitarian view of morals – a modern avatar of the Christian egalitarian view of man in Nietzsche's view. For more details on Nietzsche's scientific readings, see Richter (1911) and Andler (1958).

Nietzsche's interest in Darwinism thus came from two sources: One consisted of more or less popular books that informed him about Darwin's general theory of evolution. The other was made up of a large network of authors involved in a naturalistic and utilitarian account of human cultural evolution. It was only when he discovered the discussions about moral and cultural evolution (in the mid-1870s) that Nietzsche began to polemicize against Darwin and the "Darwinians." As far as possible, I shall separate the polemic against "Darwinism" in a narrow sense (polemic against Darwin's "struggle for existence" as a major principle of organic evolution) from the polemic against "Darwinism" as an interpretation of morals, which Nietzsche loathed. In the second case, "Darwinism" was inseparable from Spencer and, more widely, from the English utilitarian philosophers.

Among other sources of scientific information that played major roles in Nietzsche's thinking, we must mention his discovery of the energetist conception of physics through the works of the eighteenth-century Croatian physicist Boscovich and William Thomson, who continued Boscovich's work in the nineteenth century. Boscovich was noted for having proposed a reconstruction of Newtonian dynamics in a language excluding any reference to material particles or atoms. Instead, he postulated nonextended and inert points subject to the actions of forces. Nietzsche took Boscovich's nonmaterialist physics very seriously, and after discovering Boscovich's work in 1872, he referred to it relatively frequently. I mention this because I think that there probably was a common root for Nietzsche's interest in Darwin and Boscovich. Nietzsche recognized in both scientists a view of science that attributed an essential role to the notion of "force." For Boscovich, that was through his explicit use of the physical concept of the composition of forces, and for Darwin it was through his principle of the struggle for existence. No wonder, then, that Nietzsche began to comment on the two scientists at approximately the same time (1872–5), as shown by the posthumous fragments (see volumes II* and II** of the Colli-Montinari critical edition of the complete philosophical works).

Before we turn to the criticisms specifically directed at Darwin, we must deal with the embarrassing issue of whether or not Nietzsche ever read Darwin. In spite of a number of studies on this subject,[1] there is no definitive answer. It is certain that Nietzsche read "A Biographical Sketch of an Infant" (Darwin 1877), because he recommended it warmly to Paul Rée in a letter dated 1877, with a precise reference to *Mind*. But there is no direct evidence that he actually read even one book by Darwin. He never cited a precise quotation or reference, although that was his common pattern regarding almost all authors. It is likely that he consulted *The Variation of Animals and Plants under Domestication,* for Richter (1911) mentions some sentences that could hardly have come from anywhere else. It also seems hardly possible that he never read *The Descent of Man,* in view of his repeated criticisms of the Darwinian theory of the origins of moral behavior. As for *The Origin of Species,* I tend to believe that Nietzsche did not seriously read that work, but again, direct evidence is lacking. In fact, Nietzsche's knowledge of Darwin's ideas relied essentially, if not exclusively, on German accounts and/or criticisms. Rütimeyer's precise and numerous reviews of Darwin, Wallace, and Haeckel in the *Archiv für Anthropologie* clearly constituted a major source for his philosophical colleague and friend in Basel, together with a number of German popular science books found in Nietzsche's personal library (Richter 1911).

NIETZSCHE'S CRITICISM OF THE "STRUGGLE FOR LIFE"

We now come to Nietzsche's explicit discourse on Darwin. Although the earliest works and unpublished fragments (before 1872, the year of *The Birth of Tragedy*) did not mention Darwin, Nietzsche at first felt sympathy for Darwin as an author who had developed a conception of organic progress through "struggle." The struggle for existence echoed Jakob Burckhardt's reflections on the role of "agonistic" values in the Greeks' conception of existence. Burckhardt, a major philologist and historian, was a close colleague and friend of Nietzsche in Basel and had an important influence on Nietzsche's early work. A sympathy for Darwinian ideas can also be seen in the metaphorical use Nietzsche episodically made of Darwin's principle of the struggle for existence in the years 1872–6, when he wrote, for instance, that "Darwinism is also right about the production of mental images: the most vigorous image devours the weakest" (Nietzsche 1970–97, II*, fragment 19[87], 1872–73). That was one of Nietzsche's first open allusions to Darwinism. It implied that mental or moral phenomena (ideas, instincts, passions) struggle with each other just as individuals struggle with each other in Darwin's conception of evolution. Note that the sentence alludes to a struggle for existence rather than to natural selection. Note also that Nietzsche spontaneously interprets the "struggle" between organisms not as a competition but rather as a predator–prey relation, with a clear use of the idea of nutrition. As we shall see later, when Nietzsche finally rejected the Darwinian view of evolution, he did so in favor of a neo-Lamarckian conception based on the concept of nutrition.

Jean Gayon

Nietzsche began to formulate an explicit argument against Darwin's "struggle for existence" the same year he became friends with Paul Rée. Beginning in 1875, we find the first version of what I shall call the "anti-Darwin module." Let me make clear what I mean by "module." Nietzsche wrote most of his books in the form of long series of short paragraphs or aphorisms. In a number of cases we can see him coming back again and again to the same subject, repeating some sentences, but modifying and amplifying others. Sometimes the title remains the same, sometimes it changes, sometimes there is no title, sometimes one module can be integrated in another one. In the case that interests us here, a simple enumeration of the titles and dates of the "anti-Darwin module" is instructive:

- "About Darwinism" (posthumous fragment [1875], Nietzsche 1970–7, II*, 12[22])
- "Ennoblement through Degeneration" (*Human All-Too-Human,* 1878, § 224)
- "Once More the Origin of Scholars" (*The Gay Science,* 1881, § 349)
- "Against Darwinism" (*The Will to Power,* III, § 647 [1883–8])
- "Anti-Darwin" (*Twilight of the Idols,* 1888, "Expeditions of an Untimely Man," § 14)
- "Anti-Darwin" (*The Will to Power,* III, § 685 [March–June 1888])
- "Anti-Darwin" (*The Will to Power,* III, § 685 [March–June 1888])

Let us follow the evolution of the anti-Darwinian argument in this series of texts. Nietzsche's original opposition to the "struggle for existence" was related to his conception of the "intellectual progress" of peoples in the course of human history. Human cultural progress, for Nietzsche, resulted from a mixture of stability and creative change in "races." A race's ability to conserve itself depended on the existence of a majority of men "endowed with a sense of community and sympathetic feelings." In the earliest versions of the anti-Darwinian module (1875 and 1878), such individuals were said to be the "strongest." But any progress would come from "degenerated" or "weak" but creative individuals. Such individuals would introduce novelties that, by definition, would weaken the (stubborn) stability of the community: "Progress depends on unattached, much more unreliable and *weaker* individuals, who seek out novelty and explore it in various ways: many of them disappear without having had any effect, but, on the whole, they *fashion* the current order, and from time to time they *weaken* this stability" (posthumous fragment [1875], Nietzsche 1970–97, II*, 12[22]). We find the same idea three years later: "Wherever progress is to ensue, deviating natures are of greatest importance. Every progress of the whole must be preceded by a partial weakening. The strongest *retain* the type, the weaker ones help to *advance* it"

(*Human All-Too-Human,* 1878, § 224). Therefore human cultural progress implied a delicate balance between a race or a people's ability to maintain itself as such and the periodic emergence of more or less asocial but creative individuals able to introduce higher standards of culture.

This is the precise context of Nietzsche's first open criticism of Darwin: "The struggle for existence is not the key principle. . . . Progress is made possible by the weakest character, because it is the noblest or at least the freest" (posthumous fragment [1875], Nietzsche 1970–97, II*, 12[22]). The same idea can be found in *Human All-Too-Human* (1878, § 224). Nietzsche's criticism of Darwin was that in man, at least, progress was not a consequence of the better ability of the *strongest* to survive. Rather, just the opposite: Human cultural progress depended on "weak," "degenerated," "unhealthy" individuals, individuals who would perish more easily than those nicely socialized individuals who conformed to the "type." An interesting feature of that initial form of the anti-Darwinian argument is that it was restricted to the case of man: "I want to restrict my reflections to Man and to avoiding drawing conclusions about animal evolution on the basis of the laws that govern this ennoblement of man that takes place against a background of weak and degenerated characters" (posthumous fragment [1875], Nietzsche 1970–97, II*, 12[22]).

In a sense, that doctrine remained, in its most basic features, in all of Nietzsche's subsequent writings. Nietzsche always remained faithful to his heroic view of human cultural evolution, in which innumerable atypical and creative individuals would perish at the same time as they tried to bring mankind a little further in the direction of higher values. That, for instance, was the meaning of the parable of the tightrope-walker in *Zarathustra*'s Prologue. Although he fell to the ground amidst the crowd's hostility, although he failed and died, he was a symbol of man's dignity. And Zarathustra's first action was precisely to honor his memory and bury him decently, whence the famous aphorism: "Man is a rope, fastened between animal and Superman – a rope over an abyss . . . what is great in man is that he is a bridge and not a goal" (*Zarathustra,* 1883, Prologue, 4). Human cultural progress relied on innumerable individuals who, like the funambulist, tried but perished.

In the 1880s, two major modifications took place. First, Nietzsche reversed his use of "strong" and "weak." Second, he came to generalize his criticism of the "struggle for existence," extending it to all organic beings, and thus committing himself to a general attack on Darwin's theory of organic evolution.

The inversion of "weak" and "strong" can be located around 1880. Before that date, Nietzsche distinguished between "strong" and "superior." "Strong" connoted a typical (mental) constitution, firmly established and perpetuated in society through heredity and education. In contrast, independent and free individuals were said to be fragile, unhealthy, weak, both individually and in terms of descent. However, any move toward a "higher civilization" relied on those "weak" individuals. That situation generated a conflict between "superiority" (in terms of civilization) and "strength." Nietzsche's first solution was the idea that great civilizations would find a way to preserve their ability to last (by disciplining the masses) and simultaneously allow for the creativity of asocial but talented people (*Human All-Too-Human,* 1878, § 224). But his increasing obsession with "heredity" made that solution untenable. That was why he turned to a more or less deliberate Darwinian approach to the problem: The "superior" and free individuals would also be able to survive and leave progeny: "The strongest individuals will be those that are able to resist the laws of the species without perishing, the isolated individuals. The *new nobility* will be created from them: but while it is being created an infinite number of isolated individuals *will have* to perish!" (Nietzsche 1970–97, V, fragment 11[194], 1881).[2] That quotation illustrates Nietzsche's move toward a more biological vocabulary ("species" instead of "society") and toward eugenics. That probably explains the remarkable inversion of "weak" and "strong" that occurred quite suddenly in the early 1880s. That change can be clearly seen in the following unpublished fragment in which Nietzsche criticized both Spencer's conception of the evolution of altruism and all authors who made use of natural selection to explain human cultural evolution:

> Spencerian adaptation is conceivable, but only on condition that each individual becomes a useful instrument and above all feels himself to be nothing but such an instrument . . . which implies the elimination of individualism . . . !
> This transformation is possible; it perhaps corresponds to the direction taken by the whole of history! But in this case, individuals become continually *weaker* – this is the story of the decline of mankind, which is governed by the principle of unselfishness, of living for others and sociality! (posthumous fragment [1880–1], Nietzsche 1970–97, IV, 10[D60])

In that version, weakness was proportional to the degree of sociability and altruism of individuals, whereas strength characterized the individuals who resisted the process of social homogenization. In other fragments from the same period, Nietzsche criticized the idea of human evolution being driven by what he called "selection of the species," a phrase by which he meant a selection favoring the species (or the race) as such and leading to the establishment of homogeneous and altruistic types.[3] In such a conception, "the goal is to make Man as regular, as fixed as most animal species have become" (posthumous fragment [1881–2], Nietzsche 1970–97, V, 11[69]). And that was precisely what Nietzsche came to denounce more and more: In Man, anything that opposed the development of individualistic values would lead to degeneration, that is to say, an increase in the number of "meaner" individuals ("meaner" in the same sense in which Galton used the expression during the same period: the "typical" or "average" individuals). Now if we equate "weak" and "mean," on the one hand, and "strong" and "exceptional," on the other, we then have Nietzsche's anti-Darwinian leitmotiv typical of the writings of his last period. Because in Nietzsche's romantic vision of human history the immense majority of creative individuals perished, it became obvious for him that, on the whole, man's evolution did indeed follow Darwin's principle of the "struggle for existence," although seen from an apparently paradoxical point of view: Those who won were not the "strongest," but the "weakest."

Once Nietzsche had clearly adopted that formula, his comments on Darwin evolved in three directions. First, he advanced increasingly violent criticisms of all moral systems that favored the victory of the "weak" over the "strong" in the course of human history; for Nietzsche, the utilitarian-Darwinian-Spencerian view of ethics was merely the most recent avatar of that story. Second, he developed an increasingly explicit justification for intentional selection in the human species (i.e., eugenics). Third, he expanded his criticism of Darwin's principle of the "struggle for existence" in the case of man into a general criticism of Darwin's theory of evolution. Those first two aspects will be treated later in Sections 4 and 5. The remaining part of this section is devoted to the third aspect: "Darwin's error."

In the first version of the anti-Darwin module (1875), Nietzsche had explained that his reservations in regard to Darwinism did not relate to "animal evolution," but only to the application of the principle of the "struggle for existence" to man. Indeed, he never stopped emphasizing the peculiarities of human cultural evolution. However, Nietzsche went beyond his criticism of Darwin in the name of human evolution. In his last period, he tried to develop a general criticism of the principle of the struggle for existence, arguing that the principle not only was inapplicable to humans but also was doubtful as a major factor in organic evolution. What arguments did Nietzsche oppose to that principle? What precisely induced him to speak of "Darwinism's error" with respect to biological evolution in general? (*The Will to Power,* I, 395 [1888]). The reader of Nietzsche should not expect to find subtle methodological objections, but his criticisms did reflect a genuine effort to exchange views with Darwinian biologists and to challenge their most spontaneous convictions.

Although they often overlapped in the texts, his objections fell into two classes. Those in the first class challenged the outcome of the struggle for existence: It did not automatically lead to the emergence of "higher" and "stronger" types. The objections in the second class challenged the extent of the principle of competition: True, the struggle for existence went on in nature, but it was marginal; other principles came into play that also involved "struggle," but not Darwin's form of competition.

Just as he had contested the idea that the struggle for existence in societies led to the establishment of higher and stronger types, Nietzsche came to contest that principle for organic species in general:

> I see all philosophers, I see science kneeling before a reality that is the reverse of the struggle for existence as taught by Darwin's school – that is to say, I see on top and surviving everywhere those who compromise life and the value of life. – The error of the school of Darwin becomes a problem to me: how can one be so blind as to see so badly at *this* point? That *species* represent any progress is the most unreasonable assertion in the world: so far they represent one level. That the higher organisms have evolved from the lower has not been demonstrated in a single case. I see how the lower preponderate through their numbers, their shrewdness, their cunning – I do not see how an accidental variation gives an advantage, at least not for so long a period. . . . *In summa:* growth in the *power* of a species is perhaps guaranteed less by a preponderance of its children of fortune, of strong members, than by a preponderance of average and lower types. – The latter possess great fruitfulness and duration; with the former comes an increase in danger, rapid wastage, speedy reduction in numbers. (*The Will to Power,* III, § 685 [March–June 1888])

Several ideas are contained in this quotation. First, we should note that Nietzsche seemed close to contesting the very idea of common descent. To be fair, he did not openly say that "superior organisms" did not come from "inferior organisms"; he said only that it had not been demonstrated. However, Nietzsche's concern was not fixism; that would hardly have agreed with the universal mobilism he always supported. What mattered to him was to reject a naive idea of evolutionary progress. If we look carefully at the preceding quotation, it said that "inferior organisms" dominated in numbers in the living world, and it suggested that there was no sign whatsoever that what biologists called inferior organisms would ever disappear. The following passage is clearer in this respect: "The whole animal and vegetable kingdom does not evolve from the lower to the higher – but all at the same time, in utter disorder, over and against each other. The richest and most complex forms – for the expression 'higher type' means no more than this – perish more easily: only the lowest preserve an apparent indestructibility. The former are achieved only rarely and maintain their superiority with difficulty; the latter are favored by a compromising fruitfulness" (*The Will to Power,* III, § 684 [March-June 1888]). In more concrete terms, he would say that plants (vs. animals), protists (vs. multicellular organisms), invertebrates (vs. verte-brates), and inferior vertebrates (vs. superior vertebrates) did not show the slightest tendency to disappear to the benefit of higher forms. Quite the op-posite: The less complex forms were more likely to last longer, and they were more numerous. Like it or not, that was Nietzsche's argument. It was strange, but not absurd. At the level of the entire biosphere, the struggle for existence did not entail the disappearance of "inferior species." The situation was rather the opposite. To use Nietzsche's political and aristocratic metaphors: – "no-ble" species (predators) indeed required the existence of myriad "subordi-nated" species (prey and parasites).

Of course, an orthodox Darwinian would object here that that was not a fair argument against Darwin, who used the concept of the struggle for ex-istence primarily at the intra-species level. In reality, Nietzsche was aware of that, and that was why his criticism involved another dimension. Looking again at the foregoing quotation, we see that the last sentences said that any species would evolve in the direction of proliferation of the more "mediocre" types. The significance of that assertion can be gleaned by studying another passage from the same "Anti-Darwin" module:

> One credits natural selection at the same time with the power of slow and end-less metamorphosis; one wants to believe that every advantage is inherited and grows stronger with succeeding generations. . . . one observes the fortunate adaptation of certain creatures to very special conditions of life, and one ex-plains that these adaptations result from the influence of the milieu. But one nowhere finds any example of *unconscious selection* (absolutely not). The most disparate individuals unite with one another, the extremes are submerged in the mass. Everything competes to preserve its type. (*The Will to Power,* III, § 684 [March–June 1888])

That passage is interesting because it may well have been the only point at which Nietzsche attacked Darwin not only for his principle of the struggle for existence but also for his theory of natural selection. "Natural selection" was extremely rare in Nietzsche's vocabulary. That text is, in fact, the only passage where I have been able to find the term. What Nietzsche said in that exceptional fragment was that there was no selection in nature, no natural (or "unconscious") selection, because the organisms of a particular species mated indiscriminately with each other, a process that led to the conserva-tion of the mean type. That precision clarified the idea of "preponderance of mediocre types." Just as in the work of Galton, the expression "mediocre types" referred to the "meaner types" (Galton 1865), that is to say, the types that were closest to the statistical mean for the species. Nietzsche thus extended to all organisms a claim about human evolution that was quite common in the contemporary eugenic-literature: In nature, as in man, indiscriminate widespread fertility led to the domination and conservation of the more mediocre type. We see, therefore, that Nietzsche's final criticism of Darwin, with its explicit reference to natural selection, was a generalization of his former argument about the irrelevance of the struggle for existence in man. It boiled down to the paradoxical, but rather trivial, argument that in both man and nature the winners were not the "strongest" (the exceptional beings) but rather the "weakest" (the most numerous individuals) (*Twilight of the Idols,* "Expeditions of an Untimely Man," § 14). For Nietzsche, the aristocratic thinker, Darwinism, social and biological, was a "plebeian" conception: "the whole of English Darwinism breathes something like the musty air of English overpopulation, like the smell of distress and overcrowding of small people" (*The Gay Science,* § 349).[4]

We now come to the second class of objections against the idea of the struggle for existence in nature, those that challenged the extent of its application. Such objections are particularly interesting for both the historian of science and the moral philosopher, because they reveal a positive side of Nietzsche. Instead of telling us what Nietzsche did not like in the work of Darwin, they tell us what he really believed about evolution, and in relation to which philosophical theses.

The spirit of that second category of objections was that the struggle for existence, if it did occur in nature, was an exception, a marginal phenomenon, the importance of which had been exaggerated by the Darwinians. The first clear formulation of that idea can be found in *The Gay Science.* It came immediately after Nietzsche's attack on the plebeian character of Malthusianism and Darwinism in Britain (see the preceding quotation):

> In nature it is not conditions of distress that are *dominant* but overflow and squandering, even to the point of absurdity. The struggle for existence is only an *exception,* a temporary restriction of the will to life. The great and small struggle always revolves around superiority, around growth and expansion, around power – in accordance with the will to power which is the will of life.
> [*The Gay Science,* 1881, § 349]

There are similar declarations in *Twilight of the Idols* ("Untimely Meditations," § 14). What did Nietzsche mean by saying that the struggle for life was an exception? He probably did not intend to say that, empirically speaking, the phenomenon of competition was rare in nature. Nietzsche's problem was not empirical but conceptual. His idea was that the "struggle for life" (or Darwin's "struggle for existence," for Nietzsche used both expressions) was unimportant in the face of another kind of struggle, the struggle for power. Here we come close to one of the most puzzling semantic problems in the interpretation of Nietzsche's philosophy. The semantic circle was as follows: (1) The "will to power" was an essential feature of all life (posthumous fragment [September 1885], Kröner, XIII, § 591). (2) The "will to power" was not the "will to live" (an allusion to Schopenhauer; see *The Will to Power,* III, § 692 [March–June 1888]). (3) "Struggle" was essential in the concept of the "will to power." (4) What mattered was not the "struggle for life," but the "struggle for power": "'struggle for existence': that designates an exceptional circumstance. The rule is, rather, the struggle for power, for 'more' and 'better' and 'faster' and 'more often'" (posthumous fragment [April–September 1885], Kröner, XIII, § 558).

In order to fully understand those formulations, we must take into account the three levels of discourse that Nietzsche used to elucidate the concept of the "will to power": political, metaphysical, and biological. At the political level, we have the basic aristocratic schema: "Power" (or strength) always consisted of a hierarchical relation in which some entities were subordinated to others. Although not everything in the world (or even in man) was political, that schema had a metaphorical value: It was an easily understandable *model* of the general notion of "power." But it was only a model.

At the metaphysical level, the "will to power" must be understood in the context of a general "conception of the world." Nietzsche interpreted the "world" in terms of forces rather than in terms of matter: "This world: a monster of energy, without beginning, without end; a firm, iron magnitude of force that does not grow bigger or smaller, that does not expend itself but only transforms itself . . . a sea of forces flowing and rushing together, eternally changing, eternally flooding back, with tremendous years of recurrence, with an ebb and a flood of its forms. . . . – *This world is the will to power – and nothing besides*" (*The Will to Power,* IV, § 1067 [1885]). At that level, "power" connoted the idea that "centers of force" appeared at all levels of reality, generating order and complexity. That idea was quite abstract. Its function was to recall that "power" was not especially attached to a particular level of description of reality (physical, psychological, biological, political, etc.).

This brings us to the third level of elucidation of the concept of the "will to power": biology. At that level, Nietzsche thought that he had found another model that could save him from the objection of political (and psychological) anthropomorphism. That model was inspired by various German neo-Lamarckian authors, such as Rolph and Wilhelm Roux, whose work he had read in the early 1880s. For an account of those influences, see Andler (1958). From Roux, Nietzsche retained two major ideas: First, the struggle between the parts of an organism was at least as important as the struggle between organisms. Parts (cells, tissues, organs) competed against each other in the morphogenetic process, and they went on competing for nutrition over the course of life, whence the idea of an "internal adaptation" (a typical neo-Lamarckian expression) that Nietzsche believed to be much more important than Darwin's adaptation to the external milieu: "The influence of 'external circumstances' is overestimated by Darwin to a ridiculous extent: the essential thing in the life process is precisely the tremendous shaping, form-creating force working from within. . . . The new forms molded from within are not formed with an end in view; but in the struggle of the parts a new form is not left long without being related to a partial usefulness and then, according to its use, develops itself more and more completely" (*The Will to Power,* III, § 647 [1883–8]). Second, Nietzsche was enthusisastic about Roux's notion of "functional assimilation" (nutrition) as the ultimate basis of the evolutionary process. Roux thought that a change in the environment would induce a change in the nutritive regime that itself would modify the struggle between the parts of the organism. For Nietzsche, biological assimilation was a good illustration of the will to power and of his own notion of creative struggle: "The will to power can manifest itself only against resistances; therefore it seeks that which resists it – this is the primeval tendency of the protoplasm when it extends pseudopodia and feels about. Appropriation and assimilation are above all desire to overwhelm, a forming, shaping and reshaping, until at length that which has been overwhelmed has entirely gone over into the power domain of the aggressor and has increased the same" (*The Will to Power,* III, § 656 [1887]).

Those neo-Lamarckian conceptions fitted well with Nietzsche's individualism and with his idea of strength (something that originated from inside and imposed its rule throughout): "Life is not the adaptation of inner circumstances to outer ones, but will to power, which, working from within, incorporates and subdues more and more of that which is 'outside'" (*The Will to Power,* III, § 681 [1883–8]). Nietzsche generalized the notion of "assimilation," extending it to a general view of life and biological progress in terms of trophic relations between species: "A multiplicity of forces, connected by a common mode of nutrition, we call 'life'" (*The Will to Power,* III, § 641 [1883–8]). In such a conception, biological progress did not rely on a notion of struggle as competition for scarce resources, but on a notion of struggle as assimilation of certain organisms by others: "*Course of the struggle:* the fighter tries to transform his opponent into his *antithesis*" (*The Will to Power,* II, § 348 [Spring–Fall 1887]). Another aspect of that emphasis on trophic relations was Nietzsche's curious idea that species infected by many parasites were likely to be "higher species." Even though parasites would be detrimental to the host, their existence would be a sign of the host's power, or ability to structure a certain environment around it (such organisms not only "survive" but also fed other beings): "Which is the highest type of being and which is the lowest? The parasite is the lowest type; but he who is of the highest type nourishes the most parasites" (*Thus Spoke Zarathustra,* III, "Of Old and New Law-Tables," § 19).

We are now able to understand the deep philosophical roots of Nietzsche's reluctance to accept Darwin's "struggle for existence" and "natural selection." What he ultimately disliked in those principles was that they emphasized "conservation" rather than "augmentation." Indeed, in the literature of that time, both principles often were formulated in terms of "survival." Struggle for survival, in Darwin's own terms, meant that some individuals would "survive," and others not.[5] As for natural selection, it meant, in Spencer's famous phrase, "survival of the fittest." To Nietzsche, such a vocabulary evoked Spinoza's *conatus* (the effort by which each being enforces the preservation of its own being), Schopenhauer's *will to live,* and the moralist's trivial "instinct of conservation." In light of his romantic view of existence and life, that was the most miserable conception he could imagine (not to speak of the obvious religious heritage of "survival"). Nietzsche's contempt for the vocabulary of "conservation," "preservation," and "survival" pervaded all his writings. But it probably was expressed with the greatest crudity and clarity in the *Genealogy of Morals,* in the context of criticism of the "ascetic ideal": "*The ascetic ideal springs from the protecting instinct of a degenerating life* which tries by all means to sustain itself and to fight for its existence; it indicates a partial physiological obstruction and exhaustion. . . . Life wrestles in it and through it with death and *against* death. The ascetic ideal is an artifice for the *preservation* of life" (*On the Genealogy of Morals,* 1887, III, § 13). That passage suggested that the "struggle for existence" was a conception of "degenerated," "sickly," "unhappy," "exhausted," "weak" people. That was the exact opposite of what Nietzsche meant by "struggle for power": augmentation, increase, excess, prodigality. We can now understand what the philosopher meant by opposing "struggle for life" and "struggle for power." For the romantic and aristocratic thinker, life could not amount to merely "surviving": "Life itself is to my mind the instinct for growth, for durability, for an accumulation of forces, for *power*" (*The Antichrist,* 1888, § 6). In passing, that probably was why Nietzsche tended to substitute the expression "struggle for life" (the struggle of those who can do no more than just "survive," or keep alive) for Darwin's more abstract "struggle for existence."

Such a philosophical understanding of Darwinism may appear strange. Nevertheless, there is no doubt that such was the framework according to which the German philosopher received and evaluated Darwin. Like it or not, the "struggle for life" was for Nietzsche a poor, wretched, plebeian phrase:

"That our modern natural sciences have become so thoroughly entangled in this Spinozistic dogma[6] (most recently and worst of all, Darwinism with its incomprehensibly onesided doctrine of the 'struggle for existence') is probably due to the origins of most natural scientists: In this respect they belong to the 'common people'; their ancestors were poor and undistinguished people who knew the difficulties of survival only too well at firsthand" (*The Gay Science,* 1881, § 349).

Let us conclude with Nietzsche's critical observations on the concept of "struggle for life." The German philosopher certainly was not a *careful* reader of Darwin. He was not interested in the details of the English naturalist's reconstruction of the whole of natural history, but he felt it necessary to oppose his philosophical conjectures to Darwin's "struggle for existence." At the beginning, he thought that Darwin's principle did not apply to human cultural progress, which he believed to have resulted from creative but biologically and socially weak individuals. That was the first version of the "Anti-Darwin" module. Later he developed his famous interpretation of morals as a huge cultural construction directed against "vital values." That conception entailed an inversion of the meanings he had previously given to "weak" and "strong" (weak = the more "mediocre" and "numerous"). Consequently, the second version of the "Anti-Darwin" module consisted of an attack against the intuitive Darwinian idea that the strong would win. In fact, in man's recent (or "Christian") history, the winners had been the weakest, the sickly, the suffering, "the crowd." Finally, Nietzsche did not fail to generalize: He extended his pessimistic and romantic conception of culture to life in general. That final version of the "Anti-Darwin" module had two aspects: First, in nature, as well as in man, "inferior" forms of life prevailed. That theme appeared only in the later writings, under the label of "Darwinism's error." Second, Nietzsche thought that the principle of the struggle for existence was not so much false as unimportant. Hence his final formulation: "struggle for power" rather than "struggle for life," "will to power" rather than "will to live," "power" rather than "survival," "augmentation" rather than "conservation" (or "preservation"), "nutrition" rather than "competition," Lamarckism rather than Darwinism.

In this discussion I have deliberately not discussed the misunderstandings, errors, and contradictions that marked Nietzsche's knowledge of Darwin. My objective has been to describe as comprehensively as possible how Nietzsche appropriated, interpreted, and discussed Darwin's theory of evolution (or what he had heard about it). He certainly did not know much about Darwin, but he took Darwinism seriously. Why? The explanation does not need to be complicated. Nietzsche's attitude toward Darwin was deeply ambivalent. In his last period, as he became increasingly violent in his criticisms, he also declared that he would have liked for Darwin to have been right: "[This struggle for existence] exists . . . its outcome is the reverse of that desired by the school of Darwin, of that which one *ought* perhaps to desire with them: namely, the defeat of the stronger, the more privileged, the fortunate exception" (*Twilight of the Idols,* "Untimely Meditations," 1889, § 14; see also *On the Genealogy of Morals,* Foreword, § 7). That was a perfect expression of what Nietzsche had in mind: Darwin's "struggle for life" was both false and desirable. Desirable: One must help the "strong" to develop and dominate; whence Nietzsche's unambiguous commitment in favor of eugenics. The following section is devoted to this topic.

"SELECTION" (NIETZSCHE THE EUGENIST)

Before World War II, Nietzsche's commentators had not the slightest doubt about Nietzsche's support for eugenic ideas. Most recent specialists, however, seem to have ignored that aspect of his thinking. In reality, there is ample evidence that deliberate human selection was one of his most persistent ideas. It was first outlined in 1874 in *Untimely Meditations,* III ("Schopenhauer as educator"). It was episodically evoked over the next ten years, and was systematically developed in *Beyond Good and Evil* (1886) and in the posthumous fragments contained in *The Will to Power* (essentially 1881–8). From a chronological point of view, it should be noted that Nietzsche's thoughts about human selection (1874–88) developed in the same period as his comments on Darwin (1875–88). In both cases, "selection" was a key concept. But the two lines of thinking went in opposite directions: Whereas Nietzsche's criticisms of Darwin resulted in a general refusal of Darwin's evolutionary theory, Nietzsche committed himself more and more openly in favor of eugenics.

An interesting clue as to how those two aspects interacted is seen in Nietzsche's use of the word "selection." In the following paragraph, I provide a list of all occurrences of that word that I have been able to find in the complete works (excluding the correspondence, however). I do not pretend to absolute exhaustiveness; involuntary occasional omissions are possible. I must mention here a linguistic difficulty concerning the precise German term used by Nietzsche that has been translated into foreign languages (English, French) as "selection." In German, several words had been used for the translation of Darwin's "selection": *Selektion, Zuchtwahl, Züchtung.* In the nineteenth century, the first translators did not use *Selektion.* The first translation, by H. G. Bronn in 1860, used *Züchtung,* a word that, rigorously speaking, means "breeding."[7] *Züchtung* and the related substantive *Zucht* and the verb *züchten* are as ambiguous in German as "breeding" is in English or "élevage" is in French. In Carus's numerous revised editions of Bronn's translation, *Zucht-wahl* was substituted for *Züchtung* and became the conventional translation for Darwin's "selection."[8] However, even in the last edition of that translation (1920, recently reissued in 1992), *Züchtung* still appears occasionally as the translation of "selection," in the context of artificial selection (e.g., Darwin 1992, pp. 47–51). Today, *Züchtung* (or *Zucht*) would not be identified with "selection," but in Nietzsche's time, that was a common way of expressing Darwin's notions of natural and artificial selection. For that reason, I have listed a certain number of occurrences of *Zucht* and *Züchtung* as obvious evocations of Darwin's "selection."[9]

"Selection" and the related terms "select" and "selector" were rare in the books published by Nietzsche during his lifetime. When present (*Beyond Good and Evil*, 1886, §§ 61, 62, 201, 251, 262; *Ecce Homo*, 1888, § 20), "selection" was always in the sense of deliberate selection of humans. That is a good indication of what Nietzsche meant when he publicly used the word "selection." However, the posthumous fragments provide a different picture. They show that Nietzsche's uses of the term "selection" fall into three categories, corresponding to three characteristic contexts of thought. The first context consisted in critical developments concerning Darwin's theory of evolution (*The Will to Power*, I, § 54; II, § 246; III, §§ 684, 685; Kröner, XII-1, §§ 235, 243, 443; XVI, § 1073). In that context, "selection" meant "natural selection." Note, however, that the latter expression is almost never used (exception: *The Will to Power*, III, § 684; see the preceding section). The second context consisted in discussions about more or less deliberate selection (or "rearing") of human types in past or even contemporary civilizations (the Greeks, Christianity, the Hindus, modern democracy, etc.) (Kröner, X, § 287; XII-1, §§ 238, 408; XIV-1, § 151; XIV-2, § 287). Nietzsche's idea was that civilizations shaped individuals not only through education but also through the effects of their life-styles on the survival of individuals. *Beyond Good and Evil* (§ 262) provides a good example. There Nietzsche said that during the Renaissance, Venice functioned as "an institute of voluntary or involuntary selection" that favored the establishment of "a species of severe, warlike, prudently taciturn men, close-mouthed and closely linked." More generally, Nietzsche thought that all civilizations "selected" particular moral types. In that second context, "selection" corresponded to Darwin's "unconscious [artificial] selection" (Darwin 1868). Several texts suggest that Nietzsche knew the technical distinction between "unconscious" (in fact, more or less voluntary) selection, and "methodical" selection (compare, for instance, Nietzsche's allusions to "unconscious selection" and "conscious breeding" in *The Will to Power*, III, § 684, and IV, § 954 [both fragments written in March–June 1888]). The third context for his use of the term "selection" consisted in a number of fragments devoted to the "great selective [or breeding] thinking," that is, the project of "methodical, artificial and conscious selection [or breeding] of man" (*The Will to Power*, IV, § 954 [1885–6]). A large number of aphorisms in *The Will to Power* bear on this question. Some of them made explicit use of the word *Zuchtwahl*, the word that finally emerged in German as a standard equivalent for the English "selection" (I, § 134; IV, 1058). Quite often, however, Nietzsche used *Züchtung* (or a similar word with the similar root *Zucht*), but the meaning is obvious (I, § 108; II, § 398; III, § 732; IV, §§ 862, 898, 903, 954, 960, 964, 980, 1053, 1056).

Thus our quantitative inquiry as to the occurrences of "selection" or related words in Nietzsche's writings leads to the following conclusions. On the one hand, Nietzsche's first use of "selection" (fragments of 1881–2) referred to "natural selection," and although he continued using the term in that original Darwinian sense, he did so only rarely, in unpublished fragments, and always critically. On the other hand, most of his allusions referred to human selection, and most of the examples in that category were about methodical selection. In that context, Nietzsche was never critical, but rather the opposite. Therefore Nietzsche's allusions to "selection" all went in the same direction: *against* Darwin and *for* eugenics.

Now we shall deal with Nietzsche's eugenics thinking in greater detail. We must first ask whether or not it is historically correct to use the word "eugenics" here. That neologism was introduced by Francis Galton in 1883 in *Inquiries into Human Faculty and Its Development*.[10] Etymologically, "eugenics" meant the science of the "well-born" people (εὐ-γενής). According to Richter (1911), Nietzsche never used the term "eugenics," but his correspondence shows that he borrowed and read Galton's book in 1884, and probably acquired it later (the book was even found in his personal library). He may then have noted Galton's neologism and its obvious correspondence with his own ideas about hereditary aristocracy, but in the absence of any explicit allusion, we do not know that to have been the case. Nevertheless, though Nietzsche did not make use of "eugenics," he did occasionally use a term very close to it. In 1883, the year he was informed of the existence of Galton's book,[11] he wrote the following aphorism: "Natural aristocracy (γενναίος means naive!): an instinctive judgement and manner of acting belongs to the good race" (posthumous fragment [1883], Kröner, XIV-1, § 237). That aphorism was later inserted as an incidental observation in *The Genealogy of Morals* (1887, I, 10). What makes the aphorism interesting is the term γενναίος (*genaios*), a term that means primarily "well-born" or "noble." In other words, γενναίος (*genaios*) was an obvious synonym for

175

εὐ-γενής (*eu-genes*) and furthermore had the same root. Nietzsche, a first-rate Greek philologist, could not have failed to notice that. Perhaps that was no more than a coincidence. Even so, it is indeed remarkable.

Nevertheless, what matters here is not so much the word as the eugenics ideology itself, whose history began before the emergence of the word. As an ideology, eugenics can be characterized as a reformulation of a very ancient idea in the context of nineteenth-century biology. The old idea was that of preserving and improving the qualities of a population through the control of marriages (and sometimes the elimination of ill-formed newborn children – think of the Spartans and Plato's endorsement of their practices in *The Republic* and *The Laws*). In the second half of the nineteenth century, two biological theories gave that old idea an unprecedented dramatic character. One was the theory of heredity, especially (but not necessarily) the idea of "hard inheritance" (holding that individuals did not transmit their acquired characteristics, but rather the qualities that they had been given by their parents' germinal elements). In the case of humans, hereditarian thinking convinced many people that education and hygiene were unable to counter the alleged "degeneration" of the European races. The official birth of "hereditarianism" is often located in Galton's "Hereditary Talent and Character" (1865), a text that was quoted again and again in the early eugenics literature. The second biological theory that conditioned the formation of the eugenics ideology was the theory of evolution through natural selection. Shortly after the publication of Darwin's *Origin of Species* in 1859, various authors speculated whether or not the principle of natural selection was still acting in the human species, especially in the case of mental and moral characteristics. At the end of the 1860s, quite a number of authors (e.g., Haeckel 1868; Greg 1868) argued that natural selection was unable to oppose various factors of "degeneration" among the "civilized" peoples: modern medicine, which allowed the unfit to grow up and reproduce (Haeckel 1868); national wars, which killed the most vigorous and cleverest (Haeckel 1868; Darwin 1871); and finally the supposed greater fertility of the lower classes, lower races, lower "types" of all sorts (Galton 1865; Greg 1868). In those three cases, civilization was said to develop various kinds of "social selection" ("medical selection," "military selection," "class selection")[12] that opposed the preservation and improvement of the human species through natural selection. That pessimistic view of recent human evolution, in combination with hereditarianism, constituted the scientific context for the emergence of eugenics as an ideology. Eugenics involves more than the ancient idea of preserving the quality of the stock through the control of marriages, although it also contains that idea.[13] "Eugenics" is the name that was finally given to the modern ideology of the project of improving the "hereditary" qualities of the human species through methodical "selection." That ideology had just been born when Nietzsche began to be interested in Darwin. Nietzsche was in fact one of the first philosophers to adopt that ideology and to give it an important role in his thinking. He may even have been the only major modern philosopher to have done so.

I see two aspects in Nietzsche's "eugenics." One is precisely ideological. By that I mean that a number of Nietzsche's doctrines are best interpreted as echoes, in the work of the great philosopher, of a major ideology of his time. Their removal would not deeply affect the philosophical system. The second aspect consists of particular features of Nietzsche's "selective thinking" that testify to his attempt to genuinely integrate and reshape (and not only reflect) eugenics in the context of his own philosophy.

Let us begin with the ideological or stereotypical aspects. Several repetitive doctrines in the late Nietzsche writings bear witness to his adoption of the major ideas of the eugenics movement. What is distinctive, however, is the crude form in which the philosopher put most of those doctrines.

A first stereotypical item was his adoption of trivial hereditarian thought. In the course of his work, he increasingly stressed the role of heredity, blood, lineage, ancestry for the improvement of humanity, with special regard to cultural traits. Nietzsche was convinced that biological inheritance, in addition to education, played a major role in mental and moral characters: "It is simply not possible that a human being should *not* have the qualities and preferences of his parents and ancestors in his body, whatever appearances may suggest to the contrary. This is the problem of race" (*Beyond Good and Evil*, 1886, § 264). But above all he liked to apply that nineteenth-century hereditarian commonplace to aristocratic qualities: "Men and women of blood have an advantage over others, giving them an indubitable claim to higher esteem, because they possess two arts, increasingly heightened through inheritance: the art of being able to command, and the art of proud obedience" (*Human All-Too-Human*, 1878, § 440). Like a number of modern thinkers, Nietzsche had a nostalgia for hereditary aristocracy. The biology of his time seemed to him to support that nostalgia: "Nobility comes only from birth, from blood. . . . In fact, the mind alone does not lead to nobility; on the contrary, there must be something that *ennobles mind.* – What is necessary? – Blood." (Nietzsche 1970–97, XI, fragment 41[3], 1885–7). There is no sign, however, that Nietzsche adhered to the concept of "hard inheritance."

Coming now to human selection proper, it is obvious that Nietzsche advocated both "negative" and "positive" eugenics, to use expressions that appeared only later. Negative eugenics is commonly defined as the project of reducing the diffusion of hereditary handicaps by restricting individuals' access to reproduction. Positive eugenics aims at the improvement of a population by favoring the reproduction of desired types. In both cases, the founding father of eugenics, Francis Galton, thought that his program should be voluntary. Nietzsche, perhaps inspired by Haeckel's bold declarations in his *Natural History of Creation* (Haeckel proposed to imitate the Spartans), had no similar scruple. For instance, he defended authoritarian regulation of marriages, with the possible utilization of castration:

> There are cases in which a child would be a crime. . . . Society, as the great trustee of life, is responsible to life itself for every miscarried life – it also has to pay for such lives: consequently, it ought to prevent them. In numerous cases, society ought to prevent procreation: to this end, it may hold in readiness, without regard to descent, rank, or spirit, the most rigorous means of constraint, deprivation of freedom, in certain circumstances castration. – The Biblical prohibition "thou shalt not kill" is a piece of naiveté compared with the seriousness of the prohibition of life to decadents: "thou shalt not procreate!" (*The Will to Power,* III, § 734 [1888])

(See also *The Will to Power,* III, § 733; *Thus Spoke Zarathustra,* 1883–4, "Of Marriage and Children"; *Daybreak,* 1886, § 150.) Moreover, the philosopher would justify not only control over marriages but also active euthanasia, which he considered a moral duty:

> *Holy cruelty.* – A man who held a newborn child in his hands approached a holy man. "What shall I do with this child?" he asked; "it is wretched, misshapen, and does not have life enough to die." "Kill it!" shouted the holy man with a terrible voice; "and then hold it in your arms for three days and three nights to create a memory for yourself: never again will you beget a child this way when it is not time for you to beget." – When the man had heard this, he walked away, disappointed, and many people reproached the holy man because he had counseled cruelty; for he had counseled the man to kill the child. "But is it not crueler to let it live?" asked the holy man. [*The Gay Science,* 1882, § 73]

(See also *Thus Spoke Zarathustra,* 1883–4, "Of Voluntary Death.") A few months before his death, Nietzsche penned a terribly crude reformulation of that thought:

> The highest law of life, first formulated by Zarathustra, requires us to be pitiless towards any living excrement or garbage, to destroy anything that would be an obstacle, a poison, a conspiracy, an underground hostility, for ascending life. Faced with living excrement and garbage, there is only one duty: to have no *solidarity* with them. (posthumous fragment [October 1888], Nietzsche 1970–97, XIV, 23[10])

It is difficult, when reading such declarations, not to think of what Hitler wrote in 1925 in *Mein Kampf:* "The racist state . . . must take steps such that only healthy individuals procreate; it will say that there is only one shameful action: to produce children when one is sick or diseased. . . . The racist state must declare that any individual who is clearly ill or who has hereditary weaknesses that could therefore be transmitted to its offspring, has no right to reproduce and his or her ability to do so must be removed" (Hitler 1971, p. 404). Of course, Nietzsche cannot be reproached with what the Nazi state (and also other states) actually did. But it would be quite unconvincing to say that his bold declarations should be interpreted only metaphorically and to plead for his non-responsibility in the face of history. The least that can be said is that he radicalized an ideology of his time and gave it the respectability of philosophy, or at least the prestige of his own philosophy.

What about "positive eugenics"? That was what Nietzsche called "selection." At this point it is worth examining separately the theses that Nietzsche borrowed or adopted and the theses that were specific to him (they are intimately associated in the details of the texts).

The list of Nietzsche's borrowings from the common eugenics literature of his time is impressive. Like all eugenists, Nietzsche sought to justify methodical human selection on the basis of the alleged degeneration of the human species, with particular concern for the "European peoples." Of course, he had his own view of the causes of degeneration (Christianity, democracy, nihilism). But the idea that "selection" could be the most appropriate answer to European degeneration was a commonplace of nineteenth-century *fin de siècle* culture. It probably was no accident that the philosopher who tried more than any other to understand that degeneration was also the one who took eugenics so seriously. Similarly, like all early eugenists, Nietzsche argued that education (or "domestication") alone was unable to counteract the degeneration of modern societies. One had, therefore, to make use of "selection" (or "breeding") in order to "improve human nature" (*Twilight of the Idols,* 1888, "The 'Improvers' of Mankind," §§ 2, 3; *The Will to Power,* II, § 398 [1888]). With his characteristic tendency to radicalize, Nietzsche even suggested that, after all, selection could advantageously "replace moral preaching" (posthumous fragment [1883], Kröner, XIII, § 400). But – another stereotype – such an improvement of human nature would have a cost that one had to be aware of: the unfit (the "weak," in Nietzsche's vocabulary) would suffer: "A doctrine is needed powerful enough to work as a breeding agent (*züchtend zu wirken*): strengthening the strong, paralyzing and destructive for the world-weary" (*The Will to Power,* IV, § 862 [1883–8]; see also IV, § 1055).

Another typical sign of Nietzsche's adherence to early eugenics commonplaces was his observation that the fittest tended to have no children: "Curse about the fact that the best finish life without children!" (posthumous fragment [1882–5], Kröner, XIV-1, § 523). Whence numerous developments on marriage. If mankind is to progress, marriage should not be left to chance (*Daybreak,* 1886, § 250). Just as in Galton's utopia "Kantsaywhere," but in an infinitely more authoritarian manner, Nietzsche pleaded for a political rationalization of reproduction. Reproduction had to be clearly dissociated from sexual intercourse and legally regulated as a function of the value of the parent (posthumous fragment [1881–2], Kröner, XII-1, § 403). The dissociation of reproduction from sex would have precise practical consequences: Race ("the breeding of a race"), not love, would be the goal of marriage (*The Will to Power,* III, § 732). Marriage would be contingent on presentation of a "prenuptial certificate," and bachelors in good health would pay special taxes (*The Will to Power,* IV, § 733). The best men would be allowed to reproduce with several women, and, conversely, women would be allowed to change their husbands (posthumous fragment [1881–2], Kröner, XII-1, § 402). Such formulations explain why, at the end of the nineteenth century, some German feminists claimed kinship with Nietzsche, in spite of his extreme misogyny. On that interesting posthumous appropriation of Nietzsche, see Ascheim (1993).

Nietzsche made ferocious statements against "national wars." That also was a common idea: National wars (as initiated by the French revolutionists) had the net effect of killing the fittest men: "The greatest disadvantage of the conscript army, now so widely acclaimed, consists in the squandering of men of the highest civilization. . . . It is the men of highest culture who are always sacrificed in the relatively greatest number, the men who guarantee an abundant and good posterity" (*Human All-Too-Human,* 1878, § 442). Similar declarations can be found in Darwin's *The Descent of Man* (1871) and in the work of Haeckel, who spoke of "military selection" as one of the most conspicuous causes of the degeneration of modern societies (Haeckel 1876). Nietzsche's hatred of modern *national* wars (although not war in itself), became one of the most recurrent themes of his late thinking. His last written words, a few days before he descended into madness, dealt with that subject. They consisted of more or less delirious sentences about the crazed nationalism and warmongering of the Hohenzollerns, the reigning Prussian royal family (posthumous fragments [December 1888–January 1889], Nietzsche 1970–97, XIV, 25[13] through 25[21]).

A last proof of Nietzsche's adherence to contemporary eugenics can be found in his explicit adoption of what has sometimes been called the (unconscious) "central contradiction of eugenics." In its trivial form, this contradiction can be formulated thus: "Natural selection" is a natural law stating that in all living species the fittest survive and propagate their kind; but among humans the "fittest" have failed to do so; one must therefore help the "fittest" and, with the aid of artificial selection, reestablish the natural order, which requires that the fittest will necessarily win. More sophisticated formulations of that naturalistic fallacy could easily be given. The fact is that such reasoning can be clearly identified in the work of Nietzsche. On the one hand, he admitted that natural selection deserved to be imitated, amplified, and speeded up by man in order to improve human nature (posthumous fragment [1881–2], Kröner, XII-1, § 408). But on the other hand: "In the long course of history, the fundamental law must break through, and the best triumph, provided that mankind seeks, with the absolutely greatest will, *to establish the domination of the best*" (posthumous fragment [March–December 1884], Kröner, XIV-1, § 137). Or, in other words, natural selection was necessarily the paramount power in human evolution, provided that man wanted it.

To sum up, there is plenty of evidence that Nietzsche adopted the characteristic vocabulary and theses of eugenics, as that theory existed in his lifetime. I shall conclude this section by mentioning those aspects of Nietzsche's eugenics thinking that were peculiar to him and that probably would have made his position unacceptable for most eugenists.

What was specific to Nietzsche was the ultimate justification he gave for human selection. In the ordinary eugenics literature at the end of the nineteenth century, the "degeneration" of European populations was seen as indicating that less clever, less sociable, and less moral types of individuals were tending to proliferate and to swamp the mentally and morally fitter types. As a consequence, the eugenists emphasized policies that could increase the representation of the fitter types and thus increase the mean level of fitness for a given population (or of a given race, or of the human species). That was not Nietzsche's view. For him, degeneration was a historical process of homogenization that had led to the development of gregarious features in man. Christianity, with its emphasis on protecting the weak and allowing all men to be treated as equals, had played a major role in that story, just as it had produced a nonindividualistic, altruistic, uniform type of man. I do not want here to dwell on Nietzsche's well-known criticisms of Christianity and of its modern political form – democracy. I just want to point out its articulation with "the great selective thinking." Nietzsche's idea was that the dominant cultural forces that had molded modern European culture (Christianity, and later democracy) deserved to be interpreted in terms of "selection." The schema was as follows: On the one hand, Nietzsche argued that "Christianity is the principle opposed to selection" because it attributed the same value to all men, whether they were weak and sick or healthy and vigorous (posthumous fragment [1888], Kröner, XV, § 246). Thus Christianity was a cultural form that had both suspended the action of natural selection and forbidden any kind of conscious selection (posthumous fragment [1888], Kröner, XV, §§ 54, 246). On the other hand, the success of Christianity had to be explained. In that respect, Nietzsche seems to have accepted the existence of some kind of more or less conscious selection: Christianity, he often said, had favored the development of a "gregarious" human type, based on altruistic and reciprocal assistance from all to all (posthumous fragment, Kröner, XIII; § 327; XV, § 246; XVI, § 869). And democracy, which was merely the last metamorphosis of the Christian ideal in modern times, had the same selective effect: "Christianity, as a plebeian ideal, with its morality reduced to damaging the stronger, the superior, manly types and to favoring a herd-type, is a preparation to the democratic view of thinking. . . . Democratic Europe amounts only to a sublime breeding of slavery, which must be commanded by a strong race in order to sustain itself" (posthumous fragment [1885–6], Kröner, XIV-2, § 287). Thus, for Nietzsche, Christianity was a cultural form that aimed at "conserving" everything about the human species (all human types, including the weak, the sick, etc.). It was a force that opposed natural selection as well as deliberate selection. That, I believe, was the meaning of the enigmatic phrase cited earlier: "Christianity is the principle opposed to selection." And that was precisely the way Nietzsche incorporated the contemporary eugenics obsession with degeneration into his own moral philosophy (or rather his "immoralism").

The second specific trait of Nietzsche's commitment to eugenics was, of course, his own conception of the goal of human selection. That conception was a direct consequence of his critique of altruistic and egalitarian morals: "A question constantly keeps coming back to us, a seductive and wicked question perhaps; may it be whispered into the ears of those who have a right to such questionable questions, the strongest souls of today, whose best control is over themselves: is it not time, now that the type 'herd animal' is being evolved more and more in Europe, to make the experiment of a fundamental, artificial and conscious breeding (*Zuchtung*) of the opposite type and his virtues?" (*The Will to Power,* IV, § 954). A year later, in *Beyond Good and Evil* (1886, § 251), that thought persisted: "I am beginning to touch on what is *serious* for me, the 'European problem' as I understand it, the cultivation of a new caste that will rule Europe." What did Nietzsche mean by "dominant caste"? The doctrine was extremely ambiguous, not to say contradictory. There is no doubt that in a number of texts, Nietzsche really conceived of a hereditary aristocracy that would impose its will on the masses and "dominate" them in the political sense of the term: ". . . the possibility has been established for the production of international racial unions whose task will be to rear a master race, the future 'masters of the earth'; – a new, tremendous aristocracy, based on the severest legislation, in which the will of philosophical men of power and artist-tyrants will be made to endure for millenia – a higher kind of man who, thanks to their superiority in will, knowledge, riches, and influence, employ democratic Europe as their most pliant and supple instrument for getting hold of the destiny of the earth, so as to work as artists upon 'man' himself" (*The Will to Power,* IV, § 960 [1885–6]). Nietzsche even tried to set out which "races" would be potential candidates for that role. In *Beyond Good and Evil* there is a long discussion of which race or nation should dominate the European peoples. Nietzsche concluded, not without irony, that a race resulting from hybridization of Jews (intelligence) and Brandenburgers (obstinacy) might well be the solution! (*Beyond Good and Evil,* 1886, § 242). We note, in passing, that Nietzsche did not believe in the existence of original "pure races." Against Gobineau, he saw racial purity not in the past but in the future, and thus a "purification of the race" would not be a reversion, but rather the result of a gradual progressive process (*Daybreak,* 1886, IV, § 272). Again, in view of subsequent history, that hardly makes him more sympathetic.

In other texts Nietzsche had quite a different doctrine. He declared that the "superior species" should not command the "inferior species," but rather depend on the inferior species for the accomplishment of its own mission (*The Will to Power,* IV, § 901 [Spring–Fall 1887]), and that the superior caste had better remain isolated from the rest of humanity (posthumous fragment [1883], Kröner, XIV-2, § 4). In those texts, the idea was not that of a "despotic philosopher" (or artist), but that of a moral aristocracy that would "dominate" in the sense of "distinguishing itself" and "raising itself above mediocrity." The issue, then, was not to "govern," but to facilitate the development of heroic natures that would extend the creative possibilities of mankind.

Reading all of Nietzsche's sketches on "selective thinking" (more or less the entire fourth part of *The Will to Power* in the Würzbach edition) leaves one with mixed feelings of disgust and perplexity – disgust because so many sentences simply offend the most elementary sense of human dignity. For example: "I want to teach the idea that gives many the right to erase themselves – the great *cultivating* idea (*den grossen züchtenden Gedanken*)" (*The Will to Power,* IV, § 1056 [1884]); "The great majority of men have no right to existence, but are a misfortune to higher men" (*The Will to Power,* IV, § 872 [November 1887–March 1888]). Many commentators would say that such declarations should not be taken literally and would insist on consideration of their context and their metaphorical sense. Nevertheless, the question remains: can such sentences seriously be taken as mere metaphors? In fact, because the metaphorical sense of those sentences (and many others) is as clear as mud, because their literal sense is obvious, and because a lot of them were taken literally and applied later on in history, it probably would be good philosophical strategy to follow Alain Boyer's recommendation: "We should stop interpreting Nietzsche, and simply take him at face value" (Boyer 1991). That being said, Nietzsche's "selective thinking" also produces perplexity. After all, a good deal of his philosophy boils down to an obsessional plea for a mixture of aristocratic feelings and romantic individualism, together with the most violent critique ever formulated against Christianity, democracy, morals, and absolute values in general. The question here is not whether or not that critical and provocative thinking is worthy of study. It is. The question is whether or not Nietzsche really needed to develop his philosophy in the language of human methodical selection. This question can be formulated more precisely. Most of the "eugenics" developments were concentrated in *Beyond Good and Evil* and in the fourth part of *The Will to Power* – a sketch for a book that in 1888 Nietzsche had decided, at least temporarily, not to publish. Remove those passages and you still have what is probably the best of Nietzsche: *Zarathustra* and *On the Genealogy of Morals*. In those books, "selective thinking" is virtually absent. Neither the visionary reflection on the "superman" nor the radical critiques of morals required the idea of human selection.

So why did Nietzsche commit himself so openly in favor of eugenics? My feeling is that the eugenics program seduced Nietzsche because it looked like a genuine utopia, rather than a credible political program. I can hardly imagine the author of *Zarathustra* seriously believing that a coalition of biologists and politicians could actually realize his "dominant caste" of heroic artists and philosophers. In the face of his irreducible romantic individualism, I suspect that he would have found such a realistic program ridiculous. That probably is why he sometimes viewed the "dominant caste" as "selecting itself," rather than being selected (posthumous fragment [March–December 1884], Kröner, XIV-1, § 457). In fact, Nietzsche's adoption of eugenics is worth comparing again with his attitude toward Darwin's theory of evolution. At first sight, it seems puzzling that he should have been so confident of "selection" in the context of his "selective thinking" (eugenics) and yet critical toward Darwin's principles of the struggle for existence and natural selection. Early eugenists believed that natural selection was no longer able to preserve and improve the human species. But for all that, they did not think that Darwin's theory of the modification of species through natural selection was false. Nietzsche did. As we discussed earlier, his ultimate belief was that "selection" simply did not occur in nature, which meant that the only effective selective process he recognized was human selection. What a contrast between Nietzsche's feelings about "human" selection and "natural" selection!

I think that such contrast had something to do with science and myth. No doubt, for Nietzsche, Darwin represented science, indeed the best available science on the subject of "life," a key concept in his own philosophical system. Because he tried hard to clarify that concept for himself, Nietzsche probably believed that it was essential to have a critical dialogue with Darwin: The philosopher's conception and the scientist's conception of "life" and "struggle" had to be thoroughly confronted. In the case of eugenics ("selective thinking," in Nietzsche's terms), we can detect a very different attitude: no caution, no discussion about scientific principles, no polemic against the major advocates of eugenics (e.g., Galton). Nietzsche merely *adopted* eugenics. I said earlier that there was an obvious "ideological" aspect to that attitude, meaning that even the greatest philosophers remain the children of their time – not always for the best, unfortunately. But we have seen that Nietzsche not only repeated current eugenics ideas but also gave them an important role in the final formation of his most original doctrines (about decadence, the "superhuman," and the transmutation of all values). I suspect that in Nietzsche's mind, the main interest of eugenics was in its overtly utopian or even mythical nature. But that does not make "the great selective thinking" more sympathetic. I would rather say that it makes it trivial: In the culture of his time, Nietzsche could find no clearer illustration of his apology for aristocratic values than in eugenics (the "well-born" science).

THE ORIGINS OF MORAL BEHAVIOR

The two preceding sections have been devoted to the most explicit aspects of Nietzsche's interactions with Darwinism. However, I think that Nietzsche's philosophy would not be fundamentally altered if one were to remove his polemic against Darwin's theory of evolution and his eugenics fantasies. In the final sections of this essay I consider two major dimensions of Nietzsche's thinking: the question of the origin of morals, and the "perspectivist" interpretation of knowledge and culture. In both cases, there was an important connection with Darwinism. I could have begun this essay with these well-known aspects of Nietzsche, but did not, because their relationships with Darwinism were more diffuse than in the cases already analyzed, and in the second case, the relationship was not explicit. This section is devoted to the question of the origin of morals.

Nietzsche's friendship with Paul Rée was the starting point for his interest in the question of the origin of human moral behavior. That friendship began in 1873 and led to an important correspondence (Pfeiffer 1979). Rée was a young scholar who had previously published a book entitled

Psychological Observations. In 1877 he published another book, *The Origins of Moral Feelings,* that proposed to reduce all moral facts to natural causes. Rée thought that all morality originated in two primary instincts: an egoistic instinct, which was aimed at conservation of the individual, and an altruistic instinct, which made individuals act for the interest of other people. With Darwin, he accepted the existence of altruism in animals, believing that it had been developed by natural selection because it favored the survival of groups. In man, the notions of *good* and *evil* were but special developments of that basic tendency: *good* always referring to social utility, and *bad* referring to what was harmful to other people. On the basis of that identification between morality and altruism, Rée explained the origins of punishment, remorse, moral consciousness, freedom (a necessary illusion), and responsibility. For each of those classical moral notions he proposed a functional and relativistic interpretation: Morality always boiled down to categories of actions that would maintain society as such.[14] Rée's denial of any absolute criterion in morals delighted Nietzsche. As early as 1878, in *Human All-Too-Human,* he praised Rée and began to think about the "history of moral feelings." Through Rée, he discovered Darwin's speculations on the origins of morals and also became interested in the English utilitarian philosophers (especially Bentham and Mill) and in Spencer. However, he soon became convinced that Rée's work was weak. Some commentators have argued that he believed that from the beginning. Whatever the case, he did not criticize Rée, probably because they were friends. Instead, he began attacking the English authors who had influenced Rée: Darwin, Spencer, and the utilitarian philosophers. In the writings of his last period, Nietzsche often referred to that triad and polemicized against it on the question of the origin of morals. Many of Nietzsche's allusions to "Darwinism" must be understood in this perspective: not criticisms specifically addressed to Darwin (such as those analyzed earlier in this essay), but criticisms of the collective entity "Darwin, Spencer, and the utilitarian philosophers."

On one occasion, at least, Nietzsche provided a relatively precise description of the circumstances in which he became interested in the problem of the origin of morals:

> The first impulse to publish something of my hypotheses concerning the origin of morality was given me by a clear, tidy, and shrewd – also precocious – little book in which I encountered distinctly for the first time an upside-down and perverse species of genealogical hypotheses, the genuinely *English* type, that attracted me – with that power of attraction which everything contrary, everything antipodal possesses. The title of the little book was *The Origin of the Moral Sensations;* its author Dr. Paul Rée; the year in which it appeared 1877. Perhaps I have never read anything to which I would have said to my-

self No, proposition by proposition, conclusion by conclusion. [*On the Genealogy of Morals*, 1887, Foreword, § 4]

What Nietzsche did not accept in the *English* "genealogical hypotheses" with regard to morals was not the idea of providing a naturalistic and evolutionary account of morals (a program that, in fact, he praised highly). It was "the altruistic evaluation of morals." That was the common presupposition he found in "all English genealogists of morals" (*On the Genealogy of Morals*, 1887, Foreword, § 4). Nevertheless, a few pages later, in the same Foreword, Nietzsche had an ironic comment on the equivocal role Darwin played in that kind of literature:

> [Dr. Rée] had read Darwin – so that in his hypotheses, and after a fashion that is at least entertaining, the Darwinian beast and ultramodern unassuming moral milksop who "no longer bites" politely link hands, the latter wearing an expression of a certain good-nature and refined indolence, with which is mingled even a grain of pessimism and weariness, as if all these things – the problems of morality – were really not worth taking quite so seriously. [*On the Genealogy of Morals*, 1887, Foreword, § 7]

What did the Darwin-Spencer-utilitarian association really mean for Nietzsche? Why did he need to attack them collectively? The key word, of course, is "altruism." Nietzsche was not particularly worried as to whether Darwin did or did not actually use that Spencerian term. What mattered to him was the tendency of modern English authors to root altruism in the history of life and to make it an old, natural, and widely shared feature of organisms. In that respect, all authors who emphasized "social instincts" (e.g., "sympathy" or "parental affection") as the origin of moral behavior argued that altruism did indeed exist in "nature" and that it was a major biological *phenomenon:* "One tries to reconcile the altruistic mode of action with naturalness, one seeks altruism in the foundations of life; one seeks egoism and altruism as equally founded in the essence of life and nature" (*The Will to Power,* III, § 786 [Spring–Fall 1887]). Darwin, Spencer, and Rée could serve as interchangeable examples of that philosophical thesis, which Nietzsche rejected without the slightest concession. For Nietzsche, probably the most extreme philosophical apologist for individualist and aristocratic values who ever existed, altruism could not be an essential feature of "life":

> Life itself is *essentially* appropriation, injury, overpowering of what is alien and weaker; suppression, hardness, imposition of one's own forms, incorporation and at least, at its mildest, exploitation. . . . "Exploitation" does not belong to a corrupt or imperfect and primitive society: it belongs to the *essence* of what lives, as a basic organic function; it is a consequence of the will to power, which is after all the will of life. If this should be an innovation as a

theory – as a reality it is the *primordial fact* of all history: people ought to be honest with themselves at least that far. [*Beyond Good and Evil*, 1886, § 259]

What were Nietzsche's arguments against biological altruism? One of them would look quite familiar to contemporary evolutionary biologists. Nietzsche contested the altruistic nature of reproductive behavior and parental care: "Producing offspring has no altruistic aspect. Left to itself, an animal abandons itself to reproduction, to the extent that this pleasure often causes its death. To sacrifice oneself to one's offspring is to sacrifice oneself to what is closest, to one's own production, etc., this is certainly not altruism" (posthumous fragment [1879–80], Nietzsche 1970–97, IV, 1[110]). In other words, individual organisms did not reproduce for the benefit of their "race," of their "society," or of their "species." They reproduced for their own benefit. No altruism there, only egoism.

Besides that conventional argument against the notion of biological altruism, Nietzsche used a second, puzzling argument. It consisted in the suggestion that the very concept of natural selection was built as an unconscious transposition of some sort of altruistic and egalitarian moral into biological discourse. That argument (or rather suspicion) was related to a repeated criticism that Nietzsche developed against the notion of "selection of the species" (a notion that he attributed to both the Darwinians and Spencer). Nietzsche considered it absurd to hypothesize the existence of a natural process that would aim at the *conservation* of the species: "This is a wrong position: in order to conserve the species, many individuals are sacrificed. This kind of 'in order to' simply does not exist! Neither does the species exist; there are merely many different particular beings! . . . Nature does not want to 'conserve the species'! In reality many similar individuals are more easily conserved than other, abnormal individuals, in similar conditions of existence" (Nietzsche 1970–97, V, fragment 11[289], 1881). To a modern reader, such declarations are redolent of the debates over species selection. However, one should not misinterpret Nietzsche's intention: He rejected the notion of species selection, but he did not have the Darwinian notion of "individual selection." Remember that "natural selection" was an *extremely* rare expression in Nietzsche's writings, whereas "selection" (alone) was reserved for the artificial selection of humans. When he wanted to refer to natural selection, Nietzsche generally used two strange expressions: "selection of (or in) the species" and "utilitarian selection." Those terms first appeared in the early 1880s (posthumous fragments [1881–2], Nietzsche 1970–97, V, 11[67], 11[69]); they were used again in *The Will to Power* (e.g., I, § 54; see also posthumous fragments, Kröner, XII-1, §§ 235, 243), together with other equivalent phrases, such as "selection for utilitarian purposes" (posthumous fragment, Kröner, XII-1, § 243) and "general selection" (*The Will to Power,* II, 244). All of those texts had similar contents: They argued that "utilitarian selection," because it involved adaptation of the whole species to a definite environment, would essentially be a conservative force and therefore would fail to explain evolutionary change. That argument, of course, would have been unacceptable for Darwin, who emphasized individual competition. But that was precisely what Nietzsche did not see. For him, natural selection meant selection of traits advantageous to the *species* as such, and for that reason he believed that it was a fiction. It was a fiction unconsciously invented by a category of naturalists who tended to project their idealistic moral convictions onto nature. Darwin, Spencer, and others, because they thought that altruism was the first and last word on morals, adhered to the old (Christian) ideal of a "uniform," "egalitarian," "gregarious" human species. "Selection of the species" was merely a generalization of that ideal and its symbolic projection in the field of natural history. Thus, as strange as it may seem, Nietzsche criticized the Darwinian view of nature for being "plebeian" (as discussed earlier) and "Christian."[15]

That shows us what probably was Nietzsche's deepest motive for rejecting Darwinism. As a theory of the origin of morals, Darwinism claimed that altruism (and therefore morals per se) was a natural *phenomenon,* widely spread among organisms, and deeply rooted in the general history of life. Nietzsche, of course, had another theory of the origin of morals, which he laid out in a systematic manner in *The Genealogy of Morals.* I shall not analyze that famous theory here, but I want to indicate how the key theses of that book were constructed as a deliberate rejection of the common Darwinian-Spencerian-utilitarian schema. The three "dissertations" that made up the book developed three major theses on the origin of morals, each of which corresponded to a certain stage of cultural development. As noted by Dennett (1995), the organization of the book was misleading, because the first and second stages were discussed in reverse chronological order. Following Dennett's advice, I shall restore the chronology.

According to Nietzsche, the first stage in the development of human moral behavior involved the breeding of "an animal able to keep a promise," in other words, a "responsible" being (*The Genealogy of Morals,* II, 1, 2). In that process, the key event was the emergence of a guilty conscience. The overall thesis of the second dissertation was that the guilty conscience emerged as the result of incredible amounts of violence and tyranny within human societies. The emergence of the guilty conscience could not be described as the result of a progressive and adaptive process. It did not originate as a complex development of social and altruistic instincts, but through a terrible internalization of man's aggressiveness. In other words, the guilty conscience was not an (altruistic) adaptation, but the historical effect of a fantastic increase in the "will to power" and of the hierarchical relationship between men (*The Genealogy of Morals,* II, 12 17, which refer to Spencer).

The second stage in the development of morals was the emergence of moral judgment and the correlated construction of the categories of "Good" and "Evil." Nietzsche contested the utilitarian assimilation of "Good" and "non-egoistic actions." On the basis of philological evidence (which probably should be carefully checked), he argued that "Good" originally applied to the spontaneous feeling of superiority of dominant castes (*The Genealogy of Morals,* I, 2). The rest of the second dissertation was devoted to explaining how Jewish and Christian cultures succeeded in reversing the original evaluation. However, the overall message of the dissertation was clear: The utilitarian account of the origin of moral values was plain wrong.

Finally, in the third dissertation, Nietzsche argued that the general trend of European morals had been to expand "the ascetic ideal" in all areas of human culture. In that case, again, he formulated his central thesis against the Darwinian-Spencerian-utilitarian camp, with explicit reference to Darwin's struggle for existence: "*The ascetic ideal springs from the protecting instinct of a degenerating life* which tries by all means to sustain itself and to fight for its existence; it indicates a partial physiological obstruction and exhaustion. . . . The case is therefore the opposite of what those who reverence this ideal believe: life wrestles in it and through it with death and *against* death. The ascetic ideal is an artifice for the *preservation* of life" (*The Genealogy of Morals,* III, 13). I have already discussed that remarkable passage. The key word, of course, is "conservation." Nietzsche denounced a conception of morals based on "conservation" (will to survive, strategy of the "weak") instead of "augmentation" (will to power, strategy of the "strong"). At that level of discussion, the criticism of Darwinism and the criticism of Christianity do fuse.

I do not pretend, of course, to have provided an analysis of *The Genealogy of Morals,* which contains many things that I have not mentioned. My intention has been merely to show that Nietzsche's speculation on the origin (and value) of morals was profoundly motivated and structured by a controversial debate with Darwinian evolutionary ethics.

CONCLUSION: DARWINIAN OR NOT?

In the preceding sections we have seen how profound was Nietzsche's reluctance to accept the Darwinian view of life. To sum up, that reluctance developed along three lines. First, Nietzsche tried to make his own judgment of Darwin's general theory of evolution. He was initially content to raise doubts about the application of that theory to man. But he finally developed a general criticism of Darwin's theory of organic evolution. Second, Nietzsche's rejection of Darwinian evolution in man was intimately linked to his open commitment in favor of eugenics. The only serious meaning he ever gave to the word "selection" was that of methodical human selection. In his last years, that led to the quite frightening, but also confusing, utopia of "great selective thinking." The third and most important aspect of Nietzsche's opposition to Darwinism was his total disagreement with the English evolutionists' theses on the origin and meaning of moral behavior.

The foregoing considerations should be enough, I believe, to qualify Nietzsche as a remarkable anti-Darwinian philosopher. But there is a serious reservation concerning that conclusion. A major feature of Nietzsche's philosophy was what he called "perspectivism." In fact, that may well have been the most central idea of his entire thinking. I shall consider it briefly and show its extraordinary convergence with the Darwinian view of life.

By "perspectivism," Nietzsche meant that any knowledge was always an "interpretation": "In so far as the word 'knowledge' has any meaning, the world is knowable; but it is *interpretable* otherwise, it has no meaning behind it, but countless meanings. – 'Perspectivism'" (*The Will to Power,* III, § 481 [1883–8]). That thesis meant a little more than simply saying "everything is subjective." It implied that anything significant, in any area of human experience, was always local and contextual: "Is meaning not necessarily relative meaning and perspective?" (*The Will to Power,* III, § 590 [1885–6]).

For Nietzsche, perpectivism was not only a characteristic of human knowledge but also a characteristic of all life, or, even more, "the fundamental condition of all life" (*Beyond Good and Evil,* 1886, Preface). Thus "the organic process constantly presupposes interpretation" (*The Will to Power,* III, § 643 [1885–6]). And indeed, human knowledge itself had to be evaluated in terms of its biological value or "usefulness." Nietzsche applied that idea to all traditional philosophical aspects of cognition – perceptions, beliefs of all sorts, science, consciousness. Hundreds of quotations could be cited to illustrate that idea. Let me just provide a few remarkable ones:

Perceptions. "Man's firm perceptions of distance, light, colours which have been moulded and hereditarily fixed through the elimination of people with less accurate capacities" (posthumous fragment [1881–2], Kröner, XII-1, § 75).

Cognitive faculties. "The meaning of 'knowledge': here, as in the case of 'good' or 'beautiful,' the concept is to be regarded in a strict and narrow anthropological and biological sense. In order for a particular species to maintain itself and increase its power, its conception of reality must comprehend enough of the calculable and constant for it to base a scheme of behavior on it. The utility of preservation – not some abstract-theoretical need not to be deceived – stands as the motive behind the development of the organs of knowledge – they develop in such a way that their observations suffice for our preservation" (*The Will to Power,* III, § 480 [March–June 1888]).

Beliefs (of any sort). "Human beliefs which triumphed in the past are those which favored survival; not the truer ones, but the most useful" (posthumous fragment [1881–2], Kröner, XII-1, § 44).

Science. "Scientific knowledge, either true or false, lasts because it allows the human species to survive" (posthumous fragment [1881–2], Kröner, XII-1, § 75).

Consciousness. "Consciousness exists only insofar as it is useful" (posthumous fragment [1881–2], Kröner, XII-1, § 306).

I can hardly imagine a more "Darwinian" approach to human knowledge and culture. The basic program of evolutionary epistemology and evolutionary ethics is there in its totality.[16] Note also Nietzsche's vocabulary, especially the words "useful" and "survival." Add in the fact that Nietzsche liked to insist on the total contextuality or contingency of usefulness: "'Usefulness': this only applies in terms of immediate, proximate effects: any remote consequence is uncontrollable and any action may be equally characterized as either useful or harmful" (posthumous fragment [1884], Nietzsche 1970–97, X, 25[18]; see also *On the Genealogy of Morals,* II, 12). Add also Nietzsche's total rejection of any oriented progress or purposefulness in evolution and his idea that finality was but a retrospective illusion: "To make the attempt, to grasp everything seemingly purposive as the solely *life-preserving* and consequently solely *preserved*" (posthumous fragment [1884], Kröner, XIII, § 388). All of those theses suggest the image of a very "Darwinian" Nietzsche, quite different from the one I have described in this essay.

In none of the quotations just given did Nietzsche refer to "Darwin" or "Darwinism." Was he even aware that his biological perspectivism involved a remarkable convergence with the Darwinian view of life? I suspect that he would not have seen the contradiction. He probably would have answered that "perspectivism" was another word for his openly "egoistic" (or individualistic) conception of life, morals, and human culture in general.[17] He probably would have added that "Darwinism" was not an individualistic view of life and evolution. That is what this essay has been about. That was the great paradox of Nietzsche's work on Darwinism: Because of his obsessive hatred for "altruistic" and "gregarious" values, because evolutionary biology interested him only with regard to morals, and perhaps also because of his poor scientific training, Nietzsche never understood the "individualistic" nature of Darwin's concept of natural selection.

NOTES

1. That question was commonly discussed by early twentieth-century commentators; see, for instance, Richter (1911). Since that date, it has been neglected.
2. See also *Human All-Too-Human* (1878, § 230): "*Esprit fort* [in French in the original text]. – Compared with the man who has tradition on his side and needs no reasons for his actions, the free spirit is always weak, especially in his actions. . . . By what means, then, can he be made *relatively strong,* so that he can at least assert himself effectively and not perish?"
3. For a detailed discussion of this expression, see the penultimate section of this essay.
4. See also *Beyond Good and Evil* (§ 253): "The spirit of respectable but mediocre Englishmen – I name Darwin, John Stuart Mill, and Herbert Spencer – is beginning to predominate in the middle regions of European taste. . . . European vulgarity, the plebeianism of modern ideas, that of *England.*"
5. See, for instance, the last sentence in Chapter III in *On the Origin of Species:* "When we reflect on this struggle, we may console ourselves with the full belief, that . . . the vigorous, the healthy, and the happy survive and multiply" (Darwin 1859, p. 79).
6. This is an allusion to the concept of *conatus* and the "instinct of conservation."
7. The title of that first translation was *Über die Entstehung der Arten im Thier- und Pflanzen- Reich durch natürliche Züchtung, oder Erhaltung der vervolkommneten Rassen im Kampfe um's Daseyn* [*sic*]. For complete information on the first German editions of *The Origin,* see Darwin (1992, p. 611).
8. In 1867 Carus proposed a revised edition of Bronn's translation. The title was changed to *Über die Entstehung der Arten durch natürliche Zuchtwahl oder die Erhaltung der begünstigten Rassen im Kampfe um's Dasein* (Darwin 1992, p. 611). Other revised editions appeared in 1870, 1872, 1876, 1884, and 1899, with the same title.
9. Some translators have amplified the "selective" aspect of Nietzsche's thinking by introducing "selection" (or "select") wherever Nietzsche referred to "breeding" (in the sense of plant or animal breeding). That was a common tendency in the first half of the twentieth century (see, for instance, G. Bianquis's French translation of *The Will to Power*). More recent translations have tended to minimize the "selective" aspect. That was the case with Kaufmann and Hollingdale's English translation of *The Will to Power.* Those translators restricted their use of "selection" to *Zuchtwahl* and resorted to other English words ("breed," "rear," "cultivate," and corresponding substantives) when Nietzsche used German words with the root *Zucht* (especially *Züchtung*). That choice was linguistically justified and interesting, particularly from the point of view of modern English. But in the context of the early discovery of Darwin's thinking in the second half of the nineteenth century, that approach may have hidden an important aspect of Nietzsche's thought. One example will suffice to illustrate the case. Bianquis's French translation of

The Will to Power (IV, § 1056) reads as follows: "Je veux enseigner la pensée qui apprendra à un grand homme à se supprimer – la grande pensée *sélective*" (Bianquis, in fact, used the Würzbach numbering: IV, § 231). Compare Kaufmann and Hollingdale's English translation of the same passage: "I want to teach the idea that gives many the right to erase themselves – the great *cultivating* idea." The German text says "Ich will den Gedanken lehren, welcher Vielen das Recht giebt, sich durchzustreichen, – den grossen züchtenden Gedanken." The English translation of the last part of the sentence is unconvincing. For a nineteenth-century German public, the eugenics tone (and violence) of the sentence was obvious. However, I recognize that Bianquis's systematic use of "selection" is also unsatisfying: It does not do justice to the linguistic grid through which the Germans first received the work of Darwin.

10. Galton's original definition was "the science of improving stock, which is by no means confined to questions of judicious mating, but, which, especially in the case of man, takes cognisance of all influences that tend in however remote a degree to give the more suitable races or strains of blood a better chance of prevailing speedily over the less suitable than they otherwise would have had" (Galton 1883, p. 25).

11. Nietzsche was informed of the publication of Galton's *Inquiries* through a certain Doctor Paneth in the winter of 1883–4. Paneth knew Galton personally (Richter 1911, p. 41).

12. "Medical selection" and "military selection" appear in the work of Haeckel (1868), in the chapter on "Selection." "Social selection" and "class selection" came later, for instance, in the work of Vacher de Lapouge (1896).

13. For a more analytical exposition of the differences among the idea (old), the word (introduced in 1883), and the ideology of eugenics, see Gayon (1998).

14. For an analysis of Rée's book, see Andler (1958, vol. 2, pp. 304–12).

15. "*To consider:* to what extent the fateful belief in divine providence – the most paralyzing belief for hand and reason there has ever been – still exists; to what extent Christian presuppositions and interpretations still live under the formulas 'nature', 'progress', 'perfectibility', 'Darwinism', under the superstitious belief in a certain relationship between happiness and virtue, unhappiness and guilt" (*The Will to Power*, II, § 243).

16. I use "evolutionary epistemology" and "evolutionary ethics" in the most literal sense: an investigation of how knowledge and moral behavior can be illuminated by man's evolutionary past. On the ambiguities of these terms, see the illuminating discussion of Ruse (1986). Ruse preferred to speak of "Darwinian epistemology" and "Darwinian ethics."

17. There is, in fact, an aphorism that says something similar: "Even in the domain of the inorganic an atom of force is concerned only with its neighborhood: distant forces balance one another out. Here is the germ of perspectivism. This is why living beings are 'egoistic' through and through" (posthumous fragment [*WP*], 1886, Kröner, XIV-1, § 37).

NOTES ON REFERENCES

Strictly speaking, Nietzsche should now be quoted from the reference edition published simultaneously in German, French, and Italian under the direction of Giorgio

Colli and Mazzino Montinari. But that would pose problems for the general reader in English-speaking countries. I have used the following conventions:

- Books published in Nietzsche's lifetime are cited by the number of the aphorism or paragraph.
- The posthumous fragments under the title *The Will to Power* are cited from Kaufmann and Hollingdale's English translation of the Musarion standard edition.
- For other posthumous fragments, the nonexistence of English translations of Nietzsche in French libraries raised problems that I was not able to solve in a completely satisfying manner. References to fragments taken from the Würzbach edition of *The Will to Power* that do not appear in the Kaufmann and Hollingdale translation are cited from the Kröner edition of the complete works; R. M. Burian kindly translated them into English.
- Finally, in a very few cases, when I was unable to cite references from common English or German editions, the references to some posthumous fragments are from the Colli-Montinari edition of the *Complete Works*. Quotations from that trilingual edition are translated from French. I am aware that this is not a totally satisfying solution, but I have done my best for the English reader.
- All italicized terms in quotations were so in the original.

REFERENCES

Andler, C. 1958. *Nietzsche, Sa Vie et Sa Pensée,* 3 vols. Paris: Gallimard. (Originally published 1920–31.)

Ascheim, S. E. 1993. *The Nietzsche Legacy in Germany 1890–1990.* Berkeley: University of California Press.

Boyer, A. 1991. Hiérarchie et vérité. In: *Pourquoi nous ne sommes pas nietzschéens,* ed. A. Boyer et al., pp. 11–35. Paris: Grasset.

Comte-Sponville, A. 1991. La brute, le sophiste et l'esthète. In: *Pourquoi nous ne sommes pas nietzschéens,* ed. A. Boyer et al., pp. 37–98. Paris: Grasset.

Darwin, C. 1859. *On the Origin of Species.* London: John Murray.

1868. *The Variation of Animals and Plants under Domestication,* 2 vols. London: John Murray.

1871. *The Descent of Man, and Selection in Relation to Sex,* 2 vols. London: Murray.

1877. A biographical sketch of an infant. *Mind. Quarterly Review of Psychology and Philosophy* 2: 285–94. Reproduced (1977) in *The Collected Papers of Charles Darwin,* ed. P. H. Barrett, vol. 2, pp. 191–200. University of Chicago Press.

1992. *Über die Entstehung der Arten durch natürliche Zuchtwahl oder die Erhaltung der begünstigten Rassen im Kampfe um's Dasein* (reproduction of Carus's last revised translation of the last edition of *Origin* in 1899; comments and bibliographical notes by G. H. Müller). Darmstadt: Wissenchaftliche Buchgesellschaft.

Deleuze, G. 1962. *Nietzsche et la philosophie.* Paris: Presses Universitaires de France.

Dennett, D. 1995. *Darwin's Dangerous Idea. Evolution and the Meaning of Life.* London: Allen Lane, Penguin Press.

Galton, F. 1865. Hereditary talent and character. *Macmillan's Magazine* 12: 157–66, 318–27.

——— 1883. *Inquiries into Human Faculty and Its Development.* London: J. M. Dent & Sons.

Gayon, J. 1998. L'eugénisme. In: *Précis de génétique humaine,* ed. Josué Feingold, Marc Fellous, and Michel Solignac, pp. 459–83. Paris: Hermann.

Greg, W. 1868. On the failure of "natural selection" in the case of man. *Fraser's Magazine* 68: 353–62.

Haeckel, E. 1876. *The History of Creation,* 2 vols. London: Kegan Paul. (Originally published 1868.)

Hitler, A. 1971. *Mein Kampf,* trans. Ralph Manheim. Boston: Houghton Mifflin. (Originally published 1925.)

Hollingdale, R. J. 1965. *Nietzsche: The Man and His Philosophy.* London: Routledge & Kegan Paul.

Jaspers, K. 1950. *Nietzsche, introduction à sa philosophie.* Paris: Gallimard.

Kaufmann, W. 1950. *Nietzsche: Philosopher, Psychologist, Antichrist.* Princeton, NJ: Princeton University Press.

Münster, A. 1995. *Nietzsche et le nazisme.* Paris: Kimé.

Nietzsche, F. 1910–26. *Nietzsche's Werke,* 20 vols. Leipzig: A. Kröner.

——— 1937–8. *La volonté de puissance,* 2 vols. (F. Würzbach edition), trans. G. Bianquis. Paris: Gallimard.

——— 1954. *The Antichrist.* In: *The Portable Nietzsche,* trans. Walter Kaufmann. New York: Viking Press. (Originally published 1888.)

——— 1961. *Thus Spoke Zarathustra,* trans. R. J. Hollingdale. London: Penguin Books. (Originally published 1882–3.)

——— 1968a. *Beyond Good and Evil.* In: *Basic Writings of Nietzsche,* trans. Walter Kaufmann. New York: Modern Library. (Originally published 1886.)

——— 1968b. *On the Genealogy of Morals.* In: *Basic Writings of Nietzsche,* trans. Walter Kaufmann. New York: Modern Library. (Originally published 1887.)

——— 1968c. *Twilight of the Idols,* trans. R. J. Hollingdale. London: Penguin Books. (Originally published 1888.)

——— 1968d. *Ecce Homo.* In: *Basic Writings of Nietzsche,* trans. Walter Kaufmann. New York: Modern Library. (Originally published 1888.)

——— 1968e. *The Will to Power,* trans. W. Kaufmann and R. J. Hollingdale. New York: Vintage Books.

——— 1970–97. *Oeuvres philosophiques complètes,* ed. G. Colli and M. Montinari. Paris: Gallimard.

——— 1974. *The Gay Science,* trans. Walter Kaufmann. New York: Random House. (Originally published 1881.)

——— 1982. *Daybreak, Thoughts on the Prejudices of Morality,* trans. R. J. Hollingdale. Cambridge University Press. (Originally published 1886.)

1983. *Untimely Meditations,* trans. R. J. Hollingdale. Cambridge University Press. (Originally published 1873–6.)

1984. *Human All-Too-Human,* trans. Marion Faber and Stephen Lehmann. Lincoln: University of Nebraska Press. (Originally published 1878.)

Pfeiffer, E. 1979. *Friedrich Nietzsche, Paul Rée, Lou Von Salomé. Correspondance,* trans. O. Hansen-Love and J. Lacoste. Paris: Presses Universitaires de France.

Rée, P. 1877. *Des Ursprung des moralischen Empfidungen* (The Origins of Moral Feelings).

Richter, C. 1911. *Nietzsche et les Théories Biologiques Contemporaines.* Paris: Mercure de France.

Roux, W. 1881. *Der züchtende Kampf der Teile oder die Teilauslese im Organismus, zugleich eine Theorie der funktionellen Anpassung.*

Ruse, M. 1986. *Taking Darwin Seriously.* Oxford: Basil Blackwell.

Vacher de Lapouge, G. 1896. *Les sélections sociales.* Paris: Fontemoing.

Wallace, A. R. 1898. Darwinism in sociology: Dr. Alfred Russel Wallace replies to Mr. Thomas Common. *The Eagle and the Serpent* Sept. 1: 57–9.

8

Evolutionary Ethics in the Twentieth Century: Julian Sorell Huxley and George Gaylord Simpson

Michael Ruse

> If Julian Huxley is the Herbert Spencer of twentieth-century Darwinism, George Gaylord Simpson may in some ways be considered its Thomas Henry Huxley.
>
> Greene (1981, p. 168)

Philosophers usually think that evolutionary ethics – the attempt to locate and ground morality in our biological origins – met its Waterloo in the crucial year 1903. It was then that the English philosopher G. E. Moore published his devastating critique of the ideas of the prominent nineteenth-century evolutionary ethicist Herbert Spencer (1851, 1868, 1892). In a definitive manner, Moore's *Principia Ethica* (1903) showed that Spencer and all who thought like him were guilty of that gross conceptual mistake that Moore labeled the "naturalistic fallacy." Before Moore, evolutionary ethics had flourished like the rank weed that it was. After Moore, evolutionary ethics lay smoldering on the bonfire of discarded ideas. And a good thing, too, thinks the philosopher, for there have been few excesses of nineteenth-century capitalism or twentieth-century militarism and fascism that have not had their biology-oriented partisans. Choose your vileness, and there has been someone prepared to defend it in the name of evolution.

Those whose inquiries have taken them beyond philosophical folklore will know that today there is little need to spend much time on that latter charge (Russett 1976; Kelley 1981; Ruse 1986, 1996; Richards 1987; Pittenger 1993; Crook 1994). For years, historians have been looking at the claims of the evolutionary ethicists – "social Darwinians" as they are often called – and it is clear that although some pretty dreadful things have been suggested and sometimes even perpetrated in the name of evolution, the picture is by no means uniformly black. To the contrary: Some thoroughly admirable ends

I would like to thank Paul Farber, John C. Greene, and Gregg Mitman for comments on an earlier version of this essay.

have been promoted under the same name. Even Herbert Spencer had much to commend him, such as his ardent and steadfast opposition to the militarism that engulfed Europe toward the end of the nineteenth century (Spencer 1892, 1904).

But what of the earlier claim about the force and significance of Moore's *Principia Ethica?* Here, too, the historian will point out that philosophical tradition has but a tenuous connection to reality. One does not have to deny the power of Moore's vigorous critique – although one might properly note that he was less than generous in failing to acknowledge that a century and a half earlier David Hume (1978) had made some of the same crucial points – to acknowledge that before Moore there had been very effective critics of evolutionary ethics and that after Moore there still were very energetic supporters of that doctrine. Among the former were the philosopher Henry Sidgwick (1874) and the biologist Thomas Henry Huxley (1893, 1888a). Among the latter could be found just about every evolutionist of note in the first half of the twentieth century, and there have been many more since then. They have never spoken with one voice, but they have been united in the conviction that we can and must bring evolution and ethics together in one harmonious and productive relationship (Mitman 1990, 1992). It simply has to matter that we humans are, as T. H. Huxley once wrote to a friend, modified monkeys and not modified dirt (letter to F. Dyster, January 30, 1859, Huxley Papers, 15.106, Imperial College).

Whatever the merits of the arguments for and against, anyone knowledgeable in the history of evolutionary theorizing cannot be surprised that Moore's work had so little effect and that people have gone on connecting origins and ethics. For all that the past fifty years have seen a major move toward professional science, evolution has always been more than just a scientific theory – it has ever been a philosophy, a metaphysics, a weltanschauung, a secular religion (not so secular at times), even, indeed, an eschatology (Ruse 1996). And naturally, therefore, evolutionists have continued to turn their attention to morality, its nature and foundations, whether intending to support and supplement the already existing ideas and practices, sacred and secular, or hoping to start afresh, in altogether new directions. In this dreadful era of war, pestilence, famine, overpopulation, and pollution, the wonder is that *Principia Ethica* would have had any lasting effect whatsoever – except perhaps among philosophers, who have needed little persuasion anyway.

In this discussion, I want to single out for critical examination two of the most articulate and interesting of the evolutionary ethicists of this century: the Englishman Julian Huxley and the American George Gaylord Simpson. I should explain at once that I write both as a historian and as a philosopher. *Qua* historian, I choose Huxley and Simpson because they provide good contrast:

Although they were good friends and correspondents on these matters, they held very different views on the nature of the evolutionary process and the ethics to which it gave rise (Swetlitz 1991). Nor were those idiosyncratically different views. The quotation at the beginning of this essay, from the pen of the doyen of historians of evolutionary thought, John Greene, supports my belief that Huxley and Simpson are worth considering not only in their own right but also as representatives of two separate traditions in evolutionary ethical theorizing.

Qua philosopher, I choose Huxley and Simpson because today there is renewed interest in evolutionary ethics, and even philosophers (now that we are almost a century removed from *Principia Ethica*) are starting to think that perhaps our simian ancestry does count. Convinced, as an evolutionist, that the best way to understand the present is to understand the past, I use Huxley and Simpson to that end. To lay my cards openly on the table, I believe (and I am neither alone nor original in this) that of the two traditions in evolutionary ethicizing, whereas one is productive and forward-looking, the other is not (Mackie 1978; Murphy 1982; Ruse 1986; cf. Richards 1987 for a dissenting view). My fellow philosophers have been right to be suspicious, but their skepticism has turned them away too early. I am not saying that either Huxley or Simpson was completely right or completely wrong (nor that either would agree with me on what constitutes "completely right" and "completely wrong"!), but I do say that by looking at the pertinent ideas of these two men, we can carry the discussion forward in a constructive manner.

JULIAN SORELL HUXLEY (1887–1975)

Intellectually speaking, Julian Huxley was born with a silver spoon in his mouth (Huxley 1970, 1973; Waters and van Helden 1992). Grandson of Thomas Henry Huxley (and brother of Aldous Huxley), Julian was also the great-grandson of Dr. Thomas Arnold of Rugby School (nephew of Matthew Arnold). Educated at Eton and Oxford, he worked briefly at Rice Institute (now Rice University) in Houston, Texas, before returning to Oxford and then King's College, London. Soon he turned from the academic life, serving as secretary of the London Zoo and as the first director general of the United Nations Educational, Scientific, and Cultural Organization (UNESCO). Increasingly, however, his time was spent in writing articles and books (popular and learned), as well as in being a general pundit (first on radio, later on television) on matters scientific.

Huxley published creditable work on animal behavior (ethology), as well as embryology (especially on problems of comparative growth) (Huxley

Evolutionary Ethics: Julian Huxley and George Simpson

1932), but his primary importance as a scientist was as a synthesizer, most particularly of evolutionary theory. He was one of the founders of the "synthetic theory" of evolution, which was a coalescence in the 1930s and 1940s of the selectionist ideas of Charles Darwin and the particulate theory of heredity that dated back to the Moravian monk Gregor Mendel. Huxley's *Evolution: The Modern Synthesis* (1942) was a landmark survey that integrated the field as then known, covering the range of evolutionary topics from biogeography to taxonomy, from anatomy to paleontology.

But even at his most scientific, there was always more to Huxley's work than a simple quest for an empirical understanding of the world. Although, ontologically, he was at the atheistic end of the spectrum, Huxley was ever trying to make sense of life in a deeper, more spiritual sense. For him, the key concept, which flooded his whole being and infused all of his work, was that of *progress,* the belief or doctrine that all is in a state of flux or change and that the direction of the change is from lesser to greater, from less to more, from value-free to value-loaded or worthy or improved (Huxley 1912, 1936, 1942, 1954a, 1959, 1964; Huxley and Haldane 1927; Greene 1990; Swetlitz 1991).

Of course, the notion of progress was not new with Huxley. At a general cultural level, it was the ideology of the Enlightenment, and in biology it was the very mark of evolutionary theorizing from its eighteenth-century beginnings (Bury 1924; Nitecki 1988; Spadefora 1990; Ruse 1996). The key question had always been not whether or not biological progress occurred, but what form it generally took. Much ink was spilled in defense of various candidates for the true criterion: complexity, intelligence, flexibility, and more. Huxley liked the idea of complexity – "High types *are* on the whole more complex than low" – but what really excited him were the notions of *control* and *independence* (Huxley 1942). In his opinion, the more an organism was capable of exercising control over its particular environment, and the greater its independence of that environment, the higher that organism rated on the scale of improvement. Mammals, for instance, with their various methods for maintaining relatively constant body temperatures, were more in control and more independent than reptiles, which had no such methods – and mammals clearly were higher up the scale.

Huxley saw humans as right at the pinnacle of being. More than any other organism, they enjoyed "increased control over and independence of the environment." In other words, they had raised "the upper level of all-round functional efficiency and of harmony of internal adjustment" (Huxley 1942, pp. 564–5). Seeing progress in the evolution of organisms, even extending up to our own species, and believing progress to be a social phenomenon, Huxley also saw social progress as the process that took over where biological progress ended. "True human progress consists in increases of aesthetic, intellectual, and spiritual experience and satisfaction" (1942, p. 575).

Huxley's original belief in progress probably had several sources, but an early reading of Henri Bergson's *Creative Evolution* (1907, 1911) was crucial. In *The Individual in the Animal Kingdom,* whose preface stated candidly that "it will easily be seen how much I owe to M. Bergson," Huxley wrote as follows:

> Civilized man is the most independent, in our [i.e., Bergson's] sense, of any animal: this he owes partly to his comparatively large size, more to his purely mechanical complexity of body and brain, giving him the possibility of many precise and separate actions, and most to the unique machinery of part of his brain which enables him to use his size and the smoothly-working machine-actions of his body in the most varied way. [Huxley 1912, pp. 6–7]

Indeed, there is good evidence that, like Bergson, Huxley was attracted to some form of vitalism, believing in spirit forces driving organisms up the chain of being. Certainly he had had mystical experiences that had inclined him that way:

> When I was last in New York, I went for a walk, leaving Fifth Avenue and the Business section behind me, into the crowded streets near the Bowery. And while I was there, I had a sudden feeling of relief and confidence. There was Bergson's élan vital – there was assimilation causing life to exert as much pressure, though embodied here in the shape of men, as it had ever done in the earliest year of evolution: – there was the driving force of progress. [Unpublished series of lectures given at Rice Institute, lecture 1, J. Huxley Papers, Rice University]

However, unlike Bergson, Huxley realized that naked vitalism could never pass as genuine science, and so in a way his whole career can be seen as an attempt to put a scientific gloss on the philosophy – a gloss that, by and large, Huxley thought could be concocted from Darwinian selection working on Mendelian-caused heritable organic characteristics, thus yielding ever new adaptations of a progressive kind. "Bergson's *élan vital* can serve as a symbolic description of the thrust of life during its evolution, but not as a scientific explanation. To read *L'Évolution créatrice* is to realize that Bergson was a writer of great vision but with little biological understanding, a good poet but a bad scientist" (Huxley 1942, pp. 547–8).

Ardent as he was in his progressionism, Huxley, working in the first half of the twentieth century, was hardly unique in that. Indeed, until late in life, Huxley drew heavily on the then-accepted paleontological picture of life's progression, a picture of ever increasing specialization leading to extinction, save for a few forms that remained fairly generalized and thus paved the way for an adaptive breakthrough upward – ultimately yielding *Homo sapiens*. By midcentury, however, that view had come under heavy attack (with George G. Simpson playing a major role), as it was pointed out that "special" and "general" were relative terms applied retroactively. In their way, the early mammals had been highly specialized reptiles. Huxley therefore had to revise his thinking in certain respects, taking account of current opinion (variations in populations became much more important).

But Huxley never wavered from the belief that the distinctive strength of the human species was its flexibility – the human specialization had been to become generalists. With that, Huxley associated the belief that evolution, outside the human realm, had essentially ended. Humans alone continued to have the possibility and hope of further change – further progressive change – because they had transcended their biology, and human evolution in the psychosocial realm was no longer a matter of natural selection working on Mendelian genes:

> It seemed to me that, natural selection being unable to operate only on the basis of immediate biological utility, we must accept that every tendency towards improvement of the physical machinery is bound sooner or later to reach a pitch of perfection beyond which natural selection alone cannot push it. This seems to me to apply both to "one-sided" improvements, such as those of most so-called specialists, and to general improvements in what I have called biological machinery, such as sense organs, capacity for temperature regulation, etc. The one exception to the stoppage of progressive change, namely, the human stock, no longer operates primarily by natural selection, but by the primarily psychological machinery of cumulative experience. [Letter to Simpson, December 4, 1950, Simpson Papers, American Philosophical Society, Philadelphia, PA]

Along with that belief, Huxley was steadfast in his commitment to the thesis that progress was an objective fact, something to be read from the biological record, rather than to be read into it. The world, the organic world especially, was genuinely progressive, really and truly:

> Evolution, from cosmic star-dust to human society, is a comprehensive and continuous process. It transforms the world-stuff, if I may use a term which includes the potentialities of mind as well as those of matter. It is creative, in the sense that during the process new and more complex levels of organization are progressively attained, and new possibilities are thus opened up to the universal world-stuff. [Huxley and Huxley 1947, p. 131]

That was not just a question of humanly imposed values, or, rather, it did not matter if it was a question of such values: "It is immaterial whether the human mind comes to have these values *because* they make for progress in

Michael Ruse

evolution, or whether things which make for evolutionary progress become significant *because* they happen to be considered as valuable to human mind, for both are in their degree true" (Huxley 1923, pp. 59–60).

EVOLUTIONARY HUMANISM

Because we are now into the realm of values, what of ethics? Huxley, who thought of himself as an "evolutionary humanist" and as explicitly providing a secular religion, wrote on the subject many times, although – drawing the convenient philosophical distinction between normative or substantive ethics ("What should I do?") and foundational ethics or metaethics ("Why should I do that which I should do?") – in most cases it was the former topic that was his chief concern. And there, naturally enough, Huxley's interests and targets changed somewhat over the years – something that he would have predicted and expected. Huxley always emphasized that morality was relative, meaning not that one could do as one pleased, but that particular moral norms and directives would evolve, especially as circumstances (e.g., technological circumstances) changed. (He even went so far as to argue that there had been nothing wrong with slavery in earlier centuries, because it had been needed to keep societies running smoothly.)

However, beneath the changes in circumstances, and allowing for some particular quirks and special interests (I sense that, perhaps as a result of his years in Houston, Huxley always had a somewhat condescending attitude toward Africans and their abilities) (Huxley 1924, 1931), a consistent pattern or theme emerged. Although Huxley was not uninterested in life at the personal level, it was the general domain that really excited him:

> All claims that the State has an intrinsically higher value than the individual are false. They turn out, on closer scrutiny, to be rationalizations or myths aimed at securing greater power or privilege for a limited group which controls the machinery of the State.
> On the other hand the individual is meaningless in isolation, and the possibilities of development and self-realization open to him are conditioned and limited by the nature of the social organization. The individual thus has duties and responsibilities as well as rights and privileges, or if you prefer it, finds certain outlets and satisfactions (such as devotion to a cause, or participation in a joint enterprise) only in relation to the type of society in which he lives. [Huxley and Huxley 1947, pp. 138–9]

Huxley's key moral principle would seem to have been the need for planning in running the state and, above all, the application of *scientific* principles and the use of empirical findings in such planning and its implementation.

204

One simply could not (or should not) leave things to chance or intuition (the implication being that that was precisely where the average politician did leave things), but should bring trained scientific minds to bear on life's problems.

Again and again Huxley returned to that theme. For instance, in a book he wrote during the interwar years, *If I Were Dictator* (1934b), he stressed the need for science in the running of an efficient state, and that science would need to be of the social variety, as well as the physicochemical and biological. During the Second World War he wrote a highly laudatory essay on the Tennessee Valley Authority (TVA), that marvel of the Roosevelt-era New Deal whereby the federal government built and ran a massive system of river dams and irrigation in what had previously been one of the more desolate parts of the United States (Huxley 1943). Then, after the war, it was Huxley who insisted on a science component being included in the plans for UNESCO. He wrote a vigorous polemic arguing that the organization had to be run on evolutionary lines – lines demanding lots of science (Huxley 1948). So vigorous was his polemic that he upset his superiors and was denied a full four-year term as director general of UNESCO.

Now let us turn to the question of justification, the question of metaethics. One might argue, with justice, that much of what Huxley believed and preached came straight from his personal experiences and prejudices. Enthusiasm for eugenics theories, for instance, was part of the general credo of intellectuals in the first part of the twentieth century. Likewise for birth control, another key plank in the Huxley scheme of things. Huxley had known and socialized with the leaders of such intellectual movements in the early years. Obviously, his enthusiasm for science was a major component of Julian Huxley's raison d'être. The fact that he was an evolutionist was contingent, if not irrelevant. In principle, he could have believed in Genesis as literal truth so long as he was enough of a physicist or other scientist to want to advocate the all-importance of science in running the state.

However, Huxley certainly thought that he was putting his substantive ethics on an evolutionary metaethical basis (though he did not use that language), and he advanced arguments to that effect. Most explicit was the discussion that took place in 1943 (one year after the *Modern Synthesis*) in a Romanes lecture at Oxford, which he published commercially a few years later, bound together with the Romanes lecture (also on ethics) given by his grandfather some fifty years earlier (Huxley and Huxley 1947). At that point, Huxley was rather enthusiastic about Freudian theory, so that a reader has to wade through introductory filler on the psychoanalytic development of the moral sense before getting to the biological meat. But it is there, unambiguously. And it is there that we see how terribly crucial biological

205

progressionism was for a key aspect of Huxley's conception of "religion without revelation," as in the book by that name (Huxley 1927). Simply put, Huxley argued that because evolution was progressive and because progress meant that value was ever-increasing, humans had a moral obligation to cherish and promote the evolutionary scheme of things:

> When we look at evolution as a whole, we find, among the many directions which it has taken, one which is characterized by introducing the evolving world-stuff to progressively higher levels of organization and so to new possibilities of being, action, and experience. This direction has culminated in the attainment of a state where the world-stuff (now moulded into human shape) finds that it experiences some of the new possibilities as having value in or for themselves; and further that among these it assigns higher and lower degrees of value, the higher values being those which are more intrinsically or more permanently satisfying, or involve a greater degree of perfection.
>
> The teleologically-minded would say that this trend embodies evolution's purpose. I do not feel that we should use the word purpose save where we know that a conscious aim is involved; but we can say that this is the *most desirable* direction of evolution, and accordingly that our ethical standards must fit into its dynamic framework. In other words, it is ethically right to aim at whatever will promote the increasingly full realization of increasingly higher values. [Huxley and Huxley 1947, p. 137]

We see now why it was so crucial for Huxley to maintain that evolution, except in the human species, was essentially finished. It was no part of his brief to argue that humans should promote the evolution of, say, the warthog, except insofar as such animals or plants could benefit humans. It was also crucial to maintain that in humans the mode of evolution had changed. Huxley was not about to fall into the snare of arguing for some kind of laissez-faire philosophy, neither at the biological level nor at the social level. Indeed, he ardently opposed such philosophy. He was for planning, and if he was not a socialist (certainly not a social democrat), he clearly was for the use of the state to effect major schemes and projects. Remember the TVA.

Yet, equally, a characterization of progress in terms of control and independence was a vital part of Huxley's argument, the positive part. If what was needed was promotion of progress, and such progress consisted in maximizing control and independence, what better way to do it than through state-organized schemes for implementation of the output from science and technology? Although Huxley might well have agreed that it was the spur of circumstances and acquaintances that led him initially to stress the need for such things as artificial birth control and scientific agriculture and the like, he would have argued that his justification for such norms was entirely logical, or, rather, entirely evolutionary. Without such things, real progress in the human realm would become impossible.

The argument, therefore, was that the evolutionary process itself promoted value, and therefore the human ethical duty was to work with and within that process to ensure that it would be realized as fully as possible:

> From the evolutionary point of view, the destiny of man may be summed up very simply: it is to realize the maximum progress in the minimum time. That is why the philosophy of Unesco must have an evolutionary background, and why the concept of progress cannot but occupy a central position in that philosophy. [Huxley 1948, p. 11]

"New Bottles for New Wine" Huxley labeled one of his collections of essays. In that case, however, the wine was old, however fancy and new-looking the container. One can readily link Huxley's views with a long-standing – the most long-standing – tradition in evolutionary ethics, as his critics were quick to point out. Grant, for the moment, the distinctive features of Huxley's position, such as human evolution being uniquely that which is still progressing (although more on that shortly), the underlying pattern (Greene was right about this) goes back to Herbert Spencer. He, too, was an ardent evolutionary progressivist – "monad to man" – believing that the philosophy was at one with progressionism in the sociocultural human world (Spencer 1868). He, too, wanted to replace conventional religion with some sort of evolutionary substitute – the "synthetic philosophy." And he, too, saw normative directives in evolution's methods, arguing that humans had a moral obligation to promote the ends of evolution and that the reasons for that lay in the process itself. It was precisely because evolution was progressive that humans ought to cherish it, for therein and only there was the way to promote and preserve value.

For all that Bergson had been his primary nonbiological influence, I doubt that Huxley would have been particularly perturbed by my locating him in the Spencerian tradition. There was ever a Spencerian flavor to his work – "A state of equilibrium may for a time exist, but every balanced organism is as it were pressing against every other, and a change in one means a re-arrangement of them all" (Huxley 1912, pp. 114–15) – and when it came to progress, Huxley was explicit in his praise that in seeing uniformity ("homogeneity") give way to complexity ("heterogeneity"), Spencer had hit on an idea containing "a great deal of truth" (Huxley 1964, p. 35).

Because traditions can move forward as well as backward, let me point out that Julian Huxley was not the end of the line of those who would argue in a neo-Spencerian manner. To choose but one, in recent years the Harvard

Michael Ruse

entomologist and sociobiologist Edward O. Wilson has been endorsing and promoting an evolutionary ethical picture very much in line with that of Huxley (Wilson 1975, 1978, 1984, 1990, 1994; Wilson and Peter 1986). His categorical imperative beseeches us to promote biodiversity – an issue dear to the heart of one whose specialty is tropical ants and who is keenly aware of the ongoing destruction of the Brazilian rain forest – but his justification is the upwardly progressive nature of the evolutionary process. Significantly, on the wall of Wilson's laboratory hangs a picture of Herbert Spencer, and in his recent autobiography Wilson (1994) spoke of Huxley with great respect.

CRITICAL RESPONSE

What can one say about Huxley's moral philosophy? Let me now step out of my role as disinterested historian and into my role as partisan philosopher. One thing is already obvious, if further confirmation were necessary. The actual normative directives endorsed and promoted by evolutionary ethicists were very far from the evil philosophies of legend. I am not sure that I would particularly fancy having Julian Huxley as my dictator, but far rather him than some of the real ones. It is true that today some of his prescriptions have an uneasy ring – eugenics, for instance, although my reading on that particular issue suggests that Huxley was more in favor of what today we call "genetic counseling" (that is, elimination of bad genes) than of wholesale biological redesign. Likewise, nowadays, major state-run projects have fallen out of favor, although in this case there are many (myself included) who think that there is still much to be said in their favor. We also think that a major reason that such projects are unpopular is that they are seen as having served their function, and now, with increased prosperity, they are no longer needed. This, of course, speaks only to western society. Elsewhere in the world, Huxley-like projects are both needed and still promoted.

Grant that a case can be made for Huxley-style normative prescriptions and that an even stronger case can be made for the more recent prescriptions of thinkers in the same tradition. Biodiversity and the preservation of the rain forests are close to being motherhood issues. But what of the question of justification? What about the argument that Huxley used to support his normative ethics? What about the appeal to the progressive nature of the evolutionary process? It was there that Moore (1903) faulted Spencer, and likewise it was there that he would have faulted Huxley, accusing them both of having committed the naturalistic fallacy.

In the case of Huxley, the charge would have been that control and inde-

208

pendence – together with complexity, which presumably was a condition of those – were natural properties or phenomena, things of this world, as it were. However, directives that one ought to promote control abilities and independence (through science and planning) would be nonnatural, in the sense that they would not be derived from the physical realm – although obviously humans might want to apply them to or in such a realm. Hence the former simply could not be the foundational basis for the latter. Or, if we phrase things in the manner of David Hume – who argued that there was a logical gap between claims about matters of fact ("is" statements) and claims about morality ("ought" statements) (Hudson 1983) – the charge would be that Huxley sought to go, illicitly, from the way that things were (in the evolutionary process) to the way that things should be (in the future). He tried to go from "control and independence exist" to "control and independence should be maximized."

Although I am not aware that Huxley ever addressed that criticism directly (very much unlike Simpson, as we shall learn shortly), one ready response would have been to agree that certainly there were (or seemed to be) significant differences between claims about fact and claims about morality, but that that in itself was no barrier to the former being the foundation for the latter.[1] After all, in science one was quite accustomed to "reductions": derivations or justifications of one thing in terms of another very different thing (Nagel 1961; Ayala 1985). Is the "is/ought" distinction any greater than, say, the "micro/macro" division to be found in the theory of gases? One goes, there, from temperature-free molecules buzzing in empty space to a gas occupying all of the space in a container – a gas that can go from very cold to very hot and possibly can exert tremendous pressure in the bargain. Yet the one is supposed to be the foundation for the other, and there are mathematical deductions to prove it!

Perhaps Huxley did offer some sort of response along those lines, for it is worth remembering that he stressed repeatedly that he saw the progressiveness of evolution as an entirely natural objective phenomenon. Hence, for him, the value-laden nature of independence and control was not something read into nature, but something read from it. Thus the values were given to us by the natural world, and so, at least logically, there was no reason why we should not go from an "is" statement to an "ought" statement.

I shall return to this point in a moment, but, in any case, first let me agree with Huxley if, in whatever way, he would simply have refused to bow down before the preceding critical argument. Why should it be the case that in going from natural to unnatural he was thereby committing a "fallacy"? What right have his critics to label his argument fallacious – that is, without further argument on their part? After all, we do value the things that make us distinctively human, so why should we not promote them?

To his credit (a point often missed by those who quote him favorably), Moore buttressed his contention that Spencer's ethic was inadequate by offering further detailed critique. And it is worth noting that although Huxley's arguments received no greater support among professional philosophers than had Spencer's ethic, they likewise felt it necessary to take time to refute him explicitly. In particular, the Cambridge moral philosopher C. D. Broad subjected Huxley's thinking on the matter to withering critical review. Concerning the move from "is" to "ought," he was strong in his objection. Raising the key question of "whether knowledge of the facts of evolution has any bearing on the question of what is intrinsically good or bad," Broad wrote as follows:

> It is plain that Prof. Huxley thinks that it has an important bearing on this question, but I find it extremely hard to see why he does so. Perhaps I can best bring out the difficulty that I feel in the following way. Take the things which Prof. Huxley considers to be intrinsically good, and imagine him to be confronted with an opponent who doubted or denied of any of them that it was intrinsically good. How precisely would he refute his opponent and support his own opinion by appealing to the facts and laws of evolution? Unless the notion of value is surreptitiously imported into the definition of 'evolution', knowledge of the facts and laws of evolution is simply knowledge of the *de facto* nature and order of sequence of successive phases in various lines of development. [Broad 1949, p. 585]

Broad continued:

> If then, Prof. Huxley is to support his own views about the intrinsic value of so-and-so and to refute those of an opponent by appealing to the facts and laws of evolution, there must be a suppressed premiss in the argument . . . he must use *some* 'mixed' premiss, connecting certain *purely factual* characteristics, which are all that a study of evolution can possibly reveal to us, with the *value-characteristics* of intrinsic goodness and badness. [Broad 1949, p. 586]

Broad concluded:

> Now, whatever may be the evidence for such a mixed premiss, it is quite plain that it must be something different from the evidence for the facts and laws of evolution. For the premiss required asserts a connection between certain of those facts and laws and something else, viz., intrinsic value or disvalue, which forms no part of their subject-matter. [Broad 1949, p. 586]

Presumably Huxley's response at that juncture would be a reiteration of the point made earlier: that there was no reason, in principle, why one should not sometimes (this being one of those times) go from claims of fact to claims of value. But I think that Broad was trying to dig more deeply. If Huxley raised the reduction analogy (one could legitimately go from micro to macro in gas theory), the reply would be that one certainly could do so *if* one could give good reasons why that might be allowed. In gas theory, good reasons had been given. In Huxley's case, no such reasons had been given, and there were at least two major counter-reasons to think that there were no such reasons.

On the one hand, although Broad (as a non-biologist) did not really argue that, there was the question of Huxley's picture of the evolutionary process. That was crucial, or so it should have been if one was promoting an ethics based on and informed by evolution. But whether one took Huxley's original position or his later revision, there were serious questions. It was highly doubtful that the process of evolution had come to an end, save in the human line, and that consequently the only evolution possible was a nonselective sociocultural process. More than that, there was no reason to think that evolution promoted control and independence in the manner supposed by Huxley, particularly in the manner that supposed that humans had some special status in that respect – a status that allowed (required) humans to speak of progress *as a matter of objective fact.* Humans have lost the ability to synthesize vitamin C. Does this mean that we are any less worthy than formerly, unless we want to make long sea voyages? Females have lost the ability to reproduce without the aid of males. Does this make them less valuable on the scale of being? Or, conversely, if males really do control females, are males any more worthy? Sometimes independence and control are good things; sometimes not. It all depends. Whales can dive deep and live for long periods under water; birds can fly; camels can exist in the desert without water. We can do none of those things. Although we have less control and independence, are we thereby lesser beings?

On the other hand (the point that Broad did stress), suppose that person A denies that control and independence are so all-important and valuable. There is nothing in evolution, certainly nothing in Darwinian evolution as such, to convince person B otherwise. B may, of course, challenge A's values, if, for instance, A prizes weapons of attack for use in life's struggles; but B is not doing this on the basis of evolution. Rather, these are values that B imports from elsewhere. The fact of the matter is that Huxley simply read in what he wanted to read out. Huxley wanted control and independence of a peculiarly human variety to count, so he simply used them as a lens through which to view evolution. The inescapable truth is that there was a crucial gap between premises and conclusion in Huxley's argument for the foundations of ethics. State planning may be all very well (in my opinion, it is all very well), but that has been assumed and not proved. It certainly has not been proved from evolutionary premises, even if one agrees that one can get it from control and independence.

211

So much, then, for Huxley. Almost necessarily in a discussion such as this, one comes through sounding extremely negative, so let me conclude by going back to the beginning point, a conviction shared with both Julian and his grandfather that it really must matter that we are evolved beings and not just the products of a Good God on the Sixth Day, molded in his image. Unfortunately, Julian Huxley and his tradition seem not to have been on the right track. The question is whether or not that is the end of evolutionary ethics, as people like Broad believed it to be. Simpson thought not, so let us turn now to look at his thinking.

GEORGE GAYLORD SIMPSON (1902–84)

Simpson was a genius. The best paleontologist since Cuvier, his supervisor called him. Educated at the University of Colorado and Yale, Simpson spent much of his working life at the American Museum of Natural History in New York City, moving later to Harvard and then retiring to New Mexico (Simpson 1978, 1987). He worked extensively on the mammalian record. More generally, having a command of mathematics unusual among evolutionists, it was he who was primarily responsible for bringing the findings from the Darwin-Mendel synthesis to bear on the fossil record. His masterwork, *Tempo and Mode in Evolution* (revised and retitled in 1953 as *The Major Features of Evolution*), appeared in 1944, two years after Huxley's *Modern Synthesis,* although it was completed at about the same time. A shy and difficult man, Simpson had few close friends, but in later life he and Huxley (who had been friendly to the young Simpson when he had been a postdoctoral fellow in London, after Yale) became close associates, discussing in detail questions of evolution and ethics.

The place to start is with a point made earlier: that a major mission in Simpson's life was to shatter the picture of progressive change absorbed and endorsed by Huxley. Simpson denied that the past had been anything at all like that. It had been far too prone to branching for any Huxley-like scenario; the talk of the specialized forms being always dead ends was simply not true; and natural selection had been too opportunistic for anyone to suppose that independence and control could have the privileged positions presumed by Huxley (Simpson 1928, 1949, 1951, 1964). Simpson could never have written lines such as this: "One somewhat curious fact emerges from a survey of biological progress as culminating for the evolutionary moment in the dominance of *Homo sapiens.* It could apparently have pursued no other general course than that which it has historically followed" (Huxley 1942, p. 569).

For the American, evolution had been quite undirected, driven only by whatever had been immediately adaptively advantageous. And it was simply false to think that evolution had stopped or that it had come to be confined to the human realm. There were all sorts of possibilities for further change:

> . . . one cannot doubt that a *particular channel* of specialization may reach its limit (the horse cannot have less than one toe), but then progress (or change, at least) shifts to some other channel through, as a rule, some other group of animals. Even if, as you suggest, all systems except the nervous system were near the limit of biological efficiency by the end of the Miocene, the nervous system did, in our time, go on from there, which was hardly predictable at the time. I cannot feel any confidence in prediction that the nervous system is now at its limit, or even that quite other possible channels of progress, in the nervous system or in other systems, are really non-existent. [Simpson to Huxley, August 20, 1950, Huxley Papers, Rice University]

However, that much being said (and it is saying a lot), in Simpson's writings there certainly were strong echoes of Huxley's thought, which, Simpson candidly admitted, had acted as a major stimulus to his own thinking. Most particularly, Simpson was ardently and unambiguously a progressionist, and he would have thought it simply silly not to agree that humans were, in basic respects, at the top of the mountain (or "pinnacle" as he and everyone else tended to call it, following Sewall Wright). He believed that there were ecological niches existing objectively, waiting to be occupied, and that organisms would make adaptive breakthroughs in moving to take them over. And he certainly believed that there was some sort of hierarchy, of a not entirely unfamiliar kind.

Yet (and here we run straight into one of the most crucial aspects of Simpson's thought) he was keen to stress that although the denial of progress was silly, the ascription of progress was not something simply read from nature in an objective manner (as Huxley thought). Rather, it was something that had to be read in, and because one wanted to avoid the gross anthropomorphism of simply reading in criteria that were by definition humanlike, that meant that, taken generally, there could be no one mark of progress that would be especially privileged over all others.

For that reason, in his most extended discussion of the topic, in his popular book *The Meaning of Evolution* (1949), Simpson ran through a large number of criteria that had been proposed as showing biological improvement or worth: expansion of life, dominance, specialization, potential for future development, independence from the environment, control of the environment, complexity (Simpson was not very keen on that old favorite, believing that it probably was false that complexity increased, and it was an untestable concept anyway), general energy level, prenatal and postnatal care, sophistication of the nervous system, individualization, and more.

Humans certainly did not always come out on top as judged by all of those criteria. For instance, Simpson was enough of a traditionalist to think that humans were not highly specialized. Overall, however, humans tended to score well – the very best in many cases, such as dominance, prenatal and postnatal care, nervous system, and individualization. And that general consilience would seem to have been enough to convince Simpson that progress, with humans at the top, was more than just a whim or conceit. "Progress" may not have had the objectivity that Huxley supposed, but it was more than just one man's yearning.

We move on now to ethics, and again we see a reaction against the kind of thinking that Huxley represented, although again it was a sympathetic reaction. Like Huxley, Simpson was absolutely and completely committed to the view that ethics was natural, in the sense of being produced by evolution. In fact, Simpson went into more detail that Huxley on the way in which biology, through the medium of selection, could produce something like an ethical sense, pointing out that success in the struggle for existence did not necessarily mean all-out warfare, but could demand sympathetic alliance with one's fellows. (It is true that Simpson was more inclined to a group perspective on the workings of evolution than evolutionists would find acceptable today.) Ethics was natural also in having no justification or sanction outside of evolution. What had evolved was what one got. Simpson, who came from a fundamentalist Presbyterian family, was by nature (like both Huxleys) always an intensely religious man. But his faith in an existent deity was nonexistent (in middle life he worshipped with the Unitarians), and he certainly thought that there could be no divine or similar support for moral belief.

At the level of normative ethics – "What should I do?" – whereas Huxley was in favor of large-scale public works and other state-funded projects, Simpson looked much more to the individual sphere. There were two major directives. First, there was the need to improve and promote knowledge – knowledge in itself, as a good:

> The most essential material factor in the new evolution seems to be just this: knowledge, together, necessarily, with its spread and inheritance. As a first proposition of evolutionary ethics derived from specifically human evolution, it is submitted that promotion of knowledge is essentially both the acquisition of new truths or of closer approximations to truth (metaphorically the mutations of the new evolution) and also its spread by communication to others and by their acceptance and learning of it (metaphorically its heredity). This ethic of knowledge is not complete and independent. In itself knowledge is necessarily good, but it is effective only to the degree that it does spread in

a population, and its results may then be turned by human choice and responsible action for either good or evil. [Simpson 1949, p. 311]

Then, secondarily, there was personal responsibility, which would lead to integrity and dignity:

> Beyond its relationship to the ethic of knowledge, the fact of responsibility has still broader ethical bearings. The responsibility is basically personal and becomes social only as it is extended in society among the individuals composing the social unit. It is correlated with another human evolutionary characteristic, that of high individualization. From this relationship arises the ethical judgment that it is good, right, and moral to recognize the integrity and dignity of the individual and to promote the realization or fulfillment of individual capacities. It is bad, wrong, and immoral to fail in such recognition or to impede such fulfillment. [Simpson 1949, p. 315]

Why Simpson's differences with Huxley? To be honest, I doubt that they had much to do with biology, despite the supposed derivation of those ideas from evolution. In Huxley's case, I have suggested that the personal factor was significant (his acquaintances and the like), although we should not forget his upbringing, the valued place of science (and of a scientific elite) in the Huxley household, and the challenges to and successes of society, particularly society in the 1930s, when Keynesian economics seemed to offer a way forward from widespread depression and poverty. In Simpson's case, the valuing of knowledge was equally personal. In the 1930s, seeking for an alternative to conventional religious faith, he had found it in the search for knowledge – something that, incidentally, helped to salvage his conscience, as he regularly deposited his four daughters (Simpson was a single father) with his parents so that he could go off on long-term fossil-hunting trips (Simpson 1987).

The valuing of responsibility and dignity, and so forth, was equally a function of the times and the society within which Simpson lived. Those were the years when the cold war was settling into its long winter, when Soviet science was suffering under influential charlatans like T. D. Lysenko, and when issues of dictatorship and totalitarianism were all too fresh in people's memories, and still present in much of the world. From a basis in dignity and responsibility, Simpson launched straight into condemnation of the oppressive regimes then flourishing, and he juxtaposed that with praise (if not uncritical reverence) for the society within which he found himself: "Democracy is wrong in many of its current aspects and under some current definitions, but democracy is the only political ideology which can be made to embrace an ethically good society by the standards of ethics here maintained" (Simpson 1949, p. 321).

SIMPSON'S METAETHICS

What of justification? Wherein lay the foundations of normative ethics for one such as Simpson? It is not easy to give a direct answer to that question, which reflects the fact that the usually very clear Simpson was himself wrestling with his response. He knew, on the one hand, that the answer had to come from evolution. That was the whole point of naturalism. He knew, on the other hand (and, unlike Huxley, Simpson took this very seriously) that one simply could not go blithely from statements about facts, "is" statements, to statements about obligations, "ought" statements. The question was how to reconcile those demands and constraints.

At times, Simpson sounded almost like an existentialist in his thinking. Could it not be that the very fact of responsibility, of the need to make a choice in one's own right, was in some way self-validating? Ethics certainly was relative to the material and social situation in which one found oneself. Could it not be that they were relative also to the individual – the evolved individual, that is?

> It should, finally, again be emphasized that these ethical standards are relative, not absolute. They are relative to man as he now exists on the earth. They are based on man's place in nature, his evolution, and the evolution of life, but they do not arise automatically from these facts or stand as an inevitable and eternal guide for human – or any other – existence. Part of their basis is man's power of choice and they, too, are subject to choice, to selection or rejection in accordance with their own principles. [Simpson 1949, p. 324]

The trouble with such a position was that, ultimately, there was no arbiter between right and wrong. One was left with the subjectivism of the individual conscience. Yet, although Simpson had trouble with individual relationships, he had a burning sense of the imperatives of morality, imperatives that drove him to action. As Huxley worked for and defined UNESCO, so Simpson during the Second World War had engaged in difficult and dangerous military service at an age when he could have excused himself, and later in life he was outspoken about his moral disgust at America's involvement in Vietnam. There was not much relativism there.

Simpson clearly went on worrying, and in the 1960s he returned to the issue of foundations:

> The evolutionary process in itself is nonethical – there simply is no point in considering whether it is good, bad, a mixture of the two, or neither. . . . But man, with his ethical sense, was produced by evolution as the result of one evolutionary sequence. Human evolution continues and man is subject to all evolutionary principles, not simply as generalists but as they apply in his specific and extremely special circumstances. A rational, naturalistic system of

ethics cannot be independent of evolution, but neither can it be derived from evolution as a general or merely abstract principle. [Simpson 1969, p. 143]

Shortly, he continued:

> Biological trends are almost entirely directed by natural selection. Social trends operate within the framework and limitations of the biological factors and represent a remarkably flexible form of biological adaptation. . . . The point pertinent here . . . is that neither strictly organic nor social evolution necessarily leads to improvement in any sense of the word acceptable for the human situation. In fact either, in their ways that *in themselves* are blindly amoral, may have results extremely undesirable for mankind. But here is the saving grace: both organic and social evolution are now to some degree, even though a limited degree, within our own control. We, alone among all organisms, are aware of our own sociobiological evolution, can judge what seems to us desirable or undesirable in that evolution, and can deliberately work for the desirable. . . .
>
> Beyond that, it bears repeating that the evolutionary functioning of ethics depends on man's capacity, unique at least in degree, of predicting the results of his actions. A system of naturalistic ethics then demands acceptance of individual responsibility for those results, and this in fact is the basis for the origin and function of the moral sense. [Simpson 1969, pp. 145–6]

Ultimately, it all came down to evolution, and for that reason two sympathetic critics of Simpson have concluded that despite his intentions, ultimately he, too, went the way of Huxley, justifying morality in the name of evolution and hence, at the end, committing the naturalistic fallacy (Greene 1981; Swetlitz 1991). Yet, I have to say that I disagree: He spoke too strongly against the move of justifying morality by reference to the facts of evolution, although I would agree that Simpson gave no adequate analysis, and indeed at the end of the passage just quoted he seemed to slide back again into an unsatisfactory subjectivity. A murderer may accept responsibility, but that is no excuse.

Is there a way out of this dilemma? Speaking now as a philosopher, rather than a historian, I believe there is, although whether or not Simpson would have been happy with it is another matter. As a number of thinkers have argued in recent years, one might claim that inasmuch as one believes that the moral sense is a product of evolution, this suggests that ultimately there is *no* foundation for ethics considered at the normative level (Mackie 1977; Murphy 1982; Ruse 1986). This does not mean that ethics does not exist or that one can simply do what one pleases – at least, one cannot simply do what one pleases inasmuch as one is an evolved human being. What it means is that at the foundational, metaethical level, there is no ultimate justification, whether this be evolution or the Will of God or whatever else.

According to the "ethical skeptic," therefore, ethics – normative ethics – has evolved to make us good cooperators, because given the kinds of beings we humans are, cooperation is a good adaptive strategy in the struggle for existence. But there is nothing beyond this, and certainly no solid ground of proof. We have a moral sense because it is adaptively advantageous to have it, but ultimately (as in the case of the secondary qualities that appear so vividly to us) there is nothing that it is sensing! Hence, a causal explanation is all that we can give, and once it has been given, we see that it is impossible (although, thankfully, not necessary) to satisfy the call for a reasoned justification:

> The [evolutionist] may well agree . . . that value judgments are properly defended in terms of other value judgments until we reach some that are fundamental. All of this, in a sense, is the giving of *reasons*. However, suppose we seriously raise the question of why these fundamental judgments are regarded as fundamental. There may be only a *causal* explanation for this! We reject simplistic utilitarianism because it entails consequences that are morally *counterintuitive*, or we embrace a Rawlsian theory of justice because it systematizes (places in "reflective equilibrium") our *pretheoretical convictions*. But what is the status of those intuitions or convictions? Perhaps there is nothing more to be said for them than that they involve deep *preferences* (or patterns of preference) built into our biological nature. If this is so, then at a very fundamental point the reasons/causes (and the belief we ought/really ought) distinction breaks down, or the one transforms into the other. [Murphy 1982, p. 112, n. 21]

It is usually added, by people who argue in that way, that ethics works because we are part of the system. With a shared evolution, we humans have a shared insight (or, rather, sense of insight) into the norms of right and wrong. Some people may disagree with these norms, but just as children make mistakes in arithmetic, the disagreement is a function of inadequate training and does not point to an irresolvable subjectivity. Or it may be argued that the disagreement is biologically based. Just as some people are unable to tell red from green (though that is their misfortune and is not to be blamed on the nonexistence of a real difference between red and green), so some are innately morally handicapped. What is crucial is that although, given this picture, morality ultimately has no objective existence (in the sense of an objective referent – it is objective in that it certainly exists), our biology makes us think that it does! We believe that we are talking about something real, because if we did not, morality would collapse into a morass of subjective wishes and desires and directives, and it would entirely lose its force.

I like to think that Simpson would have responded favorably to the spirit of this solution:[2] One is a firm evolutionary naturalist. At the same time, one avoids the naturalistic fallacy because one is not justifying anything, including ethics, in terms of anything else, including evolution! And although personal responsibility and choice are important (the whole point of this view of ethics is that one does have the freedom to choose and is not locked into one course of action), and although it certainly is agreed that norms will change as circumstances change, ethics is more than just personal feelings. There are standards that are society-wide, if not humanity-wide.

However, reverting back to the historian, I cannot pretend that it is a solution that Simpson himself achieved or endorsed, certainly not in its entirety. There is nothing in the work of Simpson, for instance, about the lack of foundations for normative morality. Nor is there anything about having a sense that there were foundations because it would be adaptive to have such a sense. Indeed, to be candid, Simpson might well have pulled back from such a solution, believing that, ultimately, one would be downgrading ethics (normative ethics, that is) by claiming it to be (what I myself have elsewhere characterized as) an "illusion of the genes" (Ruse 1986). In defense of the solution, one can explain at length that the illusion comes not in the nonexistence of ethics as such but in the supposed referent. But that might not have convinced Simpson, or at least only in mind and not in heart.

Before leaving this discussion, however, let me refer back to Greene's claim, now to the part about Simpson being the latter-day representative of Thomas Henry Huxley. There is certainly good reason for agreeing with that, as well as with the implication that T. H. Huxley (the grandfather) and Simpson represent a rival tradition to that of Spencer and J. S. Huxley (the grandson). Although Simpson did not regard the products of evolution to be quite as vile as did T. H. Huxley (1893), they were in agreement that one must strive for a naturalistic approach to ethics, that that approach had to be evolutionary and nothing else, and that one could not justify ethics in the name of evolution – T. H. Huxley was eloquent on the degree to which evolution was not necessarily progressive in any conventional sense.

Given Simpson's fuzziness, can we usefully ask about T. H. Huxley's metaethical philosophy? Where did he think one should locate the foundations of substantive ethics? Unfortunately, as in the case of Simpson, there is (from another usually crystal-clear thinker) less than total clarity. But there is an interesting and informative little book that T. H. Huxley (1879) wrote on David Hume – interesting because, although Hume was no evolutionist, Hume's metaethics is a clear forerunner of the skeptical solution I have just sketched here, and informative because Huxley seems to have responded warmly to it. Toward the end of the discussion of Hume's moral theory, what starts as paraphrase seems to conclude as wholehearted agreement:

219

In whichever way we look at the matter, morality is based on feeling, not on reason; though reason alone is competent to trace out the effects of our actions and thereby dictate conduct. Justice is founded on the love of one's neighbour; and goodness is a kind of beauty. The moral law, like the laws of physical nature, rests in the long run upon instinctive intuitions, and is neither more nor less "innate" and "necessary" than they are. Some people cannot by any means be got to understand the first book of Euclid; but the truths of mathematics are no less necessary and binding on the great mass of mankind. . . . While some there may be, who, devoid of sympathy, are incapable of a sense of duty; but neither does their existence affect the foundations of morality. Such pathological deviations from true manhood are merely the halt, the lame, and the blind of the world of consciousness; and the anatomist of the mind leaves them aside, as the anatomist of the body would ignore abnormal specimens. [T. H. Huxley 1879, pp. 239–40]

Even if Huxley was speaking entirely for himself – and I am inclined to think that he was, because a few years later he wrote of "a keen innate sense of moral beauty" (L. Huxley 1900, pp. 2, 306) that was in itself the means and the end of morality – that was hardly a full-blooded exposition and endorsement of ethical skepticism. But it was in that spirit, not just in being naturalistic but in so firmly putting morality in the domain of feelings – something that an evolutionist like Simpson would relate back to natural selection and adaptation.[3] So, without pretending that a definitive case has been made, it seems not too unfair to suggest that just as the tradition of Spencer and Julian Huxley has been taken up today by people like Wilson, so the tradition of T. H. Huxley and Simpson has been taken up today by those evolutionists who think their theory points to ethical skepticism.

CONCLUSION

My conclusions follow quickly. As a historian I find that Julian Huxley and George Gaylord Simpson were united in believing that morality must be natural and that the science of key significance was evolution. They represented, however, two different traditions, going back (at least) to Herbert Spencer and Thomas Henry Huxley.[4] Julian Huxley wanted to justify normative ethics by reference to the fact and processes of evolution, and to do that he had to argue that progress was an objective fact of nature. He was therefore prepared to barge right through the naturalistic fallacy. Simpson wanted to explain normative ethics in terms of the fact and processes of evolution, and in doing that, although he was a progressionist, he wanted to avoid claiming that it was an objective fact of nature. He took the naturalistic fallacy seriously, as something to be avoided, and he talked much of the significance of individual responsibility as a foundation for morality.

As a philosopher, I applaud the spirits and intentions of both men. I intend this applause as showing an agreement stronger than any differences that we may have. I am happy to be set off with the evolutionists against the tradition of twentieth-century analytic philosophy, typified by G. E. Moore. However, I do agree that Julian Huxley's position was inadequate, and I would argue with the critics (in this essay, with C. D. Broad, but ultimately with David Hume) that there is a key difference between statements of fact and statements of morality and that Huxley and his tradition have failed to bridge it. Progress is not an objective fact of nature and cannot therefore be used to justify a normative ethic.

I find more encouraging Simpson's thinking about morality, and that is true more generally of his science and the interpretations he would put into it. I do not think that Simpson himself offered a full and satisfying meta-ethics, although I would defend him against critics who claim that he committed the same fallacies as did Huxley. Whether or not Simpson would appreciate the directions in which I would point him is another matter. That, however, must be left to the judgment of the reader.

NOTES

1. I confess that I am somewhat surprised at how little interest Huxley seems to have had in philosophical criticism. He must have known of it, but does not seem to have thought it worth refuting, nor his grandfather's ideas, either, for that matter.
2. Although Simpson was a little more sensitive to philosophy than Huxley, again I cannot but regret that Simpson paid so little attention to it. Had he done so, he might have made use of the then-dominant moral philosophy, emotivism, which likewise denied that morality had any objective referent.
3. Greene (1981) spoke of Huxley as being an intuitionist. I am uncomfortable with that interpretation, although it is certainly the case that if one objectifies ethics (thinks it has an objective referent), one acts and sounds like an intuitionist, and there is language by Huxley that sounds that way.
4. A question we all face is to which tradition to assign Charles Darwin. In my opinion today – a change from earlier work (Ruse 1986), where I put him in the Huxley/Simpson tradition – I think he belongs to both, especially on the basis of the *Descent of Man.*

REFERENCES

Ayala, F. J. 1985. Reduction in biology: a recent challenge. In: *Evolution at a Crossroads,* ed. D. J. Depew and B. Weber, pp. 65–80. Cambridge, MA: MIT Press.

Bergson, H. 1907. *L'Évolution créatrice.* Paris: Alcan.

1911. *Creative Evolution.* London: Macmillan.

Bradie, M. 1994. *The Secret Chain: Evolution and Ethics.* Albany: State University of New York Press.

Broad, C. D. 1949. Review of Julian S. Huxley's "Evolutionary Ethics." In: *Readings in Philosophical Analysis,* ed. H. Feigl and W. Sellars, pp. 564–86. New York: Appleton-Century-Crofts. (Originally published 1944: *Mind,* vol. 53.)

Bury, J. B. 1924. *The Idea of Progress: An Inquiry into Its Origin and Growth.* London: Macmillan. (Originally published 1920.)

Crook, P. 1994. *Darwinism: War and History.* Cambridge University Press.

Darwin, C. 1859. *On the Origin of Species.* London: John Murray.

1871. *The Descent of Man.* London: John Murray.

Dobzhansky, T. 1956. *The Biological Basis of Human Freedom.* New York: Columbia University Press.

1962. *Mankind Evolving.* Yale University Press.

1967. *The Biology of Ultimate Concern.* New York: New American Library.

Dobzhansky, T., and Wallace, B. 1959. *Radiation, Genes and Man.* New York: Henry Holt.

Farber, P. L. 1994. *The Temptations of Evolutionary Ethics.* Berkeley: University of California Press.

Greene, J. 1981. From Huxley to Huxley: transformations in the Darwinian credo. In: *Science, Ideology and World View: Essays in the History of Evolutionary Ideas,* pp. 158–93. Berkeley: University of California Press.

1990. The interaction of science and world view in Sir Julian Huxley's evolutionary biology. *Journal of the History of Biology* 23: 39–55.

Hudson, W. D. 1983. *Modern Moral Philosophy,* 2nd ed. London: Macmillan.

Hume, D. 1978. *A Treatise of Human Nature.* Oxford University Press.

Huxley, J. S. 1912. *The Individual in the Animal Kingdom.* Cambridge University Press.

1923. *Essays of a Biologist.* London: Chatto & Windus.

1924. The negro problem. *The Spectator* 133 (November 29): 821–2.

1926a. *Essays in Popular Science.* London: Chatto & Windus.

1926b. *The Stream of Life.* London: Watts.

1927. *Religion without Revelation.* London: Ernest Benn.

1931. *Africa View.* New York: Harper.

1932. *Problems of Relative Growth.* London: Methuen.

1934a. *Bird-Watching and Bird Behaviour.* London: Chatto & Windus.

1934b. *If I Were Dictator.* London: Methuen.

1935. *Science and Social Needs.* New York: Harper.

1936. Natural selection and evolutionary progress. In: *The Advancement of Science, 1936.* Addresses delivered at the annual meeting of the British Association for the Advancement of Science, pp. 81–100. London: BAAS.

1942. *Evolution: The Modern Synthesis.* London: Allen & Unwin.

1943. *TVA: Adventure in Planning.* London: Scientific Book Club.

222

1948. *UNESCO: Its Purpose and Its Philosophy.* Washington, DC: Public Affairs Press.

1954a. The evolutionary process. In: *Evolution as a Process,* ed. J. Huxley, A. C. Hardy, and E. B. Ford, pp. 1–23. London: Allen & Unwin.

1954b. Scientific humanism, evolution, and human destiny. Lecture given October 16, 1954, in Los Angeles (HP 70.9).

1955. Morphism and evolution. *Heredity* 9: 1–52.

1959. Introduction to Teilhard de Chardin's *The Phenomenon of Man,* pp. 11–28. London: Collins.

1964. *Essays of a Humanist.* London: Chatto & Windus.

1970. *Memories.* London: Allen & Unwin.

1973. *Memories II.* London: Allen & Unwin.

Huxley, J. S., and Haddon, A. C. 1935. *We Europeans: A Survey of "Racial" Problems.* London: Cape.

Huxley, J. S., and Haldane, J. B. S. 1927. *Animal Biology.* Oxford University Press.

Huxley, L. 1900. *The Life and Letters of Thomas Henry Huxley.* London: Macmillan.

Huxley, T. H. 1865. Emancipation – black and white. In: *Science and Education,* pp. 66–75. London: Macmillan.

1879. *Hume.* London: Macmillan.

1888a. Evolution and ethics. In: *T. H. Huxley's "Evolution and Ethics": with New Essays on Its Victorian and Sociobiological Context,* ed. J. Paradis and G. C. Williams, pp. 57–174. Princeton, NJ: Princeton University Press.

1888b. The struggle for existence in human society. *Nineteenth Century* 23: 161–80.

1893. Evolution and ethics. In: *Evolution and Ethics,* pp. 46–116. London: Macmillan.

Huxley, T. H., and Huxley, J. S. 1947. *Evolution and Ethics 1893–1943.* London: Pilot.

Kelley, A. 1981. *The Descent of Darwin: The Popularization of Darwinism in Germany, 1860–1914.* Chapel Hill: University of North Carolina Press.

Mackie, J. 1977. *Ethics.* Harmondsworth: Penguin.

1978. The law of the jungle. *Philosophy* 53: 553–73.

1979. *Hume's Moral Theory.* London: Routledge & Kegan Paul.

Mayr, E. 1988. *Towards a New Philosophy of Biology: Observations of an Evolutionist.* Cambridge, MA: Harvard University Press (Belknap Press).

Mitman, G. 1990. Evolution as gospel: William Patten, the language of democracy and the Great War. *Isis* 81: 44–93.

1992. *The State of Nature: Ecology, Community, and American Social Thought, 1900–1950.* University of Chicago Press.

Moore, G. E. 1903. *Principia Ethica.* Cambridge University Press.

Murphy, J. 1982. *Evolution, Morality, and the Meaning of Life.* Totowa, NJ: Rowman & Littlefield.

Nagel, E. 1961. *The Structure of Science.* London: Routledge & Kegan Paul.

Nitecki, M. (ed.). 1988. *Evolutionary Progress.* University of Chicago Press.

Pittenger, M. 1993. *American Socialists and Evolutionary Thought, 1870–1920.* Madison: University of Wisconsin Press.

Richards, R. J. 1987. *Darwin and the Emergence of Evolutionary Theories of Mind and Behavior.* University of Chicago Press.

Ruse, M. 1986. *Taking Darwin Seriously.* Oxford: Blackwell.

 1995. *Evolutionary Naturalism: Selected Essays.* London: Routledge.

 1996. *Monad to Man: The Concept of Progress in Evolutionary Biology.* Cambridge, MA: Harvard University Press.

Russett, C. E. 1976. *Darwin in America: The Intellectual Response, 1865–1912.* San Francisco: Freeman.

Sidgwick, H. 1874. *Methods of Ethics.* London: Macmillan.

Simpson, G. G. 1928. *A Catalogue of the Mesozoic Mammalia in the Geological Department of the British Museum.* London: British Museum (Natural History).

 1944. *Tempo and Mode in Evolution.* New York: Columbia University Press.

 1949. *The Meaning of Evolution.* New Haven, CT: Yale University Press.

 1951. *Horses.* Oxford University Press.

 1953. *The Major Features of Evolution.* New York: Columbia University Press.

 1964. *This View of Life.* New York: Harcourt Brace & World.

 1969. Biology and ethics. In: *Biology and Man,* pp. 130–48. New York: Harcourt Brace & World.

 1978. *Concession to the Improbable: An Unconventional Autobiography.* New Haven, CT: Yale University Press.

 1987. *Simple Curiosity: Letters from George Gaylord Simpson to His Family, 1921–1970,* ed. L. F. Laporte. Berkeley: University of California Press.

Spadefora, D. 1990. *The Concept of Progress in Eighteenth Century Britain.* New Haven, CT: Yale University Press.

Spencer, H. 1851. *Social Statics; or the Conditions Essential to Human Happiness Specified and the First of Them Developed.* London: J. Chapman.

 1868. *Essays: Scientific, Political, and Speculative.* London: Williams & Norgate.

 1892. *The Principles of Ethics.* London: Williams & Norgate.

 1904. *Autobiography.* London: Williams & Norgate.

Swetlitz, M. 1991. Julian Huxley, George Gaylord Simpson and the idea of progress in twentieth century evolutionary biology. Ph.D. dissertation, University of Chicago.

Waddington, C. H. 1960. *The Ethical Animal.* London: Allen & Unwin.

Waters, C. K., and van Helden, A. (eds.). 1992. *Julian Huxley: Biologist and Statesman of Science.* Houston, TX: Rice University Press.

Wilson, E. O. 1975. *Sociobiology: The New Synthesis.* Cambridge, MA: Harvard University Press.

 1978. *On Human Nature.* Cambridge, MA: Harvard University Press.

 1984. *Biophilia.* Cambridge, MA: Harvard University Press.

 1990. *Success and Dominance in Ecosystems: The Case of the Social Insects.* Oldendorf/Luhe: Ecology Institute.

 1994. *Naturalist.* Washington, DC: Island Books/Shearwater Books.

Wilson, E. O., and Peter, F. M. (eds.). 1986. *Biodiversity.* Washington, DC: National Academy Press.

9

The Laws of Inheritance
and the Rules of Morality

Early Geneticists on Evolution and Ethics

Marga Vicedo

In analyzing the ways in which biologists have related their scientific knowledge to moral issues, historians have not paid much attention to the geneticists
working at the turn of the twentieth century. Perhaps that has been because
they have looked mainly at the influence of evolutionism on ethics, and students of heredity in the early part of the twentieth century supposedly had
little interest in evolution, at least in Darwinian evolution (Bowler 1988).
That may explain why most major works on evolutionary ethics have focused only on geneticists' interest in eugenics (Richards 1987), and some
have even argued that biologists did not pursue evolutionary ethics at the turn
of the century (Farber 1994). Whatever the reasons may have been, the existing literature suggests that geneticists were interested in the social implications of their science, but they tended to focus rather narrowly on eugenics,
the manipulation of human breeding for social improvement.

My research has suggested a very different picture of the relationship between genetics and studies of evolution in general, and geneticists' interest
in ethics in particular. Elsewhere I have argued that the apparent schism between genetics and studies of evolution in the first two decades of the twentieth century was an artifact of a historiographic approach that considered
the ideas of William Bateson and T. H. Morgan to be fair representations of
geneticists' views during that time (Vicedo 1992). Furthermore, Barbara Kimmelman and I have shown that many first-generation geneticists in the United
States were deeply concerned with evolution and the potential for studies on
heredity to illuminate evolutionary problems (Vicedo and Kimmelman 1993).

I am very grateful to Jane Maienschein, Michael Ruse, and especially Mark Solovey for their
patience, encouragement, and suggestions for improving this essay. Thanks also to Ryan Mac-
Pherson for his excellent research assistance. Finally, I gratefully acknowledge financial support from the National Science Foundation, the Arizona State University FGIA program, and
the Arizona State University West SRCA program.

When their work is seen in this light, it is not surprising to find that geneticists were indeed interested in evolutionary ethics.

The view that genetics and evolution had little to do with each other could also be the reason that hardly any attention has been paid to the connection between genetics and a philosophical tradition strongly influenced by Darwinian thought: pragmatism. It is the only notable philosophical movement that had its main roots in the United States. Pragmatism, moreover, is regarded as the most influential philosophy in America during the first quarter of the twentieth century. Historians have also argued that pragmatist thought was representative of the life of the mind in the United States during that time (Hollinger 1985, p. 25). Some historians of science have explored the extent to which pragmatic ideas have influenced scientific research. Schweber (1986) has discussed their impact on the development of physics in the United States. Kingsland (1991) has pointed out the influence of this philosophy on the work of Charles J. Herrick and Herbert S. Jennings. Because pragmatism was developed, to a great extent, as an effort to grapple with the implications of evolutionary thought for epistemology and for social and moral thought (Wiener 1949), it is thus to be expected that biologists interested in the philosophical and social implications of evolution were attuned to this intellectual current.

My work suggests that, indeed, many American biologists were influenced by the epistemological and ethical ideas of pragmatism, especially those of William James and John Dewey. Elsewhere I have analyzed the relationship of Dewey's thought to Jennings's and Conklin's views about epistemology and ethics, as well as the implications of their biological work for those two areas of inquiry (Vicedo 1998a). Here I shall discuss some ways in which geneticists were influenced by pragmatic thought. The clearest connections can be traced from the ethics of Dewey to the positions of Jennings and Conklin. There are several factors, however, that make it difficult to assess the exact nature of that influence. Often those biologists provided very few references to other works, and that was especially so in their more popular writings. In addition, although archival material sometimes provides further evidence of specific connections, such records tend to be incomplete and fragmentary. Most importantly, the pragmatists dealt with issues that were of general concern to American intellectuals of that time. Thus many biologists may have assimilated certain ways of thinking without being aware of the entrained philosophical underpinnings. As a result, sometimes when we find striking parallels between the lines of thought articulated by pragmatists and biologists, there may be no direct way of assessing the precise influence of one on the other. Nevertheless, beyond tracing specific lines of influence, I want to show how the concerns and the work of those geneticists can be illuminated when understood within the pragmatist framework of thought.

The "geneticists" to be discussed here – Charles B. Davenport (1866–1944), Edward Murray East (1879–1938), Herbert Spencer Jennings (1868–1947), and Edwin Grant Conklin (1863–1952) – belonged to the first generation of biologists in the United States who worked on the nature of inheritance after the rediscovery of Mendel's work in 1900. Of course, none of them set out to make a career as a geneticist. They all came from other biological backgrounds, such as embryology, agricultural breeding, and cytology, but later they all worked in genetics or saw their work as contributing to an understanding of the mechanisms of inheritance.

This essay will examine the relationships between biology and morality as seen by these four geneticists. To be sure, not all geneticists of that period were equally interested in ethics; important figures such as William Ernest Castle (1867–1962) and Thomas H. Morgan (1866–1945) never wrote on such topics. But the four we shall consider not only wrote on ethics but also believed that the implications of biological thought for ethics were and indeed should be central topics for biologists. They were deeply concerned, first, about the origin and foundation of ethics. By the turn of the century, several scholars had already put forward strong objections to the project of establishing an evolutionary ethics. But our four geneticists followed Charles Darwin and Herbert Spencer in believing that such a project would be the only way to place ethics on a naturalistic and therefore objective foundation.

Second, these four geneticists confronted head-on a fundamental question raised by developments in modern science: Was determinism compatible with free will? That is, was a lawful account of human behavior compatible with the belief that humans were free and therefore responsible for their actions? William James had delineated the three possible answers: indeterminism; hard-determinism, which rejected the existence of free will; and soft-determinism, which tried to make determinism compatible with human freedom. Davenport and East were hard-determinists, whereas Jennings and Conklin struggled to maintain a compatibilistic account, following the lead of John Dewey. Thus their views were representative of the different positions that scholars had developed in trying to address a problem that was and has remained unresolved, even to this day.

In what follows, I shall first briefly introduce our subject geneticists in order to highlight some common features in their careers. Then I shall analyze their views about the evolutionary origin of ethics and the implications of such views for normative ethics. In a third section, I shall examine their ideas on the problem of determinism and moral responsibility. Finally, I shall offer some conclusions and suggestions for further study.

CAREERS IN BIOLOGY AND ETHICS IN PRAGMATIST AMERICA

Although a detailed discussion of the lives of these geneticists would be beyond the scope of this essay, a brief look at their biographies will serve to introduce them and reveal some common patterns in their careers.

Charles B. Davenport, a Harvard-trained embryologist, taught for a few years at Harvard University and the University of Chicago before moving in 1904 to Cold Spring Harbor (CSH), New York, as director of the Station for Experimental Evolution, founded by the Carnegie Institution of Washington. Davenport was interested in the different factors that played roles in growth, development, and evolution. He worked in the areas of morphology, ecology, and application of statistical methods to biological problems before moving to the field of heredity. He believed that studies of heredity held great promise for revealing empirical data regarding the evolutionary process or, as he called it, "experimental evolution." After doing some pioneering work in human inheritance, he turned his attention to the implications of genetics for social issues. In 1910 he became director of the Eugenics Record Office at CSH, and thereafter eugenics became the focus of his attention. In his view, eugenics ought to study man "as the product of breeding and as the subject of an evolutionary process" (Davenport 1921, p. 391). On Davenport's life and work, see MacDowell (1946), Rosenberg (1961), Allen (1986), and Paul (1995).

Edward M. East received his doctorate in chemistry in 1907 at the University of Illinois, but as a graduate student there he also worked on agricultural breeding experiments. In 1905 he began work for the Connecticut Agricultural Station, and in 1909 he joined Harvard's Bussey Institution for Applied Research in Biology, where he spent the rest of his career engaged in plant-breeding work. He became well known for his studies of the effects of inbreeding, for experiments with hybrid corn, and for development of the multiple-factor hypothesis that provided a Mendelian explanation for the inheritance of quantitative characters. His later work focused on the roles of sexual reproduction and self-sterility in evolution. During World War I, East became concerned about the limits of agricultural improvement in the United States and their implications for the problem of overpopulation. In 1919 he began publishing on those topics, and he continued to work extensively on eugenics, population, and birth-control issues until his death in 1938. He often emphasized that "mankind is linked with its evolutionary past" (East 1925a, p. 39) and argued that "intelligence, morals, and conduct [should] be studied along with structure, function, and environment, as a prerequisite for dealing more rationally with mental, moral, and physical adjustment" (East

1931, p. 13). On the life and work of East, see Jones (1944), Provine (1986), and Vicedo (1998b).

Herbert S. Jennings finished his doctorate in zoology in 1896 working at Harvard University with E. L. Mark and under the influence of Davenport. After a series of short appointments at the University of Montana, Dartmouth College, and the University of Michigan, he was recruited in 1903 by E. G. Conklin at the University of Pennsylvania. In 1906, he moved to Johns Hopkins University, where he became director of the Zoological Laboratory a few years later. He carried out his most famous work on the reactions of lower organisms, especially *Paramecium,* to various stimuli. Later he did pioneering work on the heredity, variation, and evolution of *Protozoa,* including analyses of the effectiveness of selection and studies of mating systems. With some important exceptions in the late 1930s and early 1940s, Jennings's most important laboratory research was conducted before 1916. Thereafter, he focused on the epistemological foundations of biology and on the implications of biological knowledge for ethics. Throughout his life, Jennings was interested in the humanities, especially in ethics, an interest that first arose during his student days at Michigan in classes taught by the philosopher John Dewey. Jennings presented a complete summary of his thought on that topic in his 1933 book *The Universe and Life.* He was convinced that the study of biology could help "in the problems of managing life, the problems of conduct, and in determining our attitude toward the world" (1933, p. 7). On the work of Jennings, see Sonneborn (1975), Kingsland (1987), Pauly (1987), and Richards (1987).

Edwin G. Conklin was trained in biology at Johns Hopkins University, and he spent his career at the University of Pennsylvania and Princeton University. In his research, Conklin concentrated on cell-lineage work to address evolutionary and developmental problems. Although he is usually regarded as an embryologist, I include him here because he perceived his work as a contribution to the understanding of inheritance and evolution. He asserted that heredity was the central problem of biology in his time, and he saw statistical, experimental, and embryological methods as complementary approaches for studying it (Conklin 1915, p. 89). In his student days at Johns Hopkins, Conklin was deeply influenced by the philosophical reflections of his teacher W. K. Brooks regarding biological topics. He later repeatedly quoted Brooks's assertion that "Nature is all there is" (Conklin 1939, p. 295; 1953, p. 73). His search for compatibility between that idea and the spiritual teachings of his strong Methodist family background was a lifelong quest. By the first decade of the twentieth century, Conklin had moved from laboratory research to broader issues concerning the ethical and social implications of biology, topics on which he wrote five books and numerous articles. He often emphasized that "the essential oneness of all life" made it possible to apply the principles of biology "to every aspect of man and his institutions" (Conklin 1917, p. 476). On the life of Conklin, see Harvey (1958), Maienschein (1991), and Cooke (1994).

A first common feature in the careers of these biologists is that they all were interested in evolution. As noted earlier, none of them was trained as a "geneticist." They came from other biological backgrounds, but then became interested in breeding and inheritance studies. Moreover, they saw genetics not as an independent area of study but as an avenue through which to begin addressing the most fundamental problems of evolution. It was not a coincidence that most of them were key speakers at a symposium that Davenport organized for the 1911 meeting of the American Society of Naturalists, entitled "The Light Thrown by the Experimental Study of Heredity upon the Factors and Methods of Evolution." That they saw their work on heredity as contributing to the study of evolution is an important key to understanding their interest in evolutionary ethics.

It is more difficult to say whether or not these biologists believed in "Darwinian evolution," in part because that will depend on how one understands this rubric. It is also difficult because their views changed during their professional careers. Nevertheless, I think it is fair to say that they all saw Darwin's work as a major contribution to the understanding of evolution, and they also recognized the validity of natural selection as one of the major forces in the evolutionary process. Yet they would also have agreed with East's assertion that biologists had passed the stage of searching for one single cause of the evolutionary process; they were ready to acknowledge the existence of a plurality of mechanisms, an assertion that Darwin would have accepted. These geneticists were also influenced by Herbert Spencer and by new proposals such as the theory of "organic selection" put forward by James Mark Baldwin, Henry Fairfield Osborn, and Conwy Lloyd Morgan in the 1890s, as well as the theory of "emergent evolution" defended by C. L. Morgan, William Morton Wheeler, and others.

A second important common characteristic was that after one or two decades of experimental research, all of these biologists had shifted their focus to the implications of such research for ethical and social issues. Most importantly, the fact that they did so in the middle of their careers suggests that their concern about ethics was central in their thought. As the foregoing quotations indicate, for them, a major goal of biological studies was to illuminate the nature of human life, including the goals and rules of the "good" life.

One measure of their deep interest in moral issues and in the implications of their work for social improvement was their strong and lasting involvement in eugenics. Whereas two of them (Jennings and Conklin) adopted a critical attitude toward the radical hereditarianism of certain colleagues (East and Davenport among them), they all remained eugenicists for life. Although I shall not talk about their specific views on eugenics here, I shall suggest that their concern with eugenics can be better understood within the larger framework of their views on biology and ethics.

We should also note that all of these geneticists did most of their work on biology and ethics after World War I. It could be that the end of the Great War simply coincided with the period during which they cut back on their experimental research, but I believe that such concordance has further significance. Biologists, like so many people at the time, were deeply impressed with the terrible effects of the war, and that led many of them to reflect on the future of humanity and on the ways in which their specific fields could contribute to the improvement of society. Mitman (1990) has already pointed out the impact of the war on some biologists, such as William Patten and Conklin.

Third, as their institutional success attests, these four biologists were in the mainstream of biological thought at the turn of the century. Harvard University, the University of Chicago, Cold Spring Harbor, the University of Pennsylvania, Princeton University, and Johns Hopkins University were the main centers at which they studied and carried out their research and teaching. Thus we are talking about central figures in biological thought who were working at leading research institutions at that time. Because they were so successful professionally, it would seem almost certain that there must have been widespread interest among biologists in the relationships of biology to ethical and social issues at the time.

Fourth, their institutional ties provide a clue to a common influence in their thought: the philosophical tradition of pragmatism. Harvard, Chicago, and Johns Hopkins were some of the major centers for the development of that philosophy, whose leaders included Charles S. Peirce, William James, and John Dewey. Peirce was lecturing on logic at Johns Hopkins during 1879–84, where Dewey went for his training in philosophy in 1882. The influence of pragmatism was still important when Jennings went there in 1906, a time when the Hopkins historian and philosopher Arthur O. Lovejoy was writing to clarify the various meanings of "pragmatism" (Lovejoy 1908). We have already noted that Jennings had attended Dewey's lectures on ethics at the University of Michigan. In personal correspondence (Jennings 1895) he acknowledged his sympathies with Dewey's ethical positions (Jennings to Mary Louise Buridge). Given that background and Jennings's interest in the humanities in general, it is most likely that he maintained his interest in that philosophy during his graduate studies at Harvard. Although Davenport did not refer to any pragmatist thinkers in his work, it seems unlikely that he would not have been aware of their ideas while working at Harvard and at Chicago, where his stay overlapped with Dewey's lectures there from 1894 to 1904. A similar situation obtained for East, given that he spent most of his career at Harvard, the premier incubator for pragmatist thought. Conklin, the only one of these biologists who did not have close ties to Harvard, received his training at Johns Hopkins, and he referred to Dewey in some writings. Indeed, the influence of Dewey on Conklin is rather clear, whether it was direct or through the writings of Jennings, whom Conklin regarded as his major influence.

We now turn to their views about the evolutionary origin of the moral sentiment and its implications for normative ethics.

THE EVOLUTIONARY ORIGINS OF ETHICS
AND ITS NORMATIVE IMPLICATIONS

Neither Charles Darwin nor other nineteenth-century biologists were the first thinkers to search for a naturalistic foundation for ethics, but Darwinism and evolutionism more generally seem to have had significant implications for ethics (Ruse 1986, 1990; Richards 1987; Bradie 1994; Farber 1994; Sloan, Chapter 3, this volume). Independently of their specific views on organic and ethical evolution, both Darwin and Herbert Spencer believed that organic evolution had profound implications for human conduct and for a naturalistic account of ethics. At the same time, however, the project of developing an evolutionary ethics was encountering important criticism from two other Englishmen: Thomas Henry Huxley and George Edward Moore.

Davenport, East, Jennings, and Conklin did not seem to be aware of or convinced by those criticisms, as they followed Darwin and Spencer rather closely in their views about the evolutionary origin of morality and in their belief in its important implications for normative ethics. Given that they all wrote after those criticisms had been put forward, it is somewhat surprising that none of them saw any need to confront such criticisms or even refer to them. Their ties to American pragmatism may have diminished their interest in responding, because at least one key figure of that movement, John Dewey, was sympathetic to evolutionary ethics and had rebuffed the criticisms advanced by Huxley.

In the *Descent of Man* (1871), Darwin argued that the moral sense could be regarded as an instinct that had developed by natural selection operating at the community level. Extrapolating from his theory of organic evolution, Darwin concluded that habits that satisfied only individual desires would disappear unless those habits also benefited the common good. Thus, only traits leading to actions that would promote the general utility of the group would become entrenched as instincts and become prevalent through natural selection at the group level. As Richards (1987) has shown, Darwin thereby established an intrinsic identity between biological good and moral good, because the utilitarian criterion of the benefit of the majority was supported both by reflective moral agents and by the workings of nature. Darwin believed that the end for both the evolutionary process and morality was the health and vigor of individuals and the community.

Herbert Spencer (1893) also sought to show that social and organic processes worked together for the common good. For him, both biological evolution and moral evolution were united in a grand cosmic evolutionary process. Each worked toward a state of harmony or equilibrium between the behaviors of organisms and their environments, leading to the progressive improvement of life in both realms. Spencer defined the moral life as one in which the maintenance of equilibrium attained a state of completeness. The test for correct actions was whether or not they helped to maintain life in such a state of balance.

The biologist and staunch Darwinian supporter T. H. Huxley believed, on the other hand, that cosmic evolution and moral evolution were opposing processes. In his opinion, ethics was established expressly to counteract the tendencies of nature. As he saw it, social progress depended on "a checking of the cosmic process." According to Huxley, progress in society was not the survival of the fittest, but the survival of those who were best from a moral point of view (Huxley 1893).

The philosopher G. E. Moore also criticized the use of evolutionary principles to justify moral precepts. As he saw it, Spencer and others who maintained that because evolution "shews us the direction in which we are developing, thereby and for that reason shews us the direction in which we ought to develop" had committed a "naturalistic fallacy," because they defined moral characteristics in terms of merely natural properties (Moore 1903, p. 46). A different but related criticism of naturalized ethics had already been advanced by David Hume, who warned that one could not deduce an "ought" statement from an "is" statement; that is, one could not derive a prescriptive or normative statement from a set of descriptive or factual statements.

In a critical review of Huxley's lectures on biology and ethics, John Dewey, the American pragmatist, strongly objected to the project of developing an ethics in isolation from biology. Dewey reacted against Huxley's view that there was a struggle between the natural processes and the ethical processes. According to Dewey, although the functions selected in biological evolution and ethical evolution might be different, both processes worked in consonance. Rejecting the dualism between the natural and social realms, he ended his review of Huxley's position with the following words:

> I question whether the spiritual life does not get its surest and most ample guarantees when it is learned that the laws and conditions of righteousness are implicated in the working processes of the universe; when it is found that man in his conscious struggles, in his doubts, temptations, and defeats, in his aspirations and successes, is moved on and beyond up by the forces which have developed nature; and that in this moral struggle he acts not as a mere individual but as an organ in maintaining and carrying forward the universal process. [Dewey 1898, p. 110]

In his most mature works, Dewey maintained a naturalistic approach to ethics, a position strongly influenced by Darwin (Moore 1961, p. 216). In *Human Nature and Conduct* he argued that good and evil sprang from nature. Appealing to the continuity of nature, he concluded that "a morals based on study of human nature instead of upon disregard for it would find the facts of man continuous with those of the rest of nature and would thereby ally ethics with physics and biology" (Dewey 1922, p. 12).

Dewey also rejected the separation between facts and value judgments that had been central to the arguments of Moore and Hume. He believed that the division, introduced by modern science, between facts and values was simply a methodological tool to promote scientific investigation, but not a feature of the universe. Furthermore, he believed that value judgments were reducible to factual judgments. "The ultimate meaning and function of a value judgment must be expressible in terms of judgments about facts," as one of his commentators has put it (Moore 1961, p. 237). It was precisely because of that connection that scientific knowledge became relevant to ethics and ethics had to go hand in hand with science.

Although it is difficult to determine the influence of Dewey's thinking on geneticists, Davenport, East, Jennings, and Conklin agreed that morality had to be understood as a natural process, whose workings would be uncovered by science. By explaining the origin of moral rules in a naturalistic way, and by taking the evolutionary process as the basis for establishing an ethical system, they hoped to provide ethics an objective foundation from which one could derive substantive normative implications. Within that general framework, they each took up a variety of issues. We shall examine them individually in order to appreciate the range of their interests and views. This seems especially important, because their ideas in such areas have thus far largely been overlooked.

Charles Davenport presented an evolutionary account of the genesis of ethics in several writings. He believed that, like any other group of social animals, human beings, at a certain stage in their development, had encountered the problem of social advancement versus individual advancement. In response to their gregarious instinct, they had come together to seek support in communal life. Once a community had been constituted, society had labeled individual actions considered detrimental to the group as immoral. So, for Davenport, codes of ethics were the sets of specific rules that society sanctioned because they would benefit the evolution of the group. He offered the following definition: "Moral law is merely this: behavior that is favorable for the specific community is 'good'; behavior that is harmful for the community is 'bad'" (Davenport 1913, p. 34). Immorality therefore was equated with social deviance, and moral behavior was that which followed society's rules.

Thus, for Davenport, actions in themselves were neither good nor bad in any absolute sense, but were considered to be one or the other depending on their effects on the community. Davenport argued, for example, that because all animals fought, violence in itself was not wrong. At the same time, "society has decided that certain kinds of behavior – reactions – are, on the whole, good for the development of society, and that certain others are bad" (Davenport 1912c, p. 2141). In conclusion, crime was bad simply because it had negative consequences for society.

Possession of the elements necessary for leading a moral life, furthermore, seemed to depend on the specific genetic configuration produced by the evolutionary process. Those traits that society catalogued as "moral" because they led to the common good had been developed, Davenport believed, through a process of selection that had operated against those who had not had the natural capacity to follow society's mores. As he expressed it, "from time to time as the social ideals developed, those incapable of sharing in the new ideals became eliminated by those who were capable" (Davenport, 1912b, p. 11). That process led to the establishment of a specific hereditary constitution that determined the capacity of a subject to follow the dictates of ethics. Thus, he claimed that the capacity to act in a moral way was determined primarily by heredity, not by moral training.

Davenport emphasized that, contrary to what he considered common opinion, not all individuals had germ plasm that was receptive to moral education. In order to lead a moral life, individuals had to possess a genetic constitution that would enable them to develop characteristics that would help to restrain immoral impulses. But "a certain percentage of the children born are without the rudiments of that inhibition and self-restraint that are essential to our advanced social life." The reason for that was that such individuals had inherited from a remote ancestry "a protoplasm that has never gained those elements" (Davenport 1912b, p. 11). Using an analogy, common at that time, between the ontogenetic and phylogenetic processes, he argued that "the traits of the feeble-minded and the criminalistic are normal traits for infants and for an earlier stage in man's evolution" (1911, 262). Those traits still appeared in certain individuals as leftovers from evolutionary history. They were, therefore, testimony to the descent of human beings from the lower animals. Such traits came directly from our animal ancestry, and they led to animalistic emotions that were not tempered by reason (Davenport 1911). In another place, Davenport called them defects of the germ plasm that have come "all the way down from man's ape-like ancestors" and "animalistic blood-lines" (1912a, p. 89).

Given Davenport's strong hereditarian bent, it seemed that such a situation could hardly be remedied by education. As he often repeated in his writings, "if the seed be poor, the harvest will be deficient" (1909, p. 18; 1912b, p. 8). Davenport maintained that even specific moral traits were dependent on hereditary makeup: "Sincerity or insincerity, generosity or stinginess, gregariousness or seclusiveness, truthfulness or untruthfulness, are all qualities whose presence or absence is determined largely by the factor of heredity" (1913, p. 36). That would have important implications for social policy, for unless people had a biological constitution receptive to moral norms, education in ethics would be in vain. Thus Davenport believed that the most important measures for improving human nature and ensuring social progress would be eugenics policies.

Edward M. East's writings on ethics and biology also reveal a deep commitment to a naturalistic perspective, with a strong hereditarian emphasis. In his view, "human society is little more than a colony of living mammals" (East 1925b, p. 669). He believed that that fact should be the foundation stone for a new ethics, a project rooted in the pioneering efforts of Darwin, "the Builder," who "led the way to a new ethics, a decent ethics, scientifically conceived" (East 1929b, p. 20). East even called Darwin's theory "the greatest effort of the human mind" (1925a, p. 24; 1931, p. 22) because it had provided a general outlook for all living forms. For East, there was no doubt that evolution had profound implications for ethical thought, implications that humankind could deny only at the risk of its own destruction.

East's account of the evolutionary origin of ethics was similar to Davenport's. According to East, the process of evolution was driven by the instinct for self-preservation. Along with the other animals, humans had developed many physical aids for survival, and in accordance with their urge for self-preservation they had also developed mental life and social customs. Somewhere along the evolutionary path, people, like other animals, had come together in groups to establish a united front against other species. Such union had provided the basis for cooperation, which had taken root because of the gains it conferred upon the members of the group. In that context, moral precepts had been developed to provide "workable rules" for the community (East 1925a, p. 38). Later on, to ensure support for those rules, society had developed legal codes and beliefs in eternal compensations and atonements.

In explaining the important implications of his views, East called for acceptance of the limitations imposed by the biological nature of humans, and he hoped to build a better world upon that foundation. He maintained that altruism, for example, could develop only insofar as it did not run counter to individualism, because man "will help his neighbor [only] if in the long run he helps himself" (East 1925a, p. 38). He thus urged his readers to accept the selfishness seen in many actions, without longing for a dream world of perfection. In his opinion, there was "dignity in selfishness," because the "foundation-stone of the marvelous diversity of the whole organic world is not a matter for cheap disdain" (1925a, p. 39). In general, East argued that regardless of the particular positions one took on social and moral issues, they had to be grounded in conclusions that followed from the study of evolution and heredity, because, he warned, "nature is not to be denied" (1925a, p. 37).

But, unfortunately in East's view, many of the moral precepts of his time were inadequate, because they failed to recognize the power of biological instincts and the significance of how evolutionary forces had an influence on ethics. Regarding the bases for moral codes, people often inverted the order of cause and effect and thus proposed moral positions that ran counter to the basic survival instinct. Inevitably, such positions were doomed to failure. Instincts that had been developed during the evolutionary process, East added, could not be eradicated by simple decisions made in the present historical moment (1925a, p. 37).

In his 1933 book *The Universe and Life,* Jennings presented a fully developed evolutionary account of morality. In his view, life was a task, a problem-solving activity that required management and decisions to determine a course of action. The penalty for making wrong choices was destruction. Some actions could help to avoid destruction, and such actions gave rise, on the one hand, to specific sensations of pain and therefore to fear of what was harmful and, on the other hand, to pleasure and a corresponding desire for what was beneficial. That provided the basis for a universal love of life and a natural fear of death. The mental lives of organisms, he continued, were to a great extent concerned with determining which actions would promote the continuance, fullness, and adequacy of life and which actions would have the opposite effect. Those were the fundamental experiences that lay at the foundation of human concepts of value, of man's ideas of good and bad. Thus the basis for human concepts of right and wrong derived from the experiences of living beings (Jennings 1933, pp. 69–70).

Jennings maintained that all aspects of the development of the moral sense had to be understood within the evolutionary framework. He argued that the evolution of a regard for the interests of others, the birth of unselfishness, and the development of a feeling for justice were processes of "natural growth," in fact, "as much a natural growth as the urge to eat or to protect one's self." Though the former undoubtedly came into being after the latter, Jennings added that "the later products of evolution are as distinctly realities – are as 'natural' – as are the earlier ones" (1933, p. 75). As he put it in another passage, "the ideals of conduct, morality, [and] ethics, grow directly out of the urge to live that is common to all organisms. Their essence lies in the effort toward the advancement of life, toward the promotion of its fullness, variety, and adequacy; they are the management of life toward these ends" (1933, p. 91).

That understanding of moral evolution clearly had normative implications. According to Jennings, what promoted the advancement of life was valuable and was therefore right. What destroyed life was wrong and therefore was to be avoided. He did not specify the things that promoted life and made it full and adequate, because, he noted, those were "things ultimate that cannot be defined except in terms of experience" (Jennings 1933, p. 78). But he did suggest that his criterion of right and wrong sanctioned eugenics measures to aid the advancement of future generations. In general, he believed that the moral ideal was to make life better for the future, to help the process of progressive evolution: "such evolution as has brought man from amoeba-like ancestors" (Jennings 1933, p. 90).

In certain key respects, Jennings's ideas reflected the influences of Dewey and Spencer. As mentioned earlier, Jennings had been a student of Dewey, whom he credited with liberating him from the intellectual grasp of Spencer. But Jennings's position had been shaped by the ideas of both thinkers. Dewey argued that experience was the ultimate test of reality. Regarding ethics, he believed that human experience was the source of our notions of good and evil. Likewise, Jennings saw experience as the source of the moral sense and argued that the standards of right and wrong were to be found in human experience. What that experience revealed, according to Dewey, was that human beings found their happiness in the exercise of their most characteristic function, one they shared with all living things: being alive. So the moral good, as Jennings also saw it, consisted in living in such a way as to make fuller living possible in the future (Moore 1961, p. 228).

That position was more Spencerian than Dewey and Jennings seem to have recognized. Like Jennings, Spencer had argued that activities that led to the preservation of life created a sensation of pleasure, whereas those that were biologically harmful were associated with pain. On that experiential basis he erected a system of ethics in which human conduct would be evaluated according to whether or not it promoted the duration, extension, and perfection of life, a position very similar to those of Jennings and Dewey. Thus the three conceptions were first anchored in the concept of experience. Second, the goal of morality in all three systems was to aid in the development of life, to help the expansion, diversity, and "advancement" of life.

In addition, all three positions shared a basic belief in the progressive character of the evolutionary process. For Spencer, evolution was a teleological process in which progress was intrinsically necessary. Jennings recognized that there was no evidence of teleology in nature, but he saw evolution as a creative process in which some forms of life, despite occasional ventures down paths ending in blind alleys, reached degrees of fullness, diversification, and adequacy that could be called "higher" (Jennings 1933, pp. 57–62). Dewey, like other pragmatists, did not think that social progress was guaranteed or necessary, but he did see cosmic evolution as a process in which there was an increasing development of intelligence leading to the establishment of "higher" personal and social arrangements. As we shall see later, E. G. Conklin saw, in a similar way, that both the organic and social worlds progressed toward increasingly higher levels of organization.

Conklin, who hired Jennings at the University of Pennsylvania and later tried to hire him at Princeton, always acknowledged Jennings's strong influence on his own views. In his book *The Direction of Human Evolution,* Conklin noted the importance of analyzing the implications for ethics and religion of the view that the "human species, no less than all others, has come into existence by a process of evolution" (1922, p. xi).

Conklin argued that moral and social ideas were products of evolution, as were physical characteristics, but he established distinctions among three aspects of the evolution of humankind: physical, intellectual, and social (1919a; 1921, p. 85; 1937, p. 596). According to him, the underlying principle was the same in each case, namely, the survival of the fittest. Yet the definitions of "fitness" varied for the three processes: "Physically, the fittest is the most viable; intellectually, it is the most rational; socially, it is the most ethical" (1919a, p. 38).

Conklin argued that those were not antagonistic lines, and the common perception that the fittest was the most powerful and aggressive was therefore not justified. He reminded his readers that Darwin himself had protested against such a conception of natural selection. In Conklin's words, Darwin "showed that in social evolution the most ethical is the most fit" (Conklin 1919a, p. 38). Conklin thus reacted against the use of evolutionary theory to justify violent behavior, as it had often been used by some German thinkers during World War I. Several other biologists also tried to counteract that reading of evolution by natural selection by emphasizing the cooperative tendencies present in nature, tendencies that they saw as providing justification for democratic and peaceful social arrangements, not authoritarian and violent ones (Mitman 1990, 1992).

Conklin thought, furthermore, that biological and social processes worked in similar ways, both leading to higher levels of organization. In his biological work, Conklin argued that organization, in the sense of the progressive differentiation and integration of the functions and structures of a system, was the biological ideal. Reaching a higher level of organization was similarly the foremost social goal, for it fostered a stable and moral society. Because differentiation and integration were the means to a higher level of organization, the ideal society, for Conklin, would be a system in which individuals would exercise and expand their natural abilities for the benefit of society as a whole. He thought that could best be realized in a democracy.

Although, for Conklin, democracy was the social form that exemplified the best social organization, it was not designed for individual gain. As "the last and highest grade of organization," society was justified in imposing restrictions on the individual's freedom for the sake of the whole society (Conklin 1915, p. 482). In Conklin's view of social organization, the individual was to the species as the cell was to the body. Thus individual freedoms were properly subordinated to the good of the species. People should be like bees: each individual serving society in the capacity for which heredity had best prepared that individual. The biological world provided the guiding principles. To create such a society, Conklin added, humans would have to organize themselves so that intelligence would rule over the more basic instincts that otherwise might incline them toward selfish goals. Given that understanding of evolution, Conklin saw the perpetuation and improvement of the human race as the highest ethical obligation, an obligation that society must take as its main duty (1915, pp. 482–91).

Conklin went on to argue that there was a harmony between the ethics of social instincts (most highly exemplified by the cooperation seen among ants and bees) and the ethical command of Christianity to be a servant to all, to work for the good of the whole. Conklin, who used religious metaphors throughout his writings, made it explicit that he saw evolution as the foundation for a new religion that would deal "with this world rather than with the next" (1922, p. 246). An understanding of religion in that light did not seem to destroy its value. On the contrary, evolution provided "a rational solution to the great problem of evil" (1922, p. 239).

Again, it is easy to see in Conklin's position traces of the theories of both Spencer and Dewey. Spencer's evolutionary theory posited increasing advances in organization, with the different aspects of the evolutionary process working toward an equilibrium in which the coordinated activities of each social unit and those of the aggregate would be in balance. Conklin also conceived of the evolutionary process as working toward a higher level of biological and social organization in a rather teleological fashion. In the case of social progress, the motor would be the application of intelligence to control basic instincts. That was very close to Dewey's view that morality was largely concerned with controlling human nature and that for moral progress, intelligence had to control habits and impulses (Dewey 1922, p. 1). The concept of intelligence as the key to a moral and stable society, with democracy as the form that society would take, was a central element in the philosophies of Dewey and Conklin.

Aside from their individual nuances, all four of these biologists held the view that the goal of morality should be to assist the evolutionary process. For them, moreover, that stance implied that the individual was always subordinate to the welfare of the community. Because the traits of individuals had influences on the evolution of society, the needs and rights of society could prevail over those of individuals. These biologists held an organismic conception of society: Society was a big organism in which the behaviors of the parts were essential for the functioning of the whole. Society was a huge colonial aggregate, East claimed. That was why monitoring the values of its parts was essential for proper functioning of the organism (East 1929a, p. 296). The same reasoning led Davenport to affirm that genes could be considered common property. As he put it, "the germ-cells do not belong to a person in the same way that his hair, or his stomach do" (Davenport 1914, p. 303). He argued that heritable traits were not personal properties, and their momentary possessors could not treat them as private and personal matters (Davenport 1912c, p. 2142). Conklin even recommended following the example of ants: forming societies in which each individual would have duties, but no rights.

I want to suggest here that these implications of their evolutionary ethics can help us understand the eugenics concerns of those biologists. Indeed, those concerns seem to have been logical consequences of their ethical positions. Of the four, Davenport and East held the strongest hereditarian views. Even though both had sophisticated knowledge of genetics – they were the first American researchers to explain the multifactorial inheritance of some characteristics, thus clarifying that there was no simple one-to-one correspondence between genes and phenotypic traits – they often talked about genes being the masters of our lives. Although they gave an occasional nod to the importance of the environment, they basically believed that the environment could only harness the forces of heredity, as East put it (1929a, p. 206). In fact, Davenport and East were among the strongest eugenicists within the genetics community. Jennings and Conklin did not support the strong hereditarianism of Davenport and East. Conklin argued that heredity conferred possibilities, and the environment then created actualities, and Jennings held similar views. Nevertheless, they believed that eugenics could help further the goals of evolution and morality (Conklin 1915, 1921, 1928; Jennings 1927b, 1933). So it is against the background of their evolutionary ethics that we can better understand the grounds on which they supported eugenics. In this light, we can also better appreciate the moralistic tone of the eugenics movement. As historians who have tracked eugenics have noted, the movement was a moral crusade, and its proponents often saw eugenics as a new religion.

In sum, all of these biologists, Davenport, East, Jennings, and Conklin, emphasized the need to study ethics within an evolutionary framework. They believed that traditional problems concerning the basis for moral precepts, the existence and force of moral norms, and the criteria for distinguishing between right or wrong had to be addressed within such a scientific framework, thus establishing an objective foundation for a normative ethical system. As they saw it, the evolutionary past of human beings had to be taken into account when formulating new moral codes. As East put it, nature cannot be denied (1925a, p. 37). Their common stance was that actions should be defined as right or wrong on the basis of their contributions to an evolutionary process in which both the biological good and the moral good would be defined in terms of what would be good for the community.

Although Davenport, East, Jennings, and Conklin all agreed that ethics had to be naturalized, they had very different views on the implications of that position for the problems of freedom and moral responsibility. Richards (1987) has shown that Darwinian evolution did not imply and did not lead to the establishment of a materialistic, mechanistic, and amoral worldview in late Victorian science. But it is also true that many feared that further scientific study of human behavior would shrink the space reserved for ethics and that human actions would be embedded in a deterministic picture of the world, one that would seem to leave no place for human freedom. Richards recognized that the problem of devising a thoroughly naturalistic account of human nature that would be compatible with freedom was the major conundrum of evolutionary ethics (1987, p. 241). In the next section, we shall see how our biologists responded to that challenge. To allow a better appreciation of their positions, I shall first provide some context by stating the problem as it was seen by pragmatist philosophers and other scholars of that time.

242

ARE THE DETERMINISTIC LAWS OF NATURE
COMPATIBLE WITH HUMAN FREEDOM?

"How is science to be accepted and yet the realm of values to be conserved?" (Dewey 1960a, p. 40) – thus did Dewey state what he saw as the major problem of modern philosophy, namely, analysis of the implications of science for human conduct. Dewey was especially concerned with what the consequences of Darwinism would be for morality. As he saw it, Darwin not only had shown that the organic world could be studied with the methods of science but also had opened the door for application of those methods in the realm of ethics and values. Dewey was a moral naturalist, but he was also aware that naturalism posed significant problems for ethics. An explanation of human behavior, and especially moral conduct, from a scientific perspective that embedded behavior within a system of natural laws would raise the fear that human beings would come to be perceived as mere automatons without free will.

The worry that a certain conception of modern science would lead to a deterministic worldview, with no place for ethics, was also forcefully expressed by William James. He saw that the proponents of determinism were impressed by the phenomenal success of the scientific community in explaining the world in terms of causes and effects. Would that eventually mean that any event in the world, including human behavior, had already been determined by past events? As many other philosophers before James had pointed out, that threatened to have disastrous consequences for ethics: If one's actions had been predetermined, then one was not free to choose between alternatives, and therefore one could not be held accountable for one's behavior. People would then be victims, not autonomous agents.

In reflecting on that issue, James introduced a distinction that has remained in use to this day. He distinguished between old-fashioned determinism, which he called hard-determinism, and a new form: soft-determinism. The hard-determinists fully accepted the consequences of determinism: fatality, bondage of the will, necessity, and predetermination. Soft-determinists attempted to defend determinism while abhorring its consequences (in James's view, its necessary consequences). James was scornful of soft-determinists, who tried to have it both ways, by affirming that determinism was true but also trying to preserve traditional notions of freedom and responsibility, a strategy that he characterized as a "quagmire of evasion" (James 1979, p. 117).

James himself accepted indeterminism on pragmatic grounds; in his opinion, that was the only way to maintain hope for a better future. He called the doctrine of free will "a general cosmological theory of promise" (1978, p. 61). For James, determinism was incompatible with the reasonableness of regret,

243

and it also implied that moral striving was meaningless. The experience of moral striving and the belief that one's actions could improve the world supported a conception that he called "meliorism." According to that view, there was no guarantee of progress, but there was real experience that human actions might change the world. That conception, however, could not operate in a deterministic universe. It would require an open-ended world in which things were not predetermined to happen in specific ways. Thus, for James, free will was needed as a "doctrine of *relief*," and he accepted the indeterminism that, in his view, such a position required (1978, p. 61).

At the turn of the century, scholars in a wide variety of fields (perhaps enchanted by the progress of modern science, as James feared) found hard-determinism appealing. For example, the famous litigator Clarence S. Darrow, in his book *Crime and Criminals* (1902), claimed that people in jail were no more responsible for being there than honest people were for being outside. That line of thought obviously had some resonance in society, because Darrow used it successfully two decades later in a famous murder trial in which he argued that the two young men accused of the crime were simply victims of their heredity and their upbringing. The contention that nurture could determine our fate was also defended by some behavioral psychologists, including John B. Watson. But whether one spoke about environmental determinism or biological determinism, the basic problem remained the same: If human actions were embedded in a scientific system that presupposed universal causation, then human actions, in principle, were as predictable as any physical event. And if one could predict a person's behavior, it would be difficult to see how one could also say that the person had genuine freedom of choice.

Some leading researchers believed that experimental biology also supported such a hard-deterministic view. Jacques Loeb, for example, argued that ethics was a matter of heritable instincts and physiological states. As he saw it, ethics was rooted in biology, and only a mechanistic conception of life could provide an objective foundation for morality. Loeb never developed his conception beyond a few remarks about altruism. According to him, altruistic behavior was not comprehensible from a utilitarian viewpoint nor from the Kantian standpoint of the categorical imperative, but it might be explainable in terms of physical and chemical states (Loeb 1964, pp. 62–3). Although Loeb did not explain how that might work, the title for one of his articles, "The Mechanistic Conception of Life," became standard shorthand for a mechanistic, deterministic, and reductionistic vision of life (Loeb 1912, 1964). Although those complex terms meant different things to different people, the common thread underlying the mechanistic conception was a worldview in which social and ethical phenomena could be explained by appealing to more fundamental biological and physical events. On the work of Loeb, see Pauly (1987).

Many students of heredity also presented a rather deterministic account of human character, one that they thought led to rejecting the existence of free will. Several American geneticists, including Raymond Pearl (1908, p. 9823), approvingly quoted the view of British statistician Karl Pearson that "no man is responsible for his own being; nature and nurture, over which he had no control, have made him the being that he is, good or evil." By the mid-1930s, Pearl was happy to report that, thanks to studies on animal behavior, scientists were "beginning to understand in some detail how conduct, normal and abnormal, moral and immoral, is the expression of 'animal drives' or urges – rather than of either free will or terrestrial and heavenly precepts" (Pearl 1936, p. 234). East and Davenport similarly argued that one's acceptance of the latest biological findings about genetics and human behavior required one to give up any belief in freedom of the will.

Davenport's rejection of free will stemmed primarily from his belief that heredity determined our capacity to act morally, as well as from a rather reductionistic view of behavior, which he defined as a reaction to a stimulus. As he saw it, "those who held the hazy doctrine of freedom of the will must have postulated uniformity of capacity for discriminating between right and wrong and uniformity in responsiveness to similar stimuli." But that seemed like a false assumption. How humans responded to stimuli depended, in his words, "on the nature of our protoplasm" (Davenport 1911, p. 260). As a consequence, people committed unacceptable acts because of the inadequacy of the determiners in their protoplasm (Davenport 1911). And because people did not have control over their genetic endowment, he believed that freedom did not exist.

Given that perspective, it is not surprising to learn that, for him, our belief in free will stemmed from ignorance of the causes that determined our actions. In his words, free will "is predicated in matters of small consequence or concern to the person so that his action is determined by habit or slight stimuli whose source is unperceived" (Davenport 1911, p. 264). Consequently, the pride that people took in the freedom of their actions rested merely on an illusion.

Whether or not that deterministic perspective exonerated people from being responsible for their actions depended, for Davenport, on what was meant by "responsibility." He believed that one was responsible in the sense of having to answer to society for one's actions. But one was not responsible in the sense of "deserving" pain or punishment. The character of a person depended on the determiners in the germ plasm and also, in smaller measure, on the culture in which the person developed; and the individual could not be held responsible for either of those factors. Thus Davenport maintained that punishing to avenge was "illogical." That did not mean that punishment should not be administered; in order to modify behavior, strict and immediate punishment was necessary. Rather, his point was that it was not "deserved" in the traditional sense that people should be commended or punished for their actions – the reason being that they did not "choose" to take those actions.

East likewise believed that the notion of free will should be discarded. He often raged about how only the megalomania of individuals prevented them from recognizing that they were governed by natural law, just as was the rest of the world. In his words, "they can not bear to think of accepting a reduced cosmic rating for their egos, of modifying their ideas of personal immortality, or of relinquishing belief in free will" (East 1927, p. 535). Nevertheless, appreciating "the full significance of biological determinism" implied, among other things, that "the idea of free will must become as obsolete as belief in the reward or punishment man was to obtain by exercising it" (East 1929a, p. 15).

Although East's views were criticized as fatalistic, he rejected that charge. Some people, he noted, called his position "Calvinistic predestination in scientific guise" (East 1929a, p. 15). But in his opinion, the modern deterministic philosophy was only in part the homologue of predestination. He saw no room for fatalism, only for a sympathetic understanding of people's problems and a recognition that not all members of society were equally responsible for their actions (1929a, p. 17). As he put it, the acceptance of determinism was not a sorry state of affairs, because the new religion of science was capable of "giving us both an emotional inspiration and a practical procedure for enriching human life" (East 1927, p. 541).

In sum, for biologists such as Loeb, Pearson, Pearl, Davenport, and East, the fact that mental and behavioral traits were subject to natural laws implied that human behavior had to be understood within a deterministic and mechanistic framework. That suggested, moreover, that the idea of free will should be rejected as a mere illusion. In James's terminology, they were hard-determinists who believed that traditional notions about moral thought, such as freedom and moral responsibility, had to disappear.

But other biologists thought that to abandon belief in free will in exchange for a scientific account of human behavior was too high a price to pay. Researchers such as Jennings and Conklin tried to reconcile the determinism inherent in a scientific account of the natural world with the freedom required by a substantive conception of morality and human responsibility. Because they were the two biologists in this group who were most influenced by pragmatism, it is not surprising that they tried to find a middle way between the traditional views of ethics and the radical idea of rejecting freedom, much in the same way as the pragmatists tried to find a worldview in which science and ethics would be compatible. They did not take the road of indeterminism, as James had. Rather, they closely followed the views of Dewey, trying to show that freedom was compatible with some variant of determinism. They were soft-determinists, those who, in James's view, wanted to defend a position – determinism – without having to accept its implications.

As he explicitly stated in his disputes with Loeb, Jennings strongly opposed mechanistic and deterministic views of behavior (Pauly 1987). As Jennings saw it, the doctrine of mechanism meant that all phenomena were determined, and therefore they could be predicted from the configuration and state of antecedent conditions (1918, pp. 593–4). That, for Jennings, led to a picture of human action in which the behavior of agents could be predicted from causal antecedents. Mechanism thus led to fatalism in the human realm. Furthermore, when mechanism was combined with a certain reading of Darwinian evolution, it supported a position that had unacceptable consequences for human behavior:

> Mingle this perfect doctrine of mechanism, as has been done, with equal parts of the perfect doctrine of natural selection, and you get a potion, a cocktail, with a kick that is warranted to knock out ethics and civilization. . . . The only conduct that is justified is that whose powerful violence leads to triumph in the struggle for existence. Ethics does not exist in the universe of perfect evolutionary mechanism; from the latter we learn the opposite of everything by which we aspire to guide our daily lives and to organize society. [Jennings 1927a, p. 23]

But Jennings was equally dissatisfied with vitalism, the main alternative to mechanism at that time. Vitalism was the position defended by the German embryologist Hans Driesch and sympathizers J. Johnstone, H. Bergson, and Jennings's one-time Harvard classmate, H. V. Neal. It seemed to Jennings that vitalism led to "experimental indeterminism," as Driesch had admitted to him (Jennings 1912). That, Jennings feared, would inhibit progress in the biological sciences, because the pursuit of scientific knowledge seemed possible only within a deterministic framework. According to Jennings, recent developments in biology had provided evidence of the success of the experimental method, and that method was founded on the assumption of experimental determinism. Thus, for Jennings, both vitalism and mechanism would have had unacceptable consequences for the conduct of science and the conduct of life (Vicedo, 1998a).

Looking for a way to defend evolutionary theory and allow a certain degree of determinism, and at the same time avoid the fatalism implicit in the extreme tenets of mechanism and determinism espoused by some of his colleagues, Jennings developed a new stance that he called "experimental determinism." Experimental determinism, he proposed, would not require that events be computable or predictable before they had occurred. It featured determinism only in the backward view, in hindsight, not in a forward sense. That would mean that if "what now occurs were different, the earlier conditions would have been different though what now occurs need not be predictable from nor existent in those earlier conditions" (Jennings 1919, p. 183). So, though the prior conditions would determine an effect, one could not predict that effect based only on knowledge of those prior conditions.

Jennings argued, moreover, not only that experimental determinism was compatible with freedom but also that some sort of determinism was necessary for it, because a reasonable theory of human conduct would not be possible if it was assumed that actions were not determined in any way. One could not be held responsible for actions that were independent of what one was, independent of one's character and personality. In a completely underdetermined world, responsibility, praise, blame, reward, and punishment would be as meaningless as in a fully deterministic world (Jennings 1919, p. 182).

Because any behavior is produced by causes, one could separate compulsive actions and free actions only by identifying some difference between the types of causes that led to those different forms of behavior. Following a long tradition of philosophical thought going back to Aristotle, Jennings understood behavioral freedom not as uncaused behavior but as behavior that was not determined by causes outside the agent and also was not completely random: "Not irresponsibility, chance or indeterminism, but holding within one's self the determinants of action is what constitutes freedom" (Jennings 1919, p. 182).

The position defended by Jennings was almost identical with that upheld by his friend and admirer Conklin, who also saw the issue of determinism as the central problem raised by modern science. Conklin stated the crux of the issue this way: "If personality in all of its main features is fixed by heredity and environment over which the individual has little or no control, and this is certainly true, personality is as inevitably determined by its antecedents as is any other natural phenomenon. That is, I believe, a conclusion from which there is no escape. . . . How then is it possible to believe in freedom and responsibility? Is there not justification for the view so often expressed of late that man is never free and that responsibility and duty are mere delusions?" (Conklin 1915, pp. 458–9).

Like Jennings, Conklin tried to steer a middle way between two extreme views, views that he called voluntarism and mechanism, because neither one seemed to be supported by experience. The first, voluntarism, held that all phenomena were expressions of the will. The second, mechanism, was a view that saw the world in terms of causes and effects, according to which natural phenomena were the results of the eternal laws of nature. Conklin argued that those views were "unreal, unscientific, and unjustifiable," because they contradicted experience. Experience, in his opinion, was the ultimate test of reality:

> We have the assurance of experience that we are not absolutely free nor absolutely bound, but that we are partly free and partly bound; the alternatives are not merely freedom *or* determinism, but rather freedom *and* determinism. [Conklin 1915, pp. 459–60]

So, by arguing in pragmatist fashion that experience was the ultimate test of reality, Conklin defended the idea that our sense of freedom, our belief in our own freedom, allowed us to assert that freedom was a fact that strict mechanism must recognize.

To achieve a middle way between voluntarism and mechanism, Conklin offered a definition of determinism that promised to leave some room for freedom. He defined determinism as the scientific doctrine that "every effect is the resultant of antecedent causes and that identical causes yield identical results." But that, he added, did not mean predeterminism; determinism was "scientific naturalism," whereas predeterminism was "fatalism." Scientific determinism, he argued, "is not incompatible with a certain amount of freedom and responsibility" (Conklin 1915, p. 460).

Regarding human actions, Conklin argued that behavior was not wholly determined by heredity and responses to stimuli, but was also influenced by the experiences and habits of the individual. Memories of past experiences served to influence and to some extent control later reactions to similar stimuli. In that way, the intellect "inhibits the strong impulse of a hungry animal to take food by the counter impulse of unpleasant memories or of fear. Here we have the beginnings of what we call freedom, the immediate response to a stimulus is suppressed, internal stimuli are balanced against external ones and final action is determined largely by past experience" (Conklin 1915, p. 465).

On that basis Conklin constructed the following definition of freedom:

> Freedom is the more or less limited capacity of the highest organisms to inhibit instinctive and non rational acts by intellectual and rational stimuli and to regulate behavior in the light of past experience. Such freedom is not uncaused activity, but freedom from the mechanical responses to external or instinctive stimuli, through the intervention of internal stimuli due to experience and intelligence. [1915, p. 466]

The positions of Jennings and Conklin were virtually the same as that developed by Dewey in *Human Nature and Conduct* (1922) and in "Philosophies of Freedom" (1960b). In the latter piece, Dewey argued that the traditional ways of discussing the free-will problem posed a fallacious dichotomy, a "hopeless dilemma" (1960b, p. 264). If the idea of determinism was correct, then there was no moral responsibility. But if determinism were not operative, then one's acts would not depend on one's personality and character, so how could one be held responsible? Thus Dewey argued not only that the existence of laws was compatible with freedom but also that laws were necessary for freedom. Freedom could not mean "without any cause," for if it did, human behavior would be totally inexplicable.

For all three thinkers, Jennings, Conklin, and Dewey, the uniqueness of the experiences of the self and the impossibility of subsuming them under general laws offered an escape from the dictates of determinism. Dewey claimed that "the essential [constituent] of choice as freedom [is] the factor of individual participation" (1960b, p. 267). Thus the existence of natural laws was not an impediment to the existence of freedom, for one used one's knowledge of those laws to foresee consequences and to consider how various possible consequences might be averted or secured. When intelligent beings began to act in that manner, then freedom began, claimed Dewey (1922, p. 312).

This is why the concept of intelligence was central to their positions. Dewey argued that intelligence was the key to freedom of action. According to him, knowledge and intelligence, rather than will, constituted freedom (1922, pp. 304, 311). By attributing to intelligence a creative and active role that would allow people to modify their natural tendencies and in that way open up new possibilities for the future, Conklin and Jennings, like Dewey, saw that interaction of people with their environment was not merely a result of a process, but was actively created, modified, and controlled by humans. Among the factors contributing to the effort to reach a Spencerian balance, a scientifically adjusted harmony, Jennings and Conklin placed strong emphasis on behavior as an active factor in the evolutionary process and on scientific knowledge. The resulting "adjustment" thus was not "driven by" a mechanistic process, but was directed by human intelligence. That emphasis on the creative and active aspects of behavior in evolution also led both Jennings and Conklin toward versions of evolutionism – organic evolution and emergent evolution – that were in accord with that outlook (Vicedo 1998a).

Their ideas also reflected the melioristic hope of the pragmatists, an attitude that was closely connected to the social worries attending the dawn of the twentieth century. The hope that people could improve the future was a core belief of the Progressive Era. It also reflected deep insecurities about the future. As James beautifully expressed it,

> our interest in religious metaphysics arises in the fact that our empirical future feels to us unsafe, and needs some higher guarantee. If the past and present

were purely good, who could wish that the future might possibly not resemble them? Who could desire free-will? Who would not say, with Huxley, "let me be wound up every day like a watch, to go right fatally, and I ask no better freedom." 'Freedom' in a world already perfect could only mean freedom *to be worse,* and who could be so insane as to wish that? [James 1978, p. 61]

But neither philosophers nor biologists at the turn of the twentieth century believed that they lived in a perfect world (Diggins 1994; Croce 1995). Thus they struggled to offer an account of the biological world that would support their hopes for improvement in the social and moral realms. As scientists, Jennings and Conklin needed a certain degree of determinism as a basis for experimental science, but they found the philosophical implications of a mechanistic account "unlivable" (Conklin 1943; 1953, p. 65). In a deeply pragmatist way, they wanted to combine the epistemology of modern science with a genuine moral life, and they struggled doggedly to find a middle way to construct a life without illusions, but with hope.

But even for those biologists, such as Davenport and East, who believed that biological determinism implied that freedom was an illusion, the recognition of humans' biological nature did not imply a passive or fatalistic position. In fact, they stressed the fact that humankind could direct the process of evolution. They did not defend a simplistic social Darwinism in which man's animal nature would have to be respected and in which even dangerous traits, such as the propensity to violence, would have to be accepted as part of the natural world. They did not believe that nature justified social practices and that we should accept its dictates or respect the "natural" direction of evolution. Neither, however, did they share Huxley's aim to subdue nature to serve human purposes. Instead, while accepting human nature as rooted in biological laws, their goal was then to use the knowledge of those laws to free humankind and to direct the future of human evolution. Not surprisingly, many of them were influenced by Francis Bacon's utopian vision of *New Atlantis,* a world in which society would have a strong "reverence" for the order of nature, but in which knowledge of causes and the "secret motions of things" would lead, nonetheless, to an "enlarging of the bounds of human empire" (Bacon 1909, pp. 25, 35; East 1929a, p. 297).

CONCLUSION

We have seen that a number of important American geneticists at the turn of the twentieth century were interested in developing an evolutionary account of ethics. In reflecting on that problem, they were influenced by an eclectic mixture of views derived from major figures in evolutionary thought, primarily Darwin and Spencer, and by the pragmatist ideas that widely permeated the society of their time. Like the pragmatists, these biologists were also interested in exploring the implications of modern science for social and moral issues. Specifically, they were concerned with the implications of a law-based description of behavior for human freedom and moral accountability. The hard-determinism defended by East and Davenport, on the one hand, and the soft-determinism upheld by Jennings and Conklin, on the other, were representative of the two major positions on the problem of determinism that any naturalistic account of morality had to address.

Although this essay has provided an overview of the thoughts of four biologists regarding the implications of biology for ethics, many issues in their positions need to be explored in greater detail. It would seem worthwhile to examine the specific ways in which their views about progress, eugenics, morality, religion, and biological knowledge were interrelated. We might also explore the specific connections between their scientific research and their moral and social views. Did they accept specific scientific views to meet the demands of their moral theories, or was it the other way around? Finally, now that we are aware of some of their important ties to pragmatism, it should be interesting to explore further their relationships to the broader social and intellectual context.

REFERENCES

Allen, G. E. 1986. The Eugenics Record Office at Cold Spring Harbor, 1910–1949: an essay in institutional history. *Osiris* (2nd ser.) 2: 225–64.
Bacon, F. 1909. *New Atlantis.* Cambridge University Press. (Originally published 1627.)
Benson, K. R., Maienschein, J., and Rainger, R. (eds.). 1991. *The Expansion of American Biology.* New Brunswick, NJ: Rutgers University Press.
Bernstein, R. J. (ed.). 1960. *On Experience, Nature, and Freedom.* Indianapolis: Bobbs-Merrill.
Bowler, P. 1988. *The Non-Darwinian Revolution: Reinterpreting a Historical Myth.* Baltimore: Johns Hopkins University Press.
Bradie, M. 1994. *The Secret Chain: Evolution and Ethics.* Albany: State University of New York Press.
Conklin, E. G. 1915. *Heredity and Environment in the Development of Men.* Princeton, NJ: Princeton University Press.
    1917. Biology and national welfare. *Yale Review* 6: 474–89.
    1919a. Has progressive evolution come to an end? *Natural History* 19: 35–9.
    1919b. Heredity and democracy. *Journal of Heredity* 10: 161–4.
    1921. Some biological aspects of immigration. *Scribner's Magazine* 69: 352–9.

1922. *The Direction of Human Evolution.* New York: Scribner. (Originally published 1921.)

1928. Some recent criticisms of eugenics. *Eugenical News* 13: 61–5.

1937. Science and ethics. *Science* 86: 595–603.

1939. Does science afford a basis for ethics? *Scientific Monthly* 49: 295–303.

1943. *Man: Real and Ideal: Observations and Reflections on Man's Nature, Development, and Destiny.* New York: Scribner.

1953. Spiritual autobiography. In: *Thirteen Americans: Their Spiritual Autobiographies,* ed. L. Finkelstein, pp. 47–76. New York: Harper & Brothers.

Cooke, K. J. 1994. A gospel of social evolution: religion, biology, and education in the thought of Edwin Grant Conklin. Ph.D. dissertation, Department of History, University of Chicago.

Croce, P. J. 1995. *Science and Religion in the Era of William James.* Chapel Hill: University of North Carolina Press.

Darrow, C. 1902. *Crime and Criminals.* New York: Charles H. Kerr & Co.

Darwin, C. 1871. *The Descent of Man and Selection in Relation to Sex,* 2 vols. London: John Murray.

Davenport, C. B. 1909. Influence of heredity on human society. *Annals of the American Academy of Political and Social Science* 34: 16–21.

1911. *Heredity in Relation to Eugenics.* New York: Henry Holt & Co.

1912a. How did feeble-mindedness originate in the first instance? *The Training School* 9: 87–90.

1912b. The relation of eugenics to religion. *The Homiletic Review* 63: 8–12.

1912c. Heredity in nervous disease and its social bearings. *Journal of the American Medical Association* 59: 2141–2.

1913. Heredity, culpability, praiseworthiness, punishment and reward. *Popular Science Monthly* 82: 33–9.

1914. Medico-legal aspects of eugenics. *The Medical Times* (October): 300–4.

1917. The effects of race intermingling. *Proceedings of the American Philosophical Society* 56: 364–8.

1921. Research in eugenics. *Science* 54: 391–7.

1928. Crime, heredity and environment. *Journal of Heredity* 19: 307–13.

Dewey, J. 1898. Evolution and ethics. *The Monist* 8: 321–41.

1922. *Human Nature and Conduct.* New York: Henry Holt & Co.

1960a. *The Quest for Certainty.* New York: Capricorn Books. (Originally published 1929.)

1960b. Philosophies of freedom. In: *On Experience, Nature, and Freedom,* ed. R. Bernstein, pp. 261–87. Indianapolis: Bobbs-Merrill. (Originally published 1928.)

Dewey, J., and Tufts, J. H. 1908. *Ethics.* New York: Henry Holt & Co.

Diggins, J. P. 1994. *The Promise of Pragmatism.* University of Chicago Press.

East, E. M. 1925a. *Mankind at the Crossroads.* New York: Scribner. (Originally published 1923.)

1925b. War or peace? *The Forum* (November 25): 668–78.

1927. Science and the new era. *The Forum* (October 27): 532–42.

1928. The man of science. *Scribner's Magazine* (December 28): 645–53.

1929a. *Heredity and Human Affairs.* New York: Scribner. (Originally published 1927.)

1929b. Darwin (book reviews). *Birth Control Review* 13: 19–20.

(ed.). 1931. *Biology in Human Affairs.* New York: McGraw-Hill.

Farber, P. L. 1994. *The Temptations of Evolutionary Ethics.* Berkeley: University of California Press.

Harvey, E. N. 1958. Edwin Grant Conklin. *National Academy of Sciences, Biographical Memoirs* 31: 54–91.

Hollinger, D. 1985. William James and the culture of inquiry. In: *In the American Province: Studies in the History and Historiography of Ideas,* pp. 3–22. Bloomington: Indiana University Press.

Huxley, T. H. 1893. *Evolution and Ethics and Other Essays.* London: Macmillan.

James, W. 1978. *Pragmatism. The Meaning of Truth.* New York. Cambridge, MA: Harvard University Press. (Originally published 1907.)

1979. The dilemma of determinism. In: *The Will to Believe and Other Essays in Popular Philosophy.* Cambridge, MA: Harvard University Press. (Originally published 1884.)

Jennings, H. S. 1895. Letter to Mrs. Mary L. Buridge, January 3, 1895. Jennings to Buridge, folder #1, Jennings Papers, American Philosophical Archives, Philadelphia.

1912. Driesch's vitalism and experimental indeterminism. *Science* 36: 434–5.

1918. Mechanism and vitalism. *Philosophical Review* 27: 577–96.

1919. Experimental determinism and human conduct. *Journal of Philosophical and Psychological Scientific Methods* 16: 180–3.

1925. *Prometheus or Biology and the Advancement of Man.* New York: Dutton.

1927a. Diverse doctrines of evolution, their relation to the practice of science and of life. *Science* 65: 19–25.

1927b. Public health progress and race progress – are they incompatible? *Science* 66: 45–50.

1930. *The Biological Basis of Human Nature.* New York: Norton.

1933. *The Universe and Life.* New Haven, CT: Yale University Press.

Jones, D. F. 1944. Edward Murray East, 1879–1938. *Proceedings of the National Academy of Sciences, Biographical Memoirs* 23: 217–42.

Kingsland, S. 1987. A man out of place: Herbert Spencer Jennings at Johns Hopkins, 1906–1938. *American Zoologist* 27: 807–17.

1991. Toward a natural history of the human psyche: Charles Manning Child, Charles Judson Herrick, and the dynamic view of the individual at the University of Chicago. In: *The Expansion of American Biology,* ed. K. R. Benson, J. Maienschein, and R. Rainger, pp. 195–230. New Brunswick, NJ: Rutgers University Press.

Loeb, J. 1912. The mechanistic conception of life. *Popular Science Monthly* 80: 5–21.

1964. *The Mechanistic Conception of Life.* Cambridge, MA: Harvard University Press (Belknap Press).

Lovejoy, A. O. 1908. The thirteen pragmatisms. *Journal of Philosophy, Psychology & Scientific Method* 5: 5–12, 29–39.

MacDowell, E. C. 1946. Charles Benedict Davenport, 1866–1944. A study of conflicting influences. *Bios* 17: 2–50.

Maienschein, J. 1991. *Transforming Traditions in American Biology, 1880–1915.* Baltimore: Johns Hopkins University Press.

Mitman, G. 1990. Evolution as gospel. William Patten, the language of democracy, and the great war. *Isis* 81: 446–63.

1992. *The State of Nature: Ecology, Community, and American Social Thought, 1900–1950.* University of Chicago Press.

Moore, E. C. 1961. *American Pragmatism: Peirce, James, and Dewey.* New York: Columbia University Press.

Moore, G. E. 1903. *Principia Ethica.* Cambridge University Press.

Osterhout, W. J. V. 1928. Jacques Loeb. *Journal of General Physiology* 8: 9–92.

Paul, D. B. 1995. *Controlling Human Heredity 1865 to the Present.* New York: Humanities Press.

Pauly, P. J. 1987. *Controlling Life: Jacques Loeb and the Engineering Ideal in Biology.* Oxford University Press.

Pearl, R. 1908. Breeding better men. *The World's Work* (January): 9818–24.

1936. Biology and human progress. *Harper's Monthly Magazine* 172: 225–35.

Provine, W. 1986. *Sewall Wright and Evolutionary Biology.* University of Chicago Press.

Richards, R. J. 1987. *Darwin and the Emergence of Evolutionary Theories of Mind and Behavior.* University of Chicago Press.

Rosenberg, C. E. 1961. Charles Benedict Davenport and the beginning of human genetics. *Bulletin of the History of Medicine* 35: 266–76.

Ruse, M. 1986. *Taking Darwin Seriously.* Oxford: Blackwell.

1990. Evolutionary ethics and the search for predecessors: Kant, Hume, and all the way back to Aristotle? *Social Philosophy and Policy* 8: 59–85.

1996. *Monad to Man: The Concept of Progress in Evolutionary Biology.* Cambridge, MA: Harvard University Press.

Schweber, S. S. 1986. The empiricist temper regnant: theoretical physics in the United States 1920–1950. *Historical Studies in the Physical and Biological Sciences* 17: 55–98.

Sonneborn, T. M. 1975. Herbert Spencer Jennings 1868–1947. *Proceedings of the National Academy of Sciences, Biographical Memoirs* 47: 143–223.

Spencer, H. 1893. *The Principles of Ethics.* London: Williams & Norgate.

Vicedo, M. 1992. Reinterpreting the relationship between Mendelian and Darwinian ideas in early genetics. Presented at the annual meeting of the History of Science Society, Washington, DC.

1998a. The conduct of science and the conduct of life: pragmatism and biology in the work of H. S. Jennings and E. G. Conklin. Unpublished manuscript.

1998b. Of maize and men: Edward Murray East's work on eugenics and the un-
derdetermination of social visions by scientific data. Unpublished manuscript.
Vicedo, M., and Kimmelman, B. 1993. Darwinism and Mendelism in early Ameri-
can genetics. Presented at a meeting of the International Society for the History,
Philosophy, and Social Studies of Biology, Brandeis University, Waltham, MA.
Wiener, P. P. 1949. *Evolution and the Founders of Pragmatism.* Cambridge University
Press.

10

Scientific Responsibility and Political Context

The Case of Genetics under the Swastika

Diane B. Paul and Raphael Falk

INTRODUCTION

Study of biology flourished under the swastika. Although we tend to dismiss Nazi science as pseudoscience and equate research in biology with racial hygiene, the history of biology during the Third Reich was in fact quite complex. Work supported by Heinrich Himmler's *Das Ahnenerbe* ("ancestral heritage"), the research and teaching arm of the Schute-Staffel (SS), was indeed racist nonsense (Deichmann 1996, pp. 251–76). But most of the science supported by the Deutsche Forschungsgemeinschaft (DFG), the major government funding agency, would have been considered mainstream science in the 1930s and 1940s. Its content and standards differed little from those of the science being pursued elsewhere in the western world.

Of course, research under the Third Reich was funded in the expectation that it would ultimately advance the aims of the regime. That fact prompts us to ask how we should think about the activities of scientists who did not engage in overtly criminal acts, but rather practiced "normal research."[1] We do not hesitate to condemn researchers who actively promoted and implemented the racial policies of the National Socialist state. We know what to think about those who produced anti-Semitic propaganda or reports on racial ancestry in connection with enforcement of the Nuremberg laws, helped formulate euthanasia policy, informed on colleagues who employed half-Jewish or politically suspect assistants, or conducted obscene experiments on human subjects. But more difficult, and more interesting, questions are raised by the behavior of scientists whose work was no more racist in either its intention or its assumptions than that of their non-German peers, but who in some way sought to profit from the National Socialist regime. About thirteen percent of German biologists lost their jobs under the Nazis, four-fifths of those for racial reasons (Deichmann 1996, p. 5). What should we think of the people who assumed their places or who in other ways seized new opportunities to advance their careers or research programs? What of those who, although not themselves Nazis, helped their more compromised colleagues reestablish their careers?

BIOLOGY DURING THE THIRD REICH AND SECOND REPUBLIC

Notwithstanding their anti-science reputation, the Nazis actually increased support of basic research in universities and in the institutes of the Kaiser Wilhelm Gesellschaft (KWG) during the war years. Funding for studies in genetics, radiobiology, and applied botany was especially generous (Macrakis 1993, pp. 123–5; Deichmann 1996, pp. 94–9, 104–10). The reasons were variously political, economic, military, medical, and ideological. Most obviously, biology provided the basis for Nazi race ideology; Hitler explicitly claimed that National Socialism was applied biology (Stein 1988). Yet what the DFG wanted from biologists was not pseudoscience, but good science. It saw the propagandistic aspect of its interest as best served by support for basic research; it rarely, if ever, funded scientifically senseless projects (Deichmann and Müller-Hill 1994, p. 160).

On the economic and military side, the DFG promoted plant-breeding research, with the aim of developing crop strains suitable for use in the occupied territories to the east, especially crops that could provide the war industry substitutes for the raw materials it lacked as a result of the Allied blockade of the Axis countries. One example was the failed effort to grow the Koksaghyz plant (*Taraxacum kok-saghyz,* or Russian dandelion) in the wide areas to the east for the essential rubber industry. The DFG also funded studies in radiation genetics for their value to the atomic-weapons program (including measures for radiation protection) and to medicine. And it supported projects whose potential benefits were much less obvious, largely out of concern that Germany would otherwise cede its preeminent position in such areas to the United States (Deichmann 1996, pp. 105–31). Given their soaring prestige and the expanded support for their field of science, it is perhaps not surprising that more than fifty percent of biologists belonged to the Nazi Party (NSDAP), over twenty percent to the Sturm-Abteilung (SA, Storm troops), and five percent to the SS – rates slightly surpassing even those for physicians (generally considered the most compromised professional group).

As noted earlier, the involvement of some biologists and medical scientists with Nazism extended far beyond joining the NSDAP (Lifton 1986; Müller-Hill 1988; Weindling 1989). Yet even scientists who were deeply implicated in the crimes of the Third Reich generally suffered only brief interruptions of their careers. Whereas it is true that in most professions the majority of "desk-bound criminals" were smoothly reintegrated into the academic, professional, and public life of Germany's Second Republic, it might be supposed that human geneticists would have been subject to special scrutiny. After all, they had played a central role in legitimating National Socialism. Thus an American occupation officer expressed the feeling that the scientists were "1,000 times more guilty than 'an idiotic SS man'" (Weindling 1989, p. 566).

However, new positions in the field of human genetics were quickly filled with old Nazis. Fritz Lenz, department head at the Kaiser Wilhelm Institute (KWI) for Anthropology, Human Heredity, and Eugenics from 1933 to 1945, and adviser on resettlement policy in the occupied east, was appointed to the first new chair in human genetics (at the University of Göttingen) in 1946. Wolfgang Lehmann, who had held the "Chair of Race Biology" at the occupied University of Strassburg, was awarded the human-genetics position at the University of Kiel in 1956 and also was appointed to the board of the Max Planck Society (MPG), the successor organization to the KWG. Most striking was the case of the virulent anti-Semite Otmar von Verschuer, director of the KWI for Anthropology, Human Heredity, and Eugenics from 1942 to 1945, editor of the Nazi medical journal *Der Erbarzt,* and mentor to Josef Mengele (the notorious "Angel of Death" at Auschwitz). As early as November 18, 1949, von Verschuer lectured on "the application of the achievements of general genetics to humans and its limits" to the Mathematics–Natural Science Section of the Academy of Sciences and Literature in Mainz. In 1951 he was awarded the new chair of human genetics at the University of Münster, and the following year he was elected head of the German Society for Anthropology. As Robert Proctor has noted, von Verschuer's "dominant position in science during the Nazi period became an asset after the war" (Proctor 1988, p. 308). It is worth noting that von Verschuer, like many highly compromised scientists, also did work that was respectable by contemporary standards; as both Robert Jay Lifton and Benno Müller-Hill have argued, it was quite possible to be simultaneously a Nazi and a good scientist or doctor.

That smooth reintegration can be partly explained by factors common to every field of science. Germans were, of course, eager to reconstruct their country and deny their past, whereas American and British reconstruction planners concluded that a strong economy would be necessary to negate the appeal of communism. Though the process of denazification was everywhere undermined by the desire to counter each move by the Soviets, the cold war in genetics was particularly fierce. In August 1948 the peasant agronomist T. D. Lysenko was made the czar of Soviet biology, and his rejection of classical genetics became official Soviet ideology. The year before, at a meeting at the Health Ministry in the Soviet occupation zone, a committee had concluded that it was too early to implement a sterilization law, but that an educational effort to inculcate the need for eugenics measures should be introduced. Soon not only eugenics but all of genetics would be under attack in Eastern Europe. Textbooks and university teaching had to follow the Lysenkoist line. In a remarkable letter, Hans Nachtsheim informed Lenz that "anyway, the Communist pressure on us is much worse than that of the Nazis" (October 18, 1948).

Lysenko's rise to power meant that genetics research would require a defense. To focus on geneticists' complicity with Nazism would be to play into the hands of the new enemy. After all, Lysenko claimed that genetics had been annexed by the Fascists. In the context of cold-war politics and Soviet denigration of genetics and persecution of geneticists, some geneticists in other nations likewise lost interest in exploring their German colleagues' activities during the Third Reich. Thus participants at the Twelfth International Congress of Genetics held in Montreal in 1958 decided to convene the next congress in Germany. Only after vigorous protests by Israeli and some American geneticists (notably, Elizabeth Goldschmidt, J. Wahrman, and L. C. Dunn) was the congress moved to The Hague in The Netherlands.

The reintegration of highly compromised scientists into postwar German academic life was frequently aided by colleagues who were judged to have "kept their vests clean" during the Nazi period. As Lenz wrote to Nachtsheim: "As I hear, especially useful for this [denazification process] are politically influential documents from unburdened personalities" (August 11, 1946). Letters from Jewish colleagues were particularly prized. Thus Ernst Rüdin, the Swiss-born wartime director of the KWI for Psychiatry in Munich (and later *Reichkommissar für Rassenhygiene*), appealed, after the war, to Franz Kallmann, who had been dismissed from his position in 1935 and had emigrated to the United States the following year (Weber 1993; Gottesman and Bertelsen 1996).[2]

Whereas it is perhaps understandable that the reconstruction of German science was a priority for Germans (both scientists and nonscientists), the responses of émigrés and other victims of the regime and members of the international scientific community are more puzzling. In the same letter, Lenz reported receiving "a wonderful certificate from a Jewish member of the Zehlendorf denazification committee." Many émigrés and other victims of Nazism accepted their former colleagues' and mentors' protestations of innocence and resumed prewar relationships (as in the case of the émigré Carl Neuberg and his successor as director of the KWI for Biochemistry, Adolf Butenandt). The Dutch ethologist Nikolaas Tinbergen, who suffered greatly during the German occupation of Holland because of his opposition to the Nazification of Leiden University, resumed a close relationship with Kon-

rad Lorenz (Deichmann 1996, pp. 199–205; Burkhardt 1996). It was difficult for anyone who had not been there to accept that respected German academics had behaved so badly, especially so for their former colleagues and
students. And it should be noted that in many cases even the worst Nazis
were capable of helping or being civil to one or more individuals with whom
they had formerly had close relationships; for those few individuals, believing the worst, or the truth, was perhaps psychologically impossible.

In the postwar period, many senior scientists alleged that their cooperation with the Nazi regime had reflected a conviction that the more "good"
people like themselves who remained, the less harm the Nazis could do. They
(or their apologists) asserted that they had joined the Nazi party or taken
other apparently compromising actions in order to resist from within – what
Michael Kater has called the classic Trojan-horse defense (Kater 1989, p. 87).
Their claim was that, like Jews and political undesirables, they had actually
been victims of the regime and had opposed it whenever they could. Thus
Ernst Rüdin was said to have tried repeatedly to convene the Society of German Neurologists and Psychiatrists to discuss the mass murders of the mentally ill, a request invariably denied by the reichsminister, but in spite of his
alleged opposition to Nazi policy, he decided not to resign his offices because "*through his remaining he would be able to prevent larger evils*" (Schulz
1953).

Postwar recriminations often were attributed to professional rivalries. In
correspondence with colleagues at home and abroad, von Verschuer bitterly
complained of persecution. Writing to the distinguished British geneticist
R. A. Fisher (whose laboratory he had once visited), he recounted having "to
fight against the denunciation of a characterless collaborator whose word
carries more weight than that of an honest man, which I think I am" (July 31,
1947). And in a second letter (October 27, 1947), he complained that "jeal-
ous former colleagues from my old Kaiser Wilhelm Institute at Dahlem have,
for purely selfish reasons, maligned me and denounced me politically. False
reports have thus been put out about me. . . . I have already received many
letters and testimonials, especially from foreign colleagues who know me
and bear witness to the fact that I am a pure scientist who has always served
the Truth."[3] In reply, Fisher offered to provide "something in the nature of a
certificate of character, i.e., an assertion that I know you to be a genuine man
of science of reputation and merit" (May 29, 1946).[4]

Von Verschuer's fervent embrace of National Socialism had been evident
in his editorials and articles in *Der Erbarzt* and other widely available journals and in his 1941 textbook *Leitfaden der Rassenhygiene,* where he paid
homage to Hitler for having based public policy on "the findings of genetics and racial hygiene" (p. 11). Even the most cursory investigation would

have exposed his lavish praise of the Nuremberg Laws for segregating Jews and Germans and his many claims to the effect that "anti-Semitism is in the first case a *völkish*-political campaign, the justification and need of which emanates from the threat that Judaism poses to our peoplehood" (von Verschuer 1937, pp. 216–17). But others were willing to take von Verschuer's word that his vest was clean. Although he was "denazified" relatively late (in 1949), he was soon thereafter lecturing abroad. In 1953 the Italian Genetic Society elected him an honorary member; a year later he was elected an honorary member of the Anthropological Society in Vienna (Weingart, Kroll, and Beyertz 1988, p. 579).

HANS NACHTSHEIM AND THE RECONSTRUCTION
OF GERMAN GENETICS PROGRAMS

In order to better understand such complex interactions in Nazi Germany, we have been exploring the life and work of Hans Nachtsheim (1890–1979), a geneticist with a reputation for probity during the Third Reich. (Our project has drawn on work by the German historian Ute Deichmann and unpublished papers at the archives of the Max Planck Society.) Nachtsheim was frequently praised by geneticist colleagues for his unusual courage, although he privately admitted regret that, in contrast to his active protests against the communists, he had remained silent during the Nazi era. Consider this typical tribute from colleagues:

> Not only did he never join the Nazi Party, not even one of its branches or departments. He openly alienated himself from the proclamation of "Einigkeit" of leading geneticists. By doing so he stressed that this "unity" was artificial and only a matter of appearances. It was achieved through the brutal exclusion, murder and incarceration of those who did not agree. . . . One must ask in surprise how it was possible that he was not arrested and could pursue his scientific work at the KWI without interference. [Gunther and Hirsch 1970]

See Vogel (1979) for similar testimony. The German Society for Anthropology and Human Genetics now awards two Hans Nachtsheim prizes annually. How did Nachtsheim acquire this enviable reputation?

Like Lenz and von Verschuer, Nachtsheim was associated with the KWI for Anthropology, Human Heredity, and Eugenics; in 1940 he was appointed head of its Department of Experimental Genetic Pathology. Unlike them, he never joined the NSDAP. Although he was (and remained) an avid eugenicist, Nachtsheim was not implicated in any racist activities. His few letters extant from the Nazi period indicate that he declined to close even official letters with the customary "Heil Hitler."

In the prewar period, Nachtsheim had achieved scientific respect and developed wide international contacts. In 1921 he was elected the first secretary of the German Society for Genetics, a position he held for ten years. He spent 1926–7 on a Rockefeller fellowship, primarily with T. H. Morgan at Columbia University and at Woods Hole, but also traveling extensively in the United States. (Some of his research projects were funded by the Rockefeller Foundation.) He served as secretary general of the Fifth International Congress of Genetics in Berlin in 1927.

Nachtsheim's wartime refusal to join the NSDAP earned him considerable moral authority in postwar Germany, as well as elsewhere. In spite of his role as department head at the KWI for Anthropology, he was viewed by foreign geneticists as a true representative of the traditional German genetics community and was seen by the Allied authorities as one who had distanced himself from the activities and ideology of the Nazi regime. The extent to which he was accepted as a representative of the "other" Germany in those years is reflected in his invitation to participate in the 1961 Conference on Human Population Genetics in Israel, held at the Hebrew University of Jerusalem. Awarded the prestigious *Grosse Bundesverdienstkreuz mit Stern* in 1955, he was also appointed expert adviser to the 1961 committee that considered whether or not to compensate those who had been sterilized against their will. (He recommended against compensation, arguing that the "Law for the Prevention of Genetically Diseased Offspring" was Nazi neither in conception nor in administration, and he noted that many U.S. states also had sterilization laws.) Nachtsheim was the only German invited to help draft the influential 1951 UNESCO statement on race. Given his reputation, he was well positioned to help rehabilitate his less-reputable colleagues, internally and internationally.

At the war's end, with von Verschuer under investigation, Nachtsheim was appointed provisional head of the KWI for Anthropology. Although that institute did not survive, his department became the Institute of Comparative Genetic Biology and Genetic Pathology under the Deutsche Akademie der Wissenschaften; he was also appointed Professor of Genetics at the Humboldt University. Both institutions were in the Soviet zone of occupation. In 1948, after a short period of struggle with Lysenkoists, he resigned from the Akademie and Humboldt University to accept a professorship in general biology and genetics at the newly founded Free University in the Western sector of Berlin.

In 1948, Nachtsheim was consulted by Gert Bonnier, secretary general of the Eighth International Congress of Genetics, regarding which Germans were known not to have been Nazis (he named eighteen, who received invitations). Within Germany, he participated in the first Max-Planck-Gesellschaft (MPG, the postwar successor to the KWI) investigation of Otmar von Verschuer – serving as von Verschuer's main defender. He also helped Lenz obtain the position in Göttingen, providing a whitewashing certificate (known as *Persilschein,* after a popular German detergent) stating that Lenz had been neither a fanatic nor a party activist and had never endorsed anti-Semitism. His career illustrates a concrete case of a scientist who never joined the Nazi party but who took advantage of opportunities afforded by the regime and who used his moral authority to help more compromised colleagues reestablish careers in postwar Germany. What does it tell us about the reasons for the aborted denazification of the genetics community?

FROM RACIAL HYGIENE TO HUMAN GENETICS

First, irrespective of their earlier attitudes toward the Nazis, German geneticists (like other scientists) shared a strong desire to see their field reestablished. That interest could only be undermined by any focus on its past association with National Socialism. The anthropologist Karl Saller lost his academic position during the Third Reich as a result of opposition to Nazi race policy; he worked as a physician and was ultimately drafted. In 1948, Nachtsheim helped him obtain a professorship in Munich. But in 1961, Saller published *Die Rassenlehre des Nationalsozialismus in Wissenschaft und Propaganda,* a book exposing the widespread involvement of geneticists with Nazism. Nachtsheim attacked Saller furiously, charging that he had "fouled his own nest," obviously fearful that Saller's book would undermine efforts to reestablish the human genetics community in Germany.

Second, members of the international community believed that the harsh treatment of Germany after its defeat in World War I had contributed to the rise of Nazism, and they wished to avoid the mistakes of the past. Less than two years after the demise of Nazism, no one in Sweden or England opposed German participation in the Eighth International Congress of Genetics in Stockholm. Only in those countries that had been occupied during the war, such as Norway, Denmark, Holland, Belgium, and France, were there demands for restrictions on German participation. When the secretary general of the congress asked Nachtsheim to recommend geneticists who "did not let themselves get confused by the Nazi teachings" (Gert Bonnier to Nachtsheim, May 12, 1947), Nachtsheim seemed to respect the reservations of colleagues from previously occupied countries, but wrote that he could not "imagine how a person like [Norway's Otto L. Mohr], even if handled especially roughly by the Nazis, would therefore wish to exclude the German geneticists from the congress" (Nachtsheim to Bonnier, May 21, 1947). Perhaps geneticists whose postwar research was being funded by the U.S. Atomic Energy Commission, the Department of Defense, and other military agencies also hesitated to condemn German scientists' former cooperation with the Nazi regime.

Personal relationships certainly help to explain the willingness to provide former colleagues with testimonials. As already noted, many émigrés wanted to believe the best about their colleagues and erstwhile mentors. But wartime associations had in some cases led to postwar moral complications. For example, in January 1944, Nachtsheim's home had been bombed, and he and his wife had moved in with von Verschuer (Nachtsheim to Geheimrat von Lewinski, September 1, 1946). The two men had already cooperated in projects supported by the Reich Research Council, and as late as August 1944 they had submitted an application for a joint research grant.

Moreover, well into the 1950s, scientists conceived their professional responsibilities narrowly. They saw science as an activity dedicated to elucidating objective truths without regard for their social or ethical consequences. That conception was crucial in providing the rationale for the status that scientists demanded during the years of the Third Reich, and it helps explain their claim of noninvolvement. Such claims did not necessarily involve deception or even self-deception (Martin 1985).

Although there were significant differences in nuances, a similar view of science and society prevailed in the western democracies. Such a conception had been nurtured for generations, reaching its highest expression in the fifty years prior to the 1930s, both with respect to conceptual scientific breakthroughs, primarily in physics, and with respect to increased autonomy and influence of scientists. Maxwell's electromagnetic theories in the second half of the nineteenth century and the theories of Rutherford, Planck, and Einstein concerning the structure of matter and the cosmos in the first decades of the twentieth convinced many people that scientists, rather than the clergy, held the key to the secrets of Nature. In biology, too, the reductionist experimentalism of the physiologists and embryologists, as well as the sweeping holistic ideas emanating from material Darwinism (eventually wedded to the new science of Mendelian genetics), seemed eminently successful.

A particularly blunt expression of the view of science that accompanied those triumphs was provided by Karl Pearson, the logical positivist and founder of biometrics, in his discussion of the "problem of alien immigration into Great Britain":

> No satisfactory conclusions can be reached by citing individual instances which may tell one way or the other. There is only one solution to a problem of this kind, and it lies in the cold light of statistical inquiry. . . . We have no axes to grind, we have no government body to propitiate by well-advertised

discoveries; we are paid by nobody to reach results of a given bias. We have
no electors, no subscribers to encounter in the marketplace. We firmly believe
that we have no political, no religious and no social prejudices, because we find
ourselves abused incidentally by each group and organ in turn. We rejoice in
numbers and figures for their own sake and, subject to human fallibility, col-
lect our data – as all scientists must do – to find out the truth that is in them.
[Pearson and Moul 1925, p. 8]

Given that concept of science, most German scientists felt that they had
properly discharged their only duty: to conduct objective research.

It is in the context of such a conceptual world that we must understand
Nachtsheim's 1956 note to the weekly *Deutsche Kommentare* in which he
characterized the Lysenko affair in the Soviet Union (not the atrocities committed during the Third Reich) as the "saddest chapter of the biology of the
twentieth century." In Nachtsheim's universe, Lysenko's crime consisted in
demolishing the rationale for scientific research and rendering worthless the
content of the Soviet genetics program. The Nazis did not undermine either
the ideology of science or the quality of research, even if they misused science for political ends.

Thus we can perhaps also understand Nachtsheim's comment that Lysen-
koism was "worse than Nazi racism." For the field of genetics, perhaps it
was: Lysenko had been responsible for only a handful of deaths, but he had
effectively destroyed genetics research in the Soviet Union. The message was
that German scientists were not to be blamed for what others had done with
their scientific findings, but rather judged only for what they had done in
their capacity as scientists. The social or political consequences that could
be drawn from scientific work, and the physical applications thereof, belonged to another category; as scientists (i.e., university professors and
research-institute investigators) they could not properly be held responsible
for such applications. There was thus no need to be apologetic about their
activities under National Socialism.

The first MPG investigation of von Verschuer illustrates the point. Josef
Mengele had been an assistant to von Verschuer. In May 1946, Robert Havemann, then director of the Soviet-dominated KWI for Physical Chemistry
in Berlin, publicly charged that their collaboration had continued when
Mengele had become camp physician at Auschwitz. The MPG appointed a
committee, consisting of Havemann, Kurt Gottschaldt, Wolfgang Heubner,
Otto Warburg, and Nachtsheim, to investigate the charge. It confirmed that
some preserved eyes of murdered gypsies had originated from Auschwitz.
We now know that Mengele sent not only the gypsy eyes but also internal
organs from murdered twins, and sera from others he had deliberately
infected with typhoid, back to the KWI for analysis (Müller-Hill 1988,

pp. 70–4; Weingart 1989, pp. 279–80). The committee concluded that the purpose had been "scientific" rather than racial, but also that von Verschuer had been aware of the mass extermination of gypsies and Jews and thus had compromised scientific standards. As a result, he was banned from office in any reopened KWI for Anthropology. (In another wrist-slapping gesture, he was classified as a "fellow traveler" and fined 600 marks by the denazification tribunal.)

In 1946, Nachtsheim defended von Verschuer, claiming that "under no circumstances" could he be considered a "racist-fanatic." He further insisted that the scientific use of organs from concentration-camp victims was not in itself a crime. Thus he noted that another professor had used a great deal of "material" from political and other inmates of prisons, resulting in valuable anatomical research that otherwise could not have been conducted. According to Nachtsheim, von Verschuer could be accused of a crime only if it could be shown that he had known that the individuals had died of other than natural causes or had been killed for the purpose of providing body parts. But von Verschuer claimed that he had not known about that and should not be denied the benefit of the doubt.

NACHTSHEIM'S SCIENCE

Nachtsheim's defense of von Verschuer becomes more understandable when we realize that a similar charge could have been leveled at him. To understand why, we need to briefly discuss the content of his scientific work.

Nachtsheim had joined Ernst Baur in the Department of Zoology at the Genetic Institute of the Landwirtschaftlichen Hochschule Berlin in 1921. Given the difficult economic situation in Germany and the need for new sources of food, Baur advised him to choose the rabbit as a research object, advice Nachtsheim accepted.

In 1933 he developed a research program to study inherited pathologic abnormalities in the rabbit as models for understanding hereditary diseases in humans, a field he called "comparative genetic pathology." It represented a major shift in the direction of his work and had been prompted, he admitted, by passage of the Nazi sterilization law. Under the July 1933 "Law for the Prevention of Genetically Diseased Offspring," individuals could be sterilized if they suffered from genetic illnesses. Believing that human genetic diseases were caused by homologous mutations in the same gene in other mammals, Nachtsheim began to investigate a number of diseases, including Parkinson's disease, Pelger anomaly, and epilepsy, in small mammals, especially rabbits. With respect to epilepsy, he hoped to find a method to distinguish between genetic and nongenetic forms so that individuals with the hereditary form of the disease could be sterilized.

As part of that work, he induced convulsions in rabbits. Pentamethylene-tetrazole, known as Cardiazol in Germany (Metrazole in the United States), was administered to rabbits to determine if it could be used for differential diagnosis of hereditary and nonhereditary epilepsy (Nachtsheim 1939, 1942). Since 1935, intravenous injections of Cardiazol had been used to produce convulsions as a "treatment" for schizophrenic patients. It was claimed that convulsions could be elicited in hereditary epileptics with lower doses of Cardiazol than were needed to elicit convulsions in nonhereditary epileptics or in nonepileptics. Nachtsheim suggested that "in the provoked Cardiazol-convulsions we have an important race-hygienic method at hand, which would allow us to determine the hereditary situation of individual epileptics" (1942, p. 23).[5]

As late as 1937, the German Society of Neurologists and Psychiatrists had rejected the use of Cardiazol for diagnosis. Yet research on the use of Cardiazol in medicine continued in Germany and elsewhere. The main conclusion from Nachtsheim's studies with rabbits was that their susceptibility to convulsions when given Cardiazol depended primarily, but not solely, on the genotype. As for humans, the mean value for the Cardiazol convulsion threshold was lowest among epileptics; for nonepileptics it was so high that with the normally used doses no seizures ever occurred. Nachtsheim obviously was among those who had no reservations about the use of Cardiazol in humans. By 1942, however, electric shock had replaced Cardiazol treatment for psychoses, and the issue of medical use of Cardiazol became irrelevant. In any event, because Cardiazol (Metrazole) was used for psychiatric diagnosis and therapeutics not only in Germany but also in Italy, Sweden, The Netherlands, the United States, and Peru, Nachtsheim's Cardiazol experiments with rabbits and its possible application to humans could, to that point, have been viewed as consistent with normal scientific practice of that time.

In the course of his rabbit research, Nachtsheim recognized that age, as well as genotype, appeared to influence the propensity for seizures. Thus young and old rabbits reacted differently to lack of oxygen, which he thought to be the proximate cause of epileptic seizures. The question was whether or not those differences would be replicated in humans. To find out, he and Gerhard Ruhenstroth-Bauer of the KWI for Biochemistry tested children ages eleven to thirteen years at low air pressure (in Luftwaffe chambers) to see if they would experience seizures. They continued those experiments with children ages five to six years, but only preliminary findings had been published before the war ended. Afterward, Nachtsheim ceased publishing on epilepsy and said nothing about those experiments. His co-author sued Benno Müller-Hill when the latter suggested (in a draft for the book *Tödliche Wissenschaft*) that the children had come from an institution where euthanasia was practiced. According to the lawyer for Professor Ruhenstroth-Bauer, the experiments had really been clinical trials, with therapeutic intent, and the children, who were not harmed, had come from an orphanage (Müller-Hill 1987, p. 10). But a letter survives from September 20, 1943, in which Nachtsheim wrote to Gerhard Koch (then a Luftwaffe doctor and SS member, and later *Dozent* in von Verschuer's institute at Münster and eventually a professor of human genetics at Erlangen) that with help from a pathologist, they had obtained six epileptic children from Görden for experiments (Koch 1993, pp. 123–5).[6] In a personal letter to one of the authors of this essay, Koch asserted that "the examined children referred to were epileptic children, and the experiments were carried out in the 'euthanasia-institute' Görden near Brandenburg" (Koch to Diane Paul, June 24, 1993). Thus, in defending von Verschuer, Nachtsheim sought to justify his own research as well.[7]

There is also strong evidence that Nachtsheim knew about the more sinister activities of his colleagues during the war. Thus in a very friendly letter sent to von Verschuer at his hideout during the last stages of the war, Nachtsheim expressed his concern about the fate of compromising documents belonging to von Verschuer that might not be destroyed in time (March 12, 1945). Likewise, he assured Lenz that his books could remain in the institute where Nachtsheim's bombed-out institute was temporarily housed, and that those books on the "black list" would be kept in a closed place (January 14, 1946).

However, there were also resentments. In particular, von Verschuer had declined to provide Nachtsheim's son with a position at the KWI that would have saved him from service on the Eastern Front, from which he never returned. At the time of the first KWI investigation, Nachtsheim could not have expected that von Verschuer would eventually regain as much influence and standing as he ultimately did; Paul Weindling has suggested that Nachtsheim may have come to view von Verschuer as a rival (1989, p. 568). In any case, it seems that envy and mistrust turned relations between Nachtsheim and von Verschuer profoundly hostile.

In 1949 von Verschuer was exonerated by a second MPG committee consisting of Adolph Butenandt, Max Hartmann, Wolfgang Heubner, and Boris Rajewsky:

> We believe that it would constitute a pharisaical attitude on our part if, in light
> of the situation today, we were to consider a few isolated events in the past as
> marks of some unpardonable moral defect in a man who, in other respects, had
> honourably and courageously pursued his difficult path, and who had often

269

Diane B. Paul and Raphael Falk

enough shown evidence of his high-minded character. We . . . unanimously
believe that Professor von Verschuer possesses all the qualities appropriate
for a scientific researcher and a teacher of academic youth. [Müller-Hill 1988,
p. 83]

By then, Nachtsheim had turned against von Verschuer, but it was too late to
matter. Complicating the moral story, Gerhard Koch, who supplied the letter
incriminating Nachtsheim, viewed von Verschuer as a "dynamic, single-
minded, far-seeing . . . and moreover certainly uncompromising personality"
and never forgave Nachtsheim for ultimately denouncing his mentor (Koch
1982).

Nachtsheim was not a Nazi in the sense of belonging to the NSDAP or,
more important, engaging in overt racist activities. Like Konrad Lorenz, he
sought to draw parallels between human cultural life and animal domestication and asserted that both had degenerating impacts. In 1940 he contributed
a paper on "Captivity-Changes in Animals – Parallel Phenomena to Those
of Civilization-Damages in Man" in a volume on "civilization damages in
man" (Zeiss and Pintschovius 1944). But it was not only Nazis (or Germans)
who subscribed to such views. It is interesting that whereas Lorenz had certainly been a Nazi and had joined the NSDAP as soon as he could, Richard
Burkhardt has noted that he had never regarded his campaign to warn of the
dangers of domestication as a Nazi project; indeed, he continued to sound
the same alarm in the immediate postwar period (Burkhardt 1996).

So did Nachtsheim. In his 1936 book *Vom Wildtier zum Haustier* he had
argued that in the wild, natural forces worked to produce an equilibrium between mutation and selection. But unlike animals in the wild, man was not
subject to selection. With respect to domesticated animals, maintaining the
standard required continuous selection. If selection was relaxed, the breeds
would rapidly begin to deteriorate. There was no difference, in that respect,
he argued, between animals and man. People needed to practice race hygiene, applying the same principles used in animal breeding to humans. In
1948 he republished the book virtually unchanged; a revised edition was
published posthumously.

In 1934, Nachtsheim applied for funds to make an instructional film about
genetic diseases in rabbits. In his application he stressed the film's relevance
to the sterilization law. In a footnote to the 1938 text that accompanied the
film, he called attention to the fact that

the concept hereditarily sick is here conceived in a wider sense than in the Law
for the Prevention of Genetically Diseased Offspring. Whereas according to
the law only individuals who phenotypically suffer from a hereditary disease
are considered hereditarily sick, here we denote all carriers of the hereditary

270

factors of the disease as hereditarily sick. The carriers of only one recessive diseased gene, the heterozygotes, though phenotypically healthy, are not hereditarily healthy. In order to free the race from a disease, one must in an animal-breeding program eliminate the apparently healthy heterozygotes from reproduction as resolutely as is done with the sick homozygotes.

Those views were so extreme that even the Nazis could not accept them. With respect to any gene that was both recessive and rare, clinically asymptomatic carriers would vastly outnumber those affected. The Nazis had no intention of sterilizing such a significant percentage of their population – especially when the targets would be mostly their prized Aryans, not political or religious minorities.

Nachtsheim's zeal to eliminate genetic defects was hardly unique in the wider scientific community. Thus the half-Jewish psychiatric geneticist Franz Kallmann, who had been forced to emigrate in 1936, had proposed sterilizing close relatives of schizophrenics in order to eliminate carriers of genetic defects. In the 1930s and 1940s, enthusiasm for eugenics ran high among geneticists; its proponents even included such Jewish victims of Nazism as Heinrich Poll, Richard Goldschmidt, and Curt Stern. Even outside of Germany, many geneticists believed that there was a desperate need to reduce the "load" of deleterious mutations. Thus Nachtsheim's views were similar to those of H. J. Muller, who argued for the need to maintain a balance between rate and selection intensity. Such widely shared assumptions help explain why geneticists who should have been seen as highly compromised were not necessarily viewed as such by either their German colleagues or their foreign peers.

It was the context rather than the content of the German scientists' views that was exceptional. And it was the denial that context mattered that allowed so many German geneticists, with apparently easy conscience, to serve the aims of the Nazi regime. Scientists believed that they did not have to concern themselves with the public's opinion. They believed that scientists were responsible only for what they (and not others) did with their work. As long as they engaged in only "pure" research, and refused to distort it for political ends, they felt morally above reproach. But as Herbert Mehrtens has pointed out, "scientific purity" is sheer fantasy. In reality, science is always carried out in a political context that inevitably shapes its meaning and consequences (Mehrtens 1994, p. 337).

The significance of context was underscored by the German psychiatrist-philosopher Alexander Mitscherlich when he responded to an essay by Nachtsheim defending eugenics (Nachtsheim 1963; Mitscherlich 1963).[8] Mitscherlich argued that problems in medicine and eugenics were not purely biological problems. To the extent that eugenics measures were needed, that could be realized only in a free society in which the political consciousness of individuals would make misuse unlikely. In Denmark, a country in which strong political consciousness was combined with courage, a sterilization law was introduced in 1929, even prior to passage of the German law. But the Danes strenuously resisted when German occupation authorities ordered them

> ... to "assemble and deliver" their Jewish citizens, that is surrender them to annihilation. To the demand to introduce the Jewish [yellow]-star the Danish officials answered the Germans, that in that case the king would be the first to carry the Jewish-star. The open resistance had the consequence that not only were the Danish Jews saved, but there followed even among the German occupiers, all the way to the "Einsatzkommandos," a remarkable change of mind, something like a stimulation of the conscience. [Mitscherlich 1963, p. 715]

Many scientists shared Nachtsheim's scientific views, some of them liberals, and some communists. Many countries passed sterilization laws, and most were social democracies. Only in Germany did sterilization become a step on the road to mass murder.

In 1942, Nikolaas Tinbergen and his colleagues at the University of Leiden responded to the dismissal of their Jewish and anti-Nazi colleagues by resigning en masse from the university (Burkhardt 1966). Their response can be contrasted with that of Ludwig Prandtl in Germany, a member of the board of directors of the Society for Applied Mathematics and Mechanics, who rejected a proposal to dissolve the society if the Jewish members of the board were dismissed, saying "this has nothing to do with considerations of what would be most honorable, since it is simply a matter of a necessity for the discipline" (Mehrtens 1994, p. 332). Likewise, Max Planck, in Germany, opposed Otto Hahn's suggestion to protest the dismissals of Jewish scientific colleagues, and he advised others to remain at their jobs "for the sake of science" (Heilbron 1986, pp. 149–51).[9] There is no reason to think that Prandtl and Planck were being disingenuous or excessively complaisant toward the Nazi regime. Their responses reflected a view of scientific responsibility with deep roots and wide resonance in western culture. In the end, that traditional commitment of scientists to the principle that they were accountable only for the quality of their "pure" research guaranteed that they would serve the political interests of whatever party was in power, and in the case of Germany that would be the National Socialist state.

NOTES

1. The phrase was coined by Benno Müller-Hill (1987, p. 9).
2. Kallmann was a confessional Christian, but considered by the Nazis to be a Jew.

According to a December 6, 1935, diary entry of Rockefeller Foundation (RF) officer Daniel P. O'Brien, Rüdin urged the RF to help Kallmann find a position outside of Germany (Records of the RF, Record Group 12.1, Box 49, Rockefeller Archive Center, North Tarrytown, New York).

3. In a passage typical of the genre, he also wrote that "I was always considered untrustworthy [by the Nazis]. I deplored many of Hitler's measures and the National Socialist system and was constantly critical and skeptically positioned." The von Verschuer–Fisher correspondence is in the papers of R. A. Fisher at the University of Adelaide. We are grateful to the archivist for making copies of the correspondence available.

4. Von Verschuer apparently had less success with H. J. Muller (Weingart et al. 1988, p. 579).

5. Ute Deichmann has commented that Nachtsheim was motivated by "an interest in diagnosing a disease as a hereditary one, not a desire to counsel sick people and develop a therapy. He sought to prevent them from reproducing in the interest of the 'Volk as a whole'" (1996, p. 234).

6. Former Nazis were particularly thick on the ground at Erlangen, whose rector deliberately made it a haven for scholars whose wartime histories precluded appointments elsewhere (Allen 1996, pp. 33–4).

7. It should be noted that in the 1940s (as well as the 1950s and 1960s), the standards for research on human subjects, in the United States as well as in Germany, were much lower than they are now (Beecher 1966; Rothman 1987).

8. As early as 1947, Mitscherlich and F. Mielke published *Das Diktat der Menschenverachtung*. It was translated by H. Norden and published in English as *Doctors of Infamy: The Story of the Nazi Medical Crimes* (New York: Henry Schuman, Inc., 1949).

9. In an undated report to Samuel Goudsmit (scientific head of the ALSOS project), the anti-Nazi mathematician Paul Rosbaud wrote as follows: "The general excuse was: 'We could not dare to protest, though the expulsion of our Jewish colleagues is completely against all our views and even against our conscience. We could not think of ourselves but of the higher purpose, the university, the academy. We had to avoid the possibility of those institutions having any trouble or their being closed. This was our first duty and so our personal views and interests, as well as those of our Jewish colleagues, had to be kept in the background'" (Goudsmit Papers, Box 37, American Physical Society).

REFERENCES

Allen, A. 1996. Open secret: a German academic hides his past – in plain sight. *Lingua Franca* (March/April): 28–41.

Beecher, H. K. 1966. Ethics and clinical research. *New England Journal of Medicine* 274: 1354–60.

Burkhardt, R. 1996. Niko Tinbergen, Konrad Lorenz, and the science and politics of the post-war reconstruction of ethology. Presented at the Population Genetics Seminar, Harvard University, November 22, 1996.

Deichmann, U. 1996. *Biologists under Hitler.* Cambridge, MA: Harvard University Press.

Deichmann, U., and Müller-Hill, B. 1994. Biological research at universities and Kaiser Wilhelm Institutes in Nazi Germany. In: *Science, Technology, and National Socialism,* ed. M. Renneberg and M. Walker, pp. 160–83. Cambridge University Press.

Gottesman, I. I., and Bertelsen, A. 1996. The legacy of German psychiatric genetics: hindsight is always 20/20. *American Journal of Medical Genetics (Neuropsychiatric Genetics).*

Gunther F. K., and Hirsch, W. 1970. Unpublished manuscript.

Heilbron, J. L. 1986. *The Dilemmas of an Upright Man: Max Planck as Spokesman for German Science.* Berkeley: University of California Press.

Kater, M. H. 1989. *Doctors under Hitler.* Chapel Hill: University of North Carolina Press.

Koch, G. 1982. Professor Otmar Freiherr von Verschuer zum Gedächtnis. In: *Inhaltsreiche Jahre eines Humangenetikers: mein Lebensweg in Bildern und Dokumenten,* pp. 249–52. Erlangen: Perimed Fachbuch. (Originally published in *Ärtzliche Praxis,* December 16, 1969.)

   1993. *Humangenetik und Neuro-Psychiatrie in meiner Zeit (1932–1978). Jahre der Entscheidung.* Erlangen: Palm & Enke.

Lifton, R. J. 1986. *The Nazi Doctors: Medical Killing and the Psychology of Genocide.* New York: Basic Books.

Macrakis, K. 1993. *Surviving the Swastika: Scientific Research in Nazi Germany.* Oxford University Press.

Martin, M. W. (ed.). 1985. *Self-deception and Self-understanding.* Lawrence: University Press of Kansas.

Mehrtens, H. 1994. Irresponsible purity: the political and moral structure of mathematical sciences in the National Socialist state. In: *Science, Technology, and National Socialism.* ed. M. Renneberg and M. Walker, pp. 324–38. Cambridge University Press.

Mitscherlich, A. 1963. Eugenik: Notwendigkeit und Gefahr. *Fortschritte der Medizin* 81: 714–15.

Müller, I. 1991. *Hitler's Justice: The Courts of the Third Reich.* Cambridge, MA: Harvard University Press.

Müller-Hill, B. 1987. Genetics after Auschwitz. *Holocaust and Genocide Studies* 2: 3–20.

   1988. *Murderous Science: Elimination by Scientific Selection of Jews, Gypsies, and Others, Germany 1933–1945,* trans. G. R. Fraser. Oxford University Press.

Nachtsheim, H. 1939. Die Bedeutung des Cardiazolkrampfes für die Diagnose der erblichen Epilepsy, Versuche am krampfbereiten und nichtkrampfbereiten Kaninchen. *Deutsche Medizinische Wochenschrift* 65: 168–71.

   1942. Krampfbereitschaft und Genotypes. III. Das Verhalten epileptischer und nichtepileptischer Kaninchen im Cardiazolkrampf. *Zeitschrift für menschliche Vererbung- und Konstitutionslehre* 26: 22–74.

   1963. Warum Eugenik? *Fortschritte der Medizin* 81: 711–13.

   1966. Eugenik im Lichte moderner Genetic. *Forschung, Praxis, Fortbildung* 17: 3–8.

Pearson, K., and Moul, M. 1925. The problem of alien immigration into Great Britain, illustrated by an examination of Russian and Polish Jewish children. *Annals of Eugenics* 1: 5–55.

Proctor, R. N. 1988. *Racial Hygiene: Medicine under the Nazis.* Cambridge, MA: Harvard University Press.

Rothman, D. J. 1987. Ethics and human experimentation: Henry Beecher revisited. *New England Journal of Medicine* 317: 1195–9.

Schulz, B. 1953. Ernst Rüdin. *Archives of Psychiatry* 190: 187–95.

Stein, G. J. 1988. Biological science and the roots of Nazism. *American Scientist* 76: 50–8.

Vogel, F. 1979. Obituary. *MPI Berichte und Mitteilungen* (3780 Sonderheft): 29.

von Verschuer, O. F. 1937. Was kann der Historiker, der Genealog und der Statistiker zur Erforschung des biologischen Problems der Judenfrage beitragen? *Forschungen zur Judenfrage* 2: 216–22.

Weber, M. 1993. *Ernst Rüdin: eine kritische Biographie.* Berlin: Springer.

Weindling, P. 1989. *Health, Race, and German Politics between National Unification and Nazism, 1870–1945.* Cambridge University Press.

Weingart, P. 1989. German eugenics between science and politics. *Osiris* 5: 260–82.

Weingart, P., Kroll, J., and Beyertz, K. 1988. *Rasse, Blut und Gene: Geschichte der Eugenik und Rassenhygiene in Deutschland.* Frankfurt am Main: Suhrkamp.

Zeiss, H., and Pintschovius, K. (eds.). 1944. *Zivilisationsschaden am Menschen,* 2nd ed. Berlin: J. F. Lehmanns.

11

The Case against Evolutionary Ethics Today

Peter G. Woolcock

INTRODUCTION

Some two thousand three hundred years ago, Plato presented us a classic formulation of an ethical problem that has disconcerted thinkers about morality ever since. In his book *The Republic,* Plato has Glaucon tell the story of a young shepherd, Gyges, who discovers a ring that enables him to become invisible at will. With the special powers the ring gives him, Gyges proceeds to seduce the queen of his city, murder the king, and seize the throne. Glaucon intends his little parable to point a lesson about human nature. All of us, he suggests, behave morally only because, unlike Gyges, we know that immoral behavior usually will be detected and punished. "There is no one," he says, "who would have such iron strength of will as to stick to what is right and keep his hands off other people's property. For he would be able to steal from the shops whatever he wanted without fear of detection, to go into any man's house and seduce his wife, to murder or release from prison anyone he felt inclined" (Plato 1955, p. 91).

What Plato has illustrated for us in this story is the ethical problem of egoism: Why, if we can get away with it, shouldn't we pursue our own self-interest at the expense of any, or even all, other people? Why should we ever behave altruistically, that is, act so as to bring about the good of others, except when we have to do so to get them to help us further our own good? Why shouldn't we all be hypocrites, pretending to care about the welfare of others, creating a reputation for reliability, benevolence, and trustworthiness only because, without such a reputation, we would be treated by other people as social pariahs and thereby would fail to maximize our own personal welfare?

Contemporary evolutionary ethicists see modern evolution theory, particularly in its sociobiological form, as providing a solution to the problem of egoism. This is because they see it as offering what I shall call the "altruism guarantee." In their view, current evolutionary theory reassures us that human nature is not essentially egoistical. Rather, they believe that natural selection has instilled in most of us a tendency to altruism, a disposition to behave in ways that will further the good of other humans even at the risk of harm to ourselves.

In this respect, modern sociobiological theorists differ from earlier evolutionary ethicists who focused on the importance of individual competition. Those earlier writers saw nature as a struggle that weeded out weak individuals, resulting in the survival of the fittest. Nature was "red in tooth and claw," a merciless process in which compassion and charity would only bring about the degeneration of the species. Spencer, for example, said that "to aid the bad in multiplying, is, in effect, the same as maliciously providing for our descendants a multitude of enemies" (1874, p. 346). Even Darwin commented to the effect that civilized innovations such as mental asylums, poor laws, hospitals, and vaccination contributed to the situation in which "the weak members of civilised societies propagate their kind. No one who has attended to the breeding of domestic animals will doubt that this must be highly injurious to the race of men" (1871, p. 168). Sociobiological theory, then, came as a relief to those who thought that evolution committed us to a pessimistic picture of the moral prospects for the human species.

There had, of course, been proposals concerning an altruism guarantee prior to modern evolutionary theory. Perhaps the most influential alternatives have been those that have featured what I shall call the "supernatural option." Various religions have postulated a number of supernatural devices that, if people in general believed them, would guarantee that most of us would behave altruistically. One such proposal was that there was an all-knowing, all-powerful God who observed our every action and who would punish or reward us in the afterlife according to our behavior in this life. Another such proposal was that the universe was governed by a moral law, the law of karma, which would either condemn us to or release us from the cycle of rebirth, depending on our behavior in this life – a law operating with all the inexorable certainty of a physical law. On the basis of either the God model or the karma model, any egoistic behavior would inevitably get its just deserts. Such supernatural forms of the altruism guarantee, however, no longer have much influence in societies whose worldviews reject supernatural explanations of phenomena in favor of purely naturalistic, scientific, empirical views.

How, then, does a purely naturalistic theory such as modern evolutionary theory deliver on the altruism guarantee? It does so, first of all, by noting that any individual human is vulnerable to attack from predators, to scarcity of food, to extremes of temperature, to disease, and so on. The survival to breeding age of the bearers of one's genetic material (namely, oneself, one's own offspring, and the offspring of one's nearest relatives) can be assisted by collective hunting, collective defense, cooperative building, skill specialization, collective contributions to medical knowledge, and so forth. That suggests that more of the offspring of individuals with a genetic disposition to behave cooperatively are likely to survive than are the offspring of those who lack such a disposition, with the eventual effect that people with an innate altruistic tendency will come to dominate in a community, that is, they will be reproductively more successful.

What, however, has that to do with ethics or morality? Morality is not just about liking your children, or your neighbors, nor about feelings of generosity and friendship. It is primarily about right and wrong, good and bad, duty, rights, and obligations versus irresponsibility. Can modern evolutionary theory tell us anything useful about such matters as these? Contemporary evolutionary ethicists believe it can. Before we look at why they think so, however, it is important to note that there are two different ways in which evolutionary theory might be said to be relevant to ethical matters.

First, one might think that all that should be expected of evolutionary theory would be that it be part of a *descriptive* account of ethics. That would include a "reportive" component. That is, it would provide a report on what the actual ethical beliefs of humans are – for example, that all but a very few humans regard incest as morally wrong. However, it would not pass judgment on such beliefs. That is, it would not aim to reach conclusions of the following kind: "Most humans believe that incest is morally wrong, whereas in fact they are mistaken and it is morally right." The other component of a descriptive account of ethics is its "explanatory" component. That is, how are we to explain the fact that most humans believe incest to be morally wrong? One kind of descriptive explanation – the one offered by the evolutionary ethicists to be considered here – would be that such a belief could best be explained in terms of reproductive success, meaning, in this case, that the offspring of those humans who found incest acceptable would, because of factors such as inbreeding, decrease in the population over time.

As part of a descriptive theory, such an explanation should be amenable to empirical testing. It should be possible to formulate predictions from the central tenets of the theory and then investigate to see whether or not the predictions can be verified. For example, current evolutionary ethicists would predict that most individuals will feel a greater moral commitment to people who are close kin than to people who are more distantly related ("kin altruism"). They would also predict that our sense of moral obligation will be proportional to the degree to which other people are in a position to reciprocate any altruistic acts we perform to benefit them. For example, they would predict that we will feel a greater obligation of charity toward the poor in our neighborhood or city or country than toward those more remote from us ("reciprocal altruism"). However, verification of such predictions is not enough to establish the truth of such theories. It will also need to be shown that there are not other, equally plausible explanations for the phenomena. (I shall sketch such an alternative theory later.) Moreover, a descriptive evolutionary explanation will need to go beyond the stage of a "just so" story. That is, it will need to do more than suggest that people with a tendency to feel moral obligation will be more reproductively successful than people without such a tendency. It will need to provide actual evidence that it has been so. The theory expects it to be so, but what evidence, independent of the theory, is there that such people are more reproductively successful? One possible way to find such evidence would be to identify people who had a disposition to altruistic feelings, but lacked a disposition to the moral point of view, and compare their reproductive success with the success rate for those people who had the moral point of view as well. As far as I know, such studies are yet to be done.

Earlier, I said that there were two ways in which evolutionary theory might be relevant to ethical matters. Besides giving a descriptive explanation of how we have come to have the ethical beliefs we have, it might also be the case that evolutionary theory could tell us what our actual obligations are, which acts are right or wrong, what our rights and duties are, which things are good and which bad. If it does that, it goes beyond *describing* what humans believe and encroaches into the arena of *prescribing* beliefs and actions. In effect, it commands or orders or directs us to act and think one way rather than another. It ceases to be merely a *descriptive theory* and becomes instead a *prescriptive theory.* For example, some contemporary evolutionary ethicists have suggested not only that evolutionary theory tells us that we shall think that we have greater obligations to children and neighbors than to other people but also that, in fact, we actually do have such greater obligations. (Prescriptive theories sometimes are also called "normative theories," because they prescribe a norm or standard that should be followed.)

It should be noted, however, that not all modern evolutionary ethicists accept the contention that evolutionary theory serves a prescriptive role as well as a descriptive role. Later we shall return to this parting of the ways. Even so, in spite of their differences over the prescriptive function of evolutionary ethics, contemporary evolutionary ethicists all accept the foregoing descriptive explanation of ethics.

When we introduced the distinction between descriptive and prescriptive ethical theories, we noted that evolutionary theory seemed to explain how we came to have altruistic feelings, but that did not amount to explaining how humans came to have moral beliefs. Moral beliefs are more than mere feelings of friendship and so forth. So what, exactly, is it that makes a belief moral? If we are to understand why evolutionary ethicists think that humans with moral beliefs are likely to be reproductively more successful than those with merely altruistic feelings, we need to clarify what it is that we mean by "moral." Much contemporary philosophical thought has gone into this matter (Hudson, 1983). The evolutionary ethicists accept its general thrust, that is, that what we mean when we call a belief "moral" is that it is a belief about how we ought to act that is prescriptive, universal, and categorical. (The term "objective" is often used instead of the term "categorical," but I shall reserve it to identify one reason why people think that moral terms are categorical, as we shall see later.) We shall now look briefly at each of these requirements that must be met for a belief to count as a "moral" belief.

First, what is involved in the "prescriptivity" requirement? We have already encountered this notion in distinguishing between descriptive and prescriptive theories. With respect to the meaning of the word "moral," it indicates the fact that we call a belief "moral" only if the person who holds it is committed to acting in accord with it. For example, if someone claims to believe that incest is wrong, then the prescriptivity requirement says that we call that belief "moral" only if the person sincerely intends not to practice incest. Second, the idea behind the requirement that a moral claim be a universal claim is that a claim is not a moral claim unless those who make it believe that it holds universally. For example, if they consider it wrong for person A to commit incest in circumstances C, then they will consider it wrong for all persons like A in circumstances similar to C. Third, the requirement that a claim be a categorical claim if it is to count as a moral claim means that a claim is not a moral claim unless those who make it believe that a person should act in accord with it regardless of his or her own desires or beliefs about the matter. A claim that incest is wrong will count as a moral claim only if those who make the claim believe that it is wrong even if the perpetrator wants to commit it or believes that it is morally right to commit it. On this understanding of morality, a belief about what people ought to do is not a moral belief if the "ought" it contains is only hypothetical. Consider, for example, "If you want to be rich, then you ought to work harder and save your money." That advice does not command all of us categorically, even if it is correct. It commands us only *if* we want to be rich, that is, hypothetically. Essential to the meaning of the moral "ought," however, is that it commands us categorically, regardless of what we happen to want or believe.

People who adopt the moral point of view, then, would appear to be much more likely to be resistant to egoism than those who merely have an inclination or preference to be altruistic. When situations arise in which there is a temptation to egoism, those people with the moral point of view who elect to give in to that temptation have to overcome more than a feeling. They have to act contrary to what they regard as an order or command that directs them to act altruistically, an order or command that applies to them regardless of any egoistic feelings they may have.

If humans with an innate tendency to the moral point of view are more reliably cooperative and altruistic than those genetically disposed to egoism or those with a mere innate disposition to behave altruistically toward their direct descendants, then, given the sociobiological story, their offspring will survive in greater numbers in the community, thereby ensuring that eventually most of the population will have an innate disposition to adopt the moral point of view. If this scenario is correct, then contemporary evolutionary ethical theory will have provided a plausible explanation for the moral behavior of humans, although, as pointed out earlier, it will not have been shown to be true. Moreover, if such an account is correct, it would appear that we are close to showing that evolutionary theory can provide us the altruism guarantee. We do not need to fear that a purely naturalistic theory, one that makes no appeal to supernatural entities or processes such as God or karma, will deprive humans of the motivation necessary to act altruistically. We need not fear Yeats's warning: "things fall apart, the centre cannot hold, mere anarchy is loosed upon the world, the blood-dimmed tide is loosed." Contrary to Dostoyevsky, God's death will not have rendered all things permissible. Humans still will have a strong inclination to act as if altruism were commanded of them, even when it turns out that such a commandment is not enforced by the threat of eternal brimstone or endless cycles of rebirth into desiring and its frustration.

However, we have not yet shown that we actually do have the obligations that evolutionary theory leads us to think we have. Does this matter? At this point, evolutionary ethicists go their different ways. As we shall see, some think that the altruism guarantee applies only if we really do have the obligations that evolution has selected us to believe that we have, that is, if it really is the case that we ought to give priority to our own offspring over the offspring of other people, and so forth. In other words, they hold that our moral claims not only have to be categorical in nature, in that they apply to us regardless of what we want or believe, but also have to be categorical for a particular kind of reason, namely, because they are objectively true, because they match the way things actually are. Consider, for example, this claim: "In 1996 the earth was an oblate spheroid." This is a categorical claim, in that it is true or false regardless of what any of us happen to believe about it. It is not hypothetically true (in the sense that its truth or falsity would depend on how different people reacted to it subjectively) as in the case of a claim like "This apple is delicious" or "Skating is fun." That is, it is not the case that how anyone felt about the earth's shape in 1996 could change its

1996 oblate spheroidicity in any way. The "deliciousness" of an apple and the "funworthiness" of an activity, by contrast, are not objective matters. They are not inherently true of apples or skating regardless of what people think or feel or want from apples or skating.

Not all evolutionary ethicists, however, think that ethical claims are objectively true in the same way that the 1996 oblate spheroidicity of the earth is objectively true, nor do they think that they need to have that kind of objective truth in order to demonstrate the applicability of the altruism guarantee. To illustrate this dispute, we shall be looking at three different positions that have been adopted on this matter. As an example of a writer who rejects the objective truth of moral claims, but who thinks that the altruism guarantee survives nonetheless, we shall consider the work of Michael Ruse. As examples of those for whom the existence of the altruism guarantee depends on the objectivity of morals, we shall look first at the work of Robert Richards and then at the work of William Rottschaefer and David Martinsen. I shall show that all of these writers have failed to establish their particular versions of evolutionary ethics. In order to assist an understanding of my criticisms of their arguments, however, I shall need to introduce a general account of the kinds of objections that, I believe, invalidate not only their proposals but also all efforts at an evolutionary ethics. I shall begin with the naturalistic fallacy.

THE PROBLEM OF THE NATURALISTIC FALLACY

A number of philosophers have argued that it is always logically improper to deduce an ethical conclusion from a "naturalistic" claim, that is, from a claim describing the natural world, the world available to us through our five senses. Naturalistic claims compose a subset of what I have previously called "descriptive" claims. Most philosophers who regard it as a fallacy to deduce ethical conclusions from naturalistic claims would also regard it as a fallacy to deduce them from any descriptive claim; so the fallacy could just as appropriately, if not more appropriately, be known as the "descriptive" fallacy. Here is an example of what they would regard as such a fallacy:

> The human species can survive only if we let severely physically and mentally handicapped infants and children die.
> Therefore: We ought to let severely physically and mentally handicapped infants and children die.

This is a case of an argument in which a value conclusion is drawn from a purely descriptive premise, one that appears to have no value component. It is what I shall call a "descripto-normative" argument. In this case, the descriptive premise is also a "naturalistic" premise in that it refers only to natural phenomena – such things as "the human species," "children." There are many problems with such an argument. An obvious one is that the premise seems to be false. Modern societies do preserve the lives of such infants and children, and yet the human species manages not only to survive but also to increase in numbers. Nonetheless, the claim could be a long-term prediction.

However, regardless of whether or not the premise is true, the argument still would be regarded by many philosophers as a failure, because the conclusion introduces an element that is not already present in the premise, namely, the notion of what we "ought" to do. An argument form is regarded as deductively invalid (that is, as not a reliable indicator that its conclusion will be true if its premises are true) if it is the case that the contents of the conclusion are not already present in the premises. We could make the argument valid by adding a second premise, namely, "We ought to do whatever is necessary to ensure the survival of the human species," but then we would no longer be deducing a value conclusion from a purely factual premise, because our new premise would have a value component. The idea that there was a gap between fact and value was first proposed by David Hume (1911, pp. 177–8).

George E. Moore (1903) thought that many writers committed the naturalistic fallacy because they implicitly identified some naturalistic property, such as the survival of the human species, with the property of what was morally good, or with the property of what ought (morally) to be done. In effect, in his view, they treated "good" as meaning "whatever leads to human survival" or some other naturalistic property such as "maximizing pleasure." To demonstrate that "good" could not be identified in meaning with any particular naturalistic property, he developed what he called the open-question argument: If "good" actually means "whatever furthers human survival," then the question "Is human survival itself good?" should amount to the question "Is human survival a case of human survival?" – just as the question "Is a bachelor unmarried?" should amount to the question "Is an unmarried man unmarried?" Now, the question "Is a bachelor unmarried?" is a closed question. We already know the answer, simply from knowing the meaning of the word "bachelor." But the answer to the question "Is human survival good?" appears not to be something we know simply from knowing the meaning of the words "human survival" and "good." It is an open question; so "human survival" cannot be part of the meaning of the word "good." A similar argument can be advanced to show that other moral words like "ought" likewise cannot be defined naturalistically. If that is so, then it seems that it must be a fallacy to deduce value claims from purely naturalistic, or descriptive, premises.

There has been a recent move to try to show that even if ethical terms cannot be defined in terms of any particular natural property, that is, are not identical in *meaning* with some natural property, nonetheless they might be identical in *fact* with some natural property, just as "water" does not mean "H₂O," even though in fact it is identical with H₂O. Now, if "goodness" is identical in fact but not in meaning with some natural property, say "maximizing pleasure," then that undermines the effectiveness of the open-question argument as a support for the existence of a naturalistic fallacy, because as long as value terms are identical in fact with some natural property, we can validly deduce value conclusions from the appropriate factual premises. Of course, supporters of that view still have to show that their favored natural properties are ontologically identical with value properties, which is no easy task.

I shall take it for granted that the naturalistic fallacy really is a fallacy, even though that is disputed by philosophers such as Boyd (1988), Harman (1977), Putnam (1981), and Scruton (1994). However, the weight of the argument at present seems to be in favor of those who support the existence of a naturalistic fallacy, the nondefinability of value words in terms of naturalistic properties, and the nonidentity, in ontological terms, of value properties and any natural properties (Hudson 1969; Ball 1988b; Horgan and Timmons 1992; Gampel 1996). Certainly, no account that defines ethical words in terms of one natural property or another or that uses inductive processes to identify ethical properties with particular natural properties has received wide support among philosophers or has even been generally regarded as plausible. An excellent survey of the issues surrounding the naturalistic fallacy as they bear on evolutionary ethics has been provided by Hughes (1986). As it happens, however, all of the writers we shall consider have denied that their own work has committed the naturalistic fallacy; so I do not need to demonstrate that it is a fallacy, but merely that their belief that they have escaped it is false.

THE PROBLEM OF DESCRIPTIVE "OUGHT"

It may well be the case that evolutionary theory can enable us to deduce a descriptive "ought" from a descriptive "is." By a descriptive "ought" I mean those cases in which an author mentions the word "ought" in order to describe what some people think ought to be done, rather than the case in which that author uses the word to recommend a course of action. Let us suppose, for example, that it is a fact that societies that do not ban incest eventually destroy themselves through inbreeding. From this, together with the hypothesis that people will tend to believe that they ought to ban things that can destroy their society, we can deduce the conclusion that people will believe that they ought to ban incest. In a sense, then, we have deduced an "ought" from an "is," but we have not ourselves endorsed that "ought" in the process of the deduction; we have merely mentioned it. We have gone from a descriptive "ought" to a descriptive "is."

That deduction, however, does not commit the naturalistic fallacy, because the naturalistic fallacy forbids only the derivation of a normative "ought" from a descriptive "is" or a normative value from a descriptive fact. If there are ways of interpreting the notion of a value such that there are descriptive values, then the naturalistic fallacy does not forbid deducing them from descriptive facts. Equally, if there are normative facts, it does not preclude them from entailing normative values. An evolutionary ethicist who claims to have shown us how to deduce a descriptive "ought" from a descriptive "is" may well have avoided the naturalistic fallacy, but will not thereby have shown us how evolutionary theory can guide our actions, will not have used it to recommend one course of action rather than another.

I suggest, then, that whenever an evolutionary theorist claims to have deduced an "ought" from an "is," we shall find, if we look closely, either that a descriptive "ought" has been deduced from a descriptive "is," and therefore we have been provided no normative advice, or that a normative "ought" has putatively been deduced from a descriptive "is," and therefore the naturalistic fallacy has been committed.

THE PROBLEM OF CAUSATION REASONS

As we shall see, Ruse has argued that our moral beliefs are ultimately based on deep preferences, and those preferences are not amenable to justification in terms of reasons, but only to explanation in terms of causes. Such a claim, however, seems to diverge radically from our normal understanding of ourselves as actors in the world. In particular, it seems to reject what I shall call the "agency" thesis, which holds that except in special cases such as kleptomania, heroin addiction, or people who fall under the M'Naghten legal rules, each of us has the capacity to resist any of the desires that overtake us. It may be that, occasionally, we shall be in abnormal states of mind that will impair our capacity to understand and control our actions. If so, then a court may judge that we have diminished responsibility, and to that extent our agency was also diminished. Most of the time, however, we regard ourselves and others as lumbered with the costs and benefits of fully fledged agency and look askance at those who too readily appeal to causality to escape the blame incurred by their actions.

In one sense, however, it can be granted that reasons can be given a causal explanation. When we act rationally, it may well be the case that a certain physical mechanism, say some kind of rationality module in our brain, is operating under causal laws, that is, deterministically. Reasons, then, can be said to be "determinism-caused." If our rationality module breaks down in some way, or is overridden by compulsions or strong emotions or whatever, then we behave irrationally. Our actions, in that case, are not only determinism-caused but also compulsion-caused, whereas our rational actions are determinism-caused but compulsion-uncaused. They are said to be "free" actions, actions in which we exercise our "free will." As a rule, we are interested only in the determinism causes of a person's actions when our purpose in seeking an explanation for those actions is not sufficiently met by an account in terms of that person's reasons or goals.

Reasons, then, can be said to be causes in the determinism-cause sense of "cause." Reasons, however, do not collapse into compulsion causes. Even if the reasons people have for their actions are determinism-caused, it is still the case that the agency thesis holds for most people most of the time. This is true even with respect to their deep preferences. For example, a woman with a deep preference for egoism might decide that her egoism was creating so many problems in her life that she would train herself to altruistic habits. Her reason for deciding to become moral may not itself be a moral reason, but nonetheless she is still able to exercise the kind of freedom necessary to change herself into a person with a deep preference for altruism.

The fact that a deep preference is resistible means that we cannot claim that it is a cause in the compulsion-cause sense. We cannot use its causal status to exonerate ourselves, to show that we are not to blame because we had no rational control over what we did. If something is a cause in a sense that relieves us of responsibility for our actions, then it cannot be offered as a reason for our actions in the sense of a justification. There are no such things as "causation reasons," that is, compulsion causes that are also explanations for the rationality of an action. This feature of reason-giving is recognized in our ordinary discourse and in the legal system. For example, a court would not permit the woman mentioned earlier to offer as a defense that because she had come to the end of whatever justifications were available for her behavior, her only remaining appeal in mitigation of her offense was that she had a deep preference for egoism. Such a plea would not be taken seriously, because a state of egoism is not normally a case of a compulsion cause and so is still subject to ordinary standards of agent evaluation. There is still a compulsion-cause/reason gap, then, for most of us most of the time, even if there is never a determinism-cause/reason gap for any of us any of the time.

It should be noted that the same point applies to the case of beliefs as well as to the case of desires. We are agents not only in the sense that we normally have the capacity to resist any desire we experience but also in the further sense that we have the capacity to change our beliefs in the light of evidence, as a consequence of argument, by the application of rules of logic, rules of inference, and so forth. In order to count as non-agents, we have to be in a state such that our rational processes have failed in some way, such that there is a breakdown in or an overriding of our rationality module. Perhaps being under hypnosis is an example, or being under the influence of certain kinds of drugs. Unless we are in such a condition, however, we cannot be taken seriously if we offer as a reason for our believing a certain claim that we were caused to believe it in the sense that we still would have continued to believe it regardless of what evidence was offered against it and what logical and inferential processes we applied to it. We standardly treat each other as epistemic agents as well as motivational agents.

If any Darwinian ethicists are committed to the view that the distinction between reasons and causes collapses in a way that really amounts to allowing causation reasons to justify actions, then either they have misunderstood what evolutionary theory entails or else evolutionary theory is mistaken. The former seems the more plausible.

THE PROBLEM OF INSTRUMENTAL JUSTIFICATION VERSUS CATEGORICAL JUSTIFICATION

One way in which altruistic behavior could be justified to people would arise if they valued the survival of the human species and altruistic behavior could be shown to further that goal. That, however, would be a hypothetical justification. What it would show is that it would be rational of them to want people to behave altruistically, in the standard, "instrumental" sense of "rational," that is, if being altruistic really was a means of instrument leading to the survival of the human species. It is what I shall call an "instrumental justification." However, it is not what I shall call a "categorical justification," that is, one that can justify people's actions regardless of what their goals might happen to be. As a consequence, it is no better as a justification than is the justification an egoist might offer for not acting altruistically. After all, not acting altruistically furthers the goal of satisfying the egoist's personal self-interest. This, too, is an instrumental justification. In fact, people who have acted malevolently may well be able to give instrumental justifications for their actions – for example, that they had the suffering of their enemies as their supreme goal, that some malicious act was the most effective means to that goal, that they could perform it with impunity, and so forth.

As the altruistic, the egoistic, and the malevolent cases are all amenable to the same kind of noncategorical, instrumental justification, the fact that the altruistic case is justifiable in this sense gives it no advantage over the others. It seems that we can show that the altruistic or moral case is better justified than the others only if it (but not the others) can be given a categorical justification. As we shall see, however, when Darwinian ethicists try to present such a justification, they end up either committing the naturalistic fallacy or falling into the causation-reasons trap.

Now to look at particular Darwinists. We shall begin with Michael Ruse, as his arguments set the agenda for much of the subsequent debate.

RUSE

Ruse postulates that there is a genetic tendency in humans to regard morality as objective and that that explains why humans by and large do not behave in a thoroughly egoistic manner, but rather are altruistically inclined, especially toward kin and toward those people who are in positions to reciprocate. Nonetheless, he argues, the objectivity of morality is an illusion: Even though we think of moral claims as not open to choice, as laid upon us from outside, in fact they are just matters of feelings and sentiments. There are no external criteria by which to judge and be judged in moral matters. There is no way of escaping from the relativity of individual inclinations (Ruse 1986, pp. 216, 277).

However, if our tendency to egoism is so strong that evolution had to develop the quite specific mechanism of a disposition to believe in the objectivity of morals to overcome it, then, surely, will not wide public dissemination of Ruse's theory undermine this disposition and thereby cast us into a Hobbesian state of nature, where each is at war against the other?

Ruse, however, offers two arguments as to why people will remain motivated to be moral, even after he has let the cat out of the bag. He suggests, first, that those of us with a genetic disposition to see morality as objective could not behave other than morally even if we tried (Ruse 1986, p. 253). Second, he suggests that morality is an evolutionary adaptation that is as useful to us as having eyes (1986, p. 253). Both of those arguments fail, however. The first fails because we can train ourselves out of any residual inclination to behave altruistically, just as atheists can train themselves out of any tendency toward religious feelings left over from their upbringing. Even if altruism is a deep preference, it is not an irresistible one. It is not a compulsion cause.

His second argument fails because our eyes are useful no matter what goals we have, even if we have egoistic goals, whereas morality or altruism is not useful to egoists as a way of better furthering their egoistic goals. It may well be that their egoistic goals will be best served by having everyone else be moral or altruistic, but that is a reason for them to be hypocritical and encourage altruism in others, not a reason to be moral or altruistic themselves. For a more detailed account of these and related criticisms, see Woolcock (1993).

Ruse, then, has failed to show that most people would continue to act morally once they realized that the objectivity of morality was an illusion. A number of possible scenarios can follow from that conclusion. If our behaving morally depends on our believing in the objectivity of morals, then one scenario is that we would all realize that that objectivity was an illusion and so would indulge in rampant egoism. Another is that we would convince ourselves of the objectivity of morality and as a consequence continue to behave morally. Another is that our behaving morally would not depend on our believing that morality was objectively true.

That third alternative, as Ruse himself recognizes, is not really open to us. Suppose, for example, we were all to agree that there were no moral claims that were true regardless of what anyone happened to feel or believe or want. The upshot of that would appear to be that we would have to surrender the notion that moral disagreements were open to rational resolution. Disputes such as whether or not the earth is round or flat are open to resolution, because the flatness or roundness of the earth is not just a function of what we want or believe. There is a fact of the matter to be found. But if there is no fact of the matter to be found in moral disputes, then there can be no categorical answer, even in principle, to the question at issue. Instead, which answer we shall find acceptable will depend on what our goals are, or on some such essentially subjective criterion. To put the point at its crudest, our abhorrence of Nazism is no more nor less rational than Hitler's adherence to it. There is no way in which we can show that our feelings are the correct ones, the ones that correspond to the truth, that match reality, and that Hitler's feelings were shockingly mistaken. We may believe that of him, but our lack of an objective reference for our beliefs renders such a belief unwarranted. If we suppress Nazism, that can occur only because we are more powerful than the Nazis, not because our exercise of power is any more justified than would be their exercise of power to suppress us.

It might be thought that this kind of moral subjectivism does not matter a great deal. Natural selection will already have instilled altruistic instincts in most of us. True, but these altruistic instincts manifest themselves in a huge variety of ways. Many people are prepared to sacrifice their own self-interest on behalf of family, tribe, nation, political party, religion, social class, race, gender, and profession. Those people who would sacrifice their self-interest on behalf of one of those causes often are equally prepared to harm those people committed to one of the other causes. How are we to determine which has priority? Even Hitler was prepared to die for the good of the German race. As we have seen, Darwinian theory would predict that most of us would sacrifice self-interest with greatest alacrity for family, then for neighbors, then for humans in general, but if there is no objective truth about morality, then that order of priority is no more susceptible to rational justification than any other, even an order that would include egoism. If that is the case, then why shouldn't a Hitler seek to change our inclinations so that our first priority in our feelings of obligation would be to the *Führer,* rather than the family? Of what moral force is the fact that feelings of obligation to the family are the ones that natural selection has favored, at least up until now? In other words, of what *normative* interest, of what help in choosing how to act, is the kind of *descriptive* theory that Ruse has offered us about how we come to have certain sorts of feelings?

The problem is that in order to have some normative relevance, a descriptive theory would seem to have to be able to leap the "is/ought" gap, or the "reasons/causes" gap discussed earlier. Ruse believes that his theory enables him to "go around" these gaps (1986, p. 256); that is, he believes that it does have ethical relevance of a kind: It does so because those who act in accordance with the dispositions that evolution has implanted in us are no worse off than those who act, say, in accordance with utilitarianism. Suppose we ask utilitarians, "Why do you believe that you ought maximize happiness?" Ultimately, they will have to ground their principle in some deep preference, in a description of the kind of people they are, in their very nature. But that is precisely what can be done by people whose feelings of obligation are the ones that Darwinism predicts – that is, for example, people who would reject the utilitarian injunction to give charity to those in greatest need, regardless of country, and who instead would give it to the poor of their own country. In the end, according to Ruse, the gap between explanation and justification, causes and reasons, collapses: When pushed to the most fundamental form of justification, all we would be able to offer would be basic intuitions, and those intuitions could not be justified, only explained. We would be able to show what caused us to feel a certain way, but we would not be able to provide reasons for feeling that way rather than another. Moreover, given that Darwinian deep preferences are the ones most of us are likely to have if Ruse's evolutionary account is right, those who think they have utilitarian or Kantian or whatever deep preferences probably have misread their own natures. In other words, if Ruse's Darwinian theory were true, it would show that a normative Darwinian ethic would be more justifiable (in that it would be better grounded causally) than its normative competitors.

However, that attempt to give a normative role to descriptive Darwinian theory fails. As we saw in the preceding section, we can never justify our actions on the grounds that they arose out of our deep preferences. Such an attempt at justification would imply that we were locked into deep preferences in some way not amenable to argument or evidence. If someone said to us, "It is just plain selfish chauvinism to give to the poor of your own country ahead of those in other countries with greater need," we would not be regarded as having justified our actions were we to say, "Well, that just happens to be our deep preference." Such an answer would suggest that we could not have acted differently, which our critics certainly would not believe, except in the rare case of the act being compulsion-caused.

What our critics will standardly want is some broader generalization from which our preference for the poor of our own nation is derived, a generalization that incorporates values crucial to all parties to the argument. In the case of Darwinism, the broader principle from which obligations of charity are derived is something like "Maximize your own reproductive success," because that is the normative version of the descriptive law at work in natural selection. As Ruse has pointed out, however, "the person who helped another, consciously intending to promote his own biological advantage, would not be moral. He would be crazy" (1986, p. 222). Such a person would be crazy for two reasons: first, because there is no automatic link between a particular altruistic act and maximization of one's reproductive success; second, because it is not rational to think that other people would regard your action as justified (from their point of view, as opposed to yours) by the fact that it was intended to maximize your reproductive success, especially when it may have harmed them or their loved ones.

Justification, then, does not collapse into explanation, nor do reasons collapse into causes. Even utilitarians seek to justify their choices of supreme principles. Like Mill, they may try to show that the only evidence we can have for what is desirable is what is actually desired, or some such argument. That particular proposal may not be successful, but that does not show that such arguments can never succeed.

Ruse, then, has not shown us that we are as justified in adopting a Darwinian normative morality as we are in adopting utilitarianism or Kantianism, nor has he shown us, at the descriptive level, that natural selection has developed in us a specific genetic disposition to believe in the objectivity of morality. Nonetheless, his emphasis on the fact that people tend to believe in the objectivity of morality has added an important element to what evolutionary ethicists need to explain, even though he thinks that such objectivity is ultimately delusory. Other evolutionary ethicists have taken up his emphasis on the objectivity of morality and have tried to show that evolutionary

291

ethics does not need to treat it as an illusion. We shall now look at two such cases: first, Robert Richards, and second, Rottschaefer and Martinsen.

RICHARDS

Richards is less optimistic than Ruse. He seems to doubt that people will continue to behave morally if they come to believe that the objectivity of morality is an illusion. He tries to use evolutionary theory to demonstrate that morality is objective, that regardless of your individual beliefs and desires it really is the case that "the community welfare is the highest good" (Richards 1986a, p. 286). Unless this can be shown to be so by an appeal to the facts, especially to facts about human nature, he seems to fear that people will not behave altruistically, that there will be no altruism guarantee. Fortunately, in his view, the evolutionary facts do reassure us that people, for the most part, either will behave altruistically or, when they do not, will feel the kind of guilt that will push them back onto the straight and narrow path of moral behavior.

Before we look at whether or not he is successful, it should be noted that his advice to pursue the community welfare is not especially determinate. Certainly it does direct us to behave in an altruistic manner, rather than an egoistic one, so it rules out some of the behavioral options available to us. However, exactly what counts as the community welfare or good is itself a function of the more general normative position that one holds. A liberal account of the community good, for example, will differ considerably from that endorsed by a Marxist or a communitarian, especially in terms of the emphasis it places on individual liberty and autonomy as components of the good of the community.

According to Richards, evolutionary theory itself tells us that we ought to pursue the community good. It does so because the claim that we ought to act altruistically is just another way of saying that we must act that way unless our altruistic tendency is overridden by some defeating condition such as a surge of jealousy or hatred. The only exceptions are people who are sociopaths or psychopaths. Evolutionary theory, then, predicts that most of us will behave altruistically, because it holds that natural selection has favored the survival of groups in which a disposition to altruism is dominant.

That contention, however, has not given us a descripto-normative deduction. It has merely deduced a descriptive "ought" from a descriptive "is." The word "ought" does not become normative simply because it is used in a moral context. To become normative, it must be used to prescribe, not merely to predict, altruistic behavior.

So, on Richards's account thus far, it still looks as if we shall have to commit the naturalistic fallacy to get from the evolutionary facts to some normative recommendation.

Richards tries to bridge the gap by introducing a new rule of inference whose essence appears to be that we should regard as sound ethical injunctions "whatever [things] we are prompted to do by those dispositions whose dominance in a group furthers the survival of its members." Such injunctions, presumably, would include "Protect your offspring," "Provide for the general well-being of members of the community (including yourself)," and "Defend the helpless against aggression," all of which Richards offers as implications of the supreme principle of promoting the community welfare (Richards 1986a, p. 286).

However, utilitarians could equally well adopt that strategy to go from descriptive fact to normative value by advocating the inference rule that we should regard as sound ethical injunctions "whatever will maximize utility." Contractarians could also do the same by proposing the inference rule that we should regard as sound ethical injunctions "whatever could be agreed to by the parties in the original position behind the veil of ignorance." Even egoists could avail themselves of it, postulating an inference rule such as that we should regard as sound ethical injunctions "whatever furthers the satisfaction of your own interests."

That suggestion, then, does not enable evolutionary theory to give Richards's normative principle any advantage over its competitors. In fact, evolutionary theory could not arbitrate between those alternative normative positions, but merely could enable anyone who agreed with Richards's supreme principle to derive more specific injunctions from it. Whether or not we ought to follow those more specific injunctions, however, would depend on whether or not we ought to follow the supreme principle from which they were derived, and Richards's argument has not demonstrated that.

Richards tries to show us why we should prefer his particular rule of inference by offering a general criterion of choice, namely, that "frameworks, their inference rules, and their principles are usually justified in terms of intuitively clear cases – i.e., in terms of matters of fact. Such justifying arguments, then proceed from what people as a matter of fact believe to conclusions about what principles would yield these matters of fact" (Richards 1986a, p. 284).

There is little point, however, in trying to discover which one of these meta-moral rules of inference matches people's moral intuitions, because these rules of inference merely parallel the different theories that have arisen from differing pre-theoretical convictions. Those whose pre-theoretical moral feelings incline them toward adopting a utilitarian theory as the best

Peter G. Woolcock

integration of their moral opinions will find the utilitarian rule of inference the most persuasive, and so on. The "appeal to rules of inference" strategy, then, fails to arbitrate between competing normative positions.

Moreover, Richards's claim that the only way to justify a first principle is to show "that it would itself justify conclusions we antecedently and clearly know to be correct but not conclusions we clearly know to be incorrect" (1989, p. 334) does simplify the situation a little. But the criterion needs to be stronger than that; otherwise it allows in what are merely entrenched prejudices in a particular society that are best explained by indoctrinatory upbringing. As it happens, logic teachers do not seek to justify a rule of inference such as "From 'argument x has the form AAA-3' conclude 'the argument is invalid'" by testing the claim against lots of examples and then making the kind of abductive leap that Richards seems to be proposing. Instead, they use Venn diagrams or truth tables to demonstrate why that rule must hold for all cases of AAA-3. Certainly intuitions are appealed to, but the ones that are relevant here are intuitions about the relationship of containment captured by the Venn diagram or the exhaustive coverage of possibilities captured by truth tables. We cannot, however, use Venn diagrams, truth tables, or similar devices to justify the rule of inference that "From 'action x promotes the community good' conclude 'x ought to be done'" unless we define "ought" in terms of whatever promotes the community good. But that is to commit the naturalistic fallacy. So Richards's rule of ethical inference does not really seem to parallel rules of deductive inference.

Perhaps the element of Richards's theory that most reveals its problems, however, is his treatment of the nonaltruistic individual. According to Richards, such a person is "hardly a man" (1986a, p. 285), "not a human being in the full sense" (1986a, p. 291), has been "like the creature born without cerebral hemispheres . . . deprived of what we have come to regard as an essential organ of humanity" (1986a, p. 291), is a sociopath or a psychopath and "cannot be held responsible for his actions, morally guilty of his crimes, since he, through no fault of his own, has not been provided with the equipment to make moral decisions" (1986a, p. 291). There are at least three serious objections to that kind of suggestion on Richards's part.

First, the class of nonaltruistic people seems much wider than the class of people we would normally call "sociopaths" or "psychopaths." We seem to have argument by persuasive definition here. "Nonaltruists" could well include benevolent egoists, that is, those who were not altruistically inclined, but who were not malevolently inclined either: They would enjoy the company of other humans; they would not be the kind of person who "robs, rapes and murders" as Richards would have it (1986a, p. 285). These people would still seem to be agents, to have the capacity to have altruistic desires; it is simply that that would not be their deep preference. They would be as capable of altruistic desires, presumably, as altruistic people are capable of egoistic desires, and so they would be just as responsible for their actions as are the altruistic.

Second, there is an important omission from Richards's account. He suggests an absence of the relevant organ as an explanation for why some people do not desire to act altruistically but the only criterion he offers for demonstration of that deprivation is that such people lack altruistic desires. However, what is needed is a criterion that would enable us to identify that lack of an organ prior to determining cases of nonaltruistic behavior, in order to enable us to use its absence to predict nonaltruistic behavior and thereby explain it. Until Richards offers such a criterion, his explanation is only the promise of an explanation.

Third, the same argument might be used to show that celibates or homosexuals were not fully human, were not responsible for their actions, and so forth. Evolution not only has constituted us as moral beings but also has constituted us as heterosexual beings, because the preponderance of such a tendency in a group is essential to its survival. Thus, by analogy with his account of morality, Richards seems committed to the view that those who lack such a tendency presumably lack an organ essential for humanity. Clearly, celibates and homosexuals have the relevant sexual organs, so the organs they lack must be as-yet-undiscovered organs that generate heterosexual desire. An absence of such organs, Richards would have to say, would mean that such people would not be responsible for their actions, just as those who lack the as-yet-undiscovered organs of moral desire are not responsible for their actions. However, we cannot conclude that a man is not responsible for his actions just because he lacks a particular kind of desire. Perhaps lack of such a desire can be correlated with a person's not being responsible for his actions, but to show that, we must be able to identify when a person is responsible for his actions independently of and prior to knowing that he lacks that desire. Richards does not do that.

In fact, there seems to be no reason to think that even most of those humans with malevolent dispositions (those who do rob and rape and murder) lack the capacities that the rest of us have to resist our desires. If they do have those capacities, then they are as responsible for their actions as the rest of us. Again, there seems to be no reason that the mere presence of a deep preference of one kind rather than another would mean that one would not be an agent, that malevolent desires would be irresistible in a way that nonmalevolent desires are not.

None of this is to deny that there are people who have desires that they cannot resist, such as kleptomaniacs, heroin addicts, and so forth. What has not been shown is that such lack of capacity to resist occurs only with non-altruistic desires. Perhaps there are "altrupaths," people who cannot help sacrificing their interests for the interests of others, who find altruism irresistible, whose behavior has to be explained in terms of compulsion causes. The existence of such people, however, would no more show that all people with altruistic desires act out of compulsion causes than that all people who act out of nonaltruistic desires act out of compulsion causes. In fact, Richards's view seems to be that, except for the few "sociopaths," all of us are altrupaths who are sometimes subject to surges of nonaltruistic emotions. Why, then, are we responsible for our actions in a way that sociopaths, as he defines them, are not? He does seem to regard us as free to use our reason to determine the best means to achieve our altruistic goals, but offers no argument to show that nonaltruists cannot equally use their reason to determine the best means to their goals.

ROTTSCHAEFER AND MARTINSEN

Rottschaefer and Martinsen claim that human survival and reproduction (S/R) constitute an important normative value (1990, p. 163). In defense of that claim, they take Ruse to task for his thesis that Darwinism shows the objectivity of ethics to be an illusion. We have already seen that what Ruse means by such a thesis is that there are, in fact, no actions we ought to take regardless of what we believe or want. There are no such things as categorical obligations, only hypothetical ones.

Rottschaefer and Martinsen want to argue, by contrast, that such things as "food, shelter, and companionship have objectively valuable properties" (1990, p. 159). They suggest that these properties can be understood using an analogy with color properties. Science has shown us that objects do not have the property of color as such, but rather have properties that cause us to see objects as being colored. Likewise, objects do not have moral properties as such, but rather have properties that cause us to perceive certain objects as valuable. Value, in effect, supervenes on these natural properties but is not identical with any of them in particular. As a result, normative properties like "goodness" cannot be defined in terms of these natural properties, any more than "malleability" can be defined in terms of a particular set of natural properties.

What Rottschaefer and Martinsen have offered us here, however, is a descriptive account of value. It is an account of the properties that objects possess that lead people to *think* that an object is valuable, not an account of what actually does make the object valuable. It is, consequently, of no normative use. If we want to know whether or not we should choose a particular object, Rottschaefer and Martinsen's theory can tell us only that the object has the properties that normally lead most people to choose it in certain circumstances. If they then were to conclude that we also should choose the object in those circumstances because it has those properties, that would be to offer us a descripto-normative deduction and, thereby, commit the naturalistic fallacy.

Alternatively, Rottschaefer and Martinsen could be telling us that what really is valuable, as opposed to merely being thought valuable by most people, is the goal of human survival and reproduction, and that we should choose whichever of two possible courses of action is most consistent with that goal. If they are doing that, then they certainly are engaging in the normative enterprise of directing choices, but their moral advice is no longer grounded in evolutionary theory. It could be the case, for example, that natural selection has locked us into thinking that loyalty to the nation is valuable, that is, that acts of loyalty to the nation contain exactly the kinds of natural properties that lead us standardly to judge them "good," and that scientists are able to identify what those natural properties are in such a way that they can predict that when most people are in the presence of an act that possesses those properties, they will judge it good. Suppose, however, that there are good reasons for thinking that evolution has led us astray here, that loyalty to humanity as a whole, not loyalty to the nation, is what is crucial for human survival and reproduction. Should that turn out to be the case, then Rottschaefer and Martinsen's normative theory would lead them to dissent from the common view that loyalty to the nation is good, just as a utilitarian might dissent from the common view that one should save the life of one's own child ahead of that of a stranger, even though evolutionary processes have led us to be inclined to make the common judgment. If the utilitarian goal of maximizing happiness or the Rottschaefer and Martinsen goal of human survival and reproduction should coincide with what natural selection has led most of us to think is good, then that is merely serendipitous, not something that either theory required *qua* normative theory.

Rottschaefer and Martinsen developed their supervenience argument in order to show that Ruse was mistaken in his rejection of the objective nature of moral properties. Their argument, however, bears on Ruse's point only if those naturalistic properties entail that there are obligations that each of us has regardless of what we believe or want. Those naturalistic properties may be objective in the sense of "out in the world" rather than in us as part of human nature and yet still not be objective in the sense that we ought (normatively) to behave in response to their presence in a particular way regardless of what we think or desire. Food, shelter, and companionship do not seem to have properties that require us, as rational agents, to value them regardless of what our goals may be. Were one to have a goal that denied the value of, say, food (e.g., if one had a painful terminal illness and wished to starve oneself to death), then what force would the objective value of food, in Rottschaefer and Martinsen's sense, have as a reason for eating? They could tell us that food had a set of natural properties that caused most people in contact with those properties to value it. That, however, would amount to the argument that we ought to value food because most people in fact value it, which again would be a descripto-normative deduction and therefore would entail the naturalistic fallacy.

Nonetheless, Rottschaefer and Martinsen may be able to engage in justification of a kind. As they say later, "a person who appeals to the ultimate Darwinian metaethical principle to justify his or her action is using that principle as a justificatory reason. That reason can move the person to act if she has adopted in a consistent manner a Darwinian ethics. And it can be used to justify her actions" (1990, p. 167). That certainly justifies her action in the sense that it shows that it is a rational act for her to perform, given her goals, but it does not justify it in the sense that it shows that it is the act that everyone ought to perform in those circumstances regardless of their wants. It remains a noncategorical justification. In exactly the same way, an egoistical man who performs some act B because his doing B will maximize the satisfaction of his self-interest has thereby justified his act. He has justified it in the sense that he has shown that he acted rationally, given his particular supreme desire. Nonetheless, it has not been shown why one should do B in those circumstances regardless of what one's desires are.

Rottschaefer and Martinsen also want to claim that the fact that causes can serve as reasons provides them an avenue of justification. They say that "we can appeal to both a well-functioning perceptual system, normal perceptual circumstances, and the causal operation of the object of perception on our perceptual system in order not only to explain a perception but to justify a perceptual claim" (1990, p. 167). Likewise, a person can use the ultimate Darwinian metaethical principle "Promote human adaptations and, thereby, fitness leading normally to S/R" as a justificatory reason in that it can move her "to act if she has adopted in a consistent manner a Darwinian ethics. And it can be used to justify her actions" (1990, p. 167).

It does seem to be the case that I can appeal to certain factual conditions, such as my attribute of a well-functioning perceptual system, and so forth, to justify some perceptual claim – for example, that I see a glass on the table. Those factual conditions justify my perceptual claim, however, because they support the view that the perceptual claim is more likely to be true than false.

In other words, they justify the perceptual claim in the sense that they support the rationality of my trust in my perception.

The parallel argument in the evolutionary case would have to show that the relevant factual conditions justified the relevant normative claim, for example, that food was good. They would need to show that such a normative claim was more likely to be true than false. Truth, however, is a property that a proposition possesses or fails to possess regardless of what any of us happen to think or want. It is a categorical property. In order for the normative claim that we ought to value food to be true, it would have to be the case that we would always value food regardless of what anyone thinks or wants, just as the proposition that I am seeing a glass on the table is true or false regardless of what anyone believes or desires about the matter. Rottschaefer and Martinsen, however, have not offered us an argument to show that we ought to value food, or anything else, quite independently of what our desires might happen to be. They have not, therefore, shown that how our moral beliefs are caused can be used to justify them, in the sense of showing that they are likely to be true. They have not, then, shown that causes can be reasons with respect to our moral perceptions in the way that they can be for our visual perceptions.

Rottschaefer and Martinsen offer as a further argument for the use of evolutionary theory as a source of normative advice that it is unlikely that we would have the moral sentiments we do if evolution had not selected out such sentiments as reliable guides to what was good. Evolutionary theory, however, tells us only that those sentiments are reliable guides to what is good from the viewpoint of human survival. It does not, of itself, tell us that human survival is a good. We may, then, be able to justify acting on our moral sentiments if our goal is human survival, but all that we shall have done will have been to offer an instrumental justification for acting on our moral sentiments, not a categorical one. If our goal were to further our own self-interest, we could give an instrumental justification for encouraging everyone else to act on the moral sentiments, while ignoring those sentiments when it would benefit us personally to do so. The altruistic and egoistic courses of action are equally well justified vis-à-vis their respective goals, but that kind of hypothetical justification has done nothing to justify altruism over egoism.

Moreover, the strategy that Rottschaefer and Martinsen have adopted seems to land them in the same kinds of difficulties as Richards's sociopath caused him. Suppose that goodness and rightness really do supervene on natural properties and that food, shelter, companionship, human survival, and reproduction are things that can be demonstrated through an appropriate theory to be good, that it is right to pursue such things. What sense, then, are we to make of people who deny that these things are good, who deny that actions conducive to them are right? How are we to deal with the case that Rottschaefer and Martinsen themselves mention – those people who would be prepared to let human beings, or even the whole of humanity, die rather than experiment on animals? Are they morally blind? Well, there is no reason to believe that they cannot see every natural property of objects that the rest of us see, nor any reason to believe that they do not possess exactly the same kind of rationality module as the rest of us. What, then, are they missing? Presumably some organ of moral insight, some device that would enable them to go from the perception of certain natural properties to a moral perception. But that, of course, is just a form of naturalistic moral intuitionism, as opposed to Moore's nonnaturalistic version. It moves moral debate out of the realm of argument and into the realm of nontestable assertion (Hudson, 1983, pp. 100–7).

EVOLUTIONARY THEORY AND CATEGORICAL JUSTIFICATION

As I suggested at the beginning of this essay, all the writers we have examined so far seem to want evolutionary theory to serve as a naturalistic altruism guarantee in much the same way as religious believers might have hoped that notions like God, karma, hell, and so forth, would have constituted a supernatural altruism guarantee. Those supernatural altruism guarantees tried to get God's law or karmic law to perform a role in the control of human actions similar to the role of scientific law. The way the world is (the 'true') does not allow us to adopt any goal we choose and successfully act on it. Our actions can be irrational in the sense that if they are inconsistent with what is true, then no matter what we want, we doom ourselves to frustration. However, if what we do is inconsistent with a particular conception of the 'right', there does not seem to be anything in the world that inevitably dooms us to frustration when we act inconsistently with that conception of the 'right'. As a consequence, religions can be interpreted as having developed models of how the universe is (the 'true') in which those who behave immorally are doomed to frustration, in the long run at least, in much the same way as is someone who disregards laws of nature.

If there is no such thing as a categorical justification, and intimidatory models like God and karma are no longer rationally persuasive, then should we fear the onset of moral chaos? The writers we have examined all reassure us that such fears are unwarranted. Evolutionary theory, they tell us, guarantees us that the vast majority of people will continue to act altruistically, will continue to value the community good and the survival and reproduction of humanity. Ruse and Richards think that this will be so because of the dispositions developed in us by natural selection. Rottschaefer and Martinsen seem to think it will be so because the natural properties of objects are such that we perceive certain of them as good, just as we perceive certain objects as colored. As we have seen, however, none of those strategies has been successful.

Does this mean, then, that the worst tendencies of human nature will run rampant? It certainly means that the egoist and the malevolent individual are no more irrational per se than are moral or altruistic individuals, that they must undermine their most fundamental goals in a way not true of the moral person. But even if it is true that no categorical justification can be provided for altruism over egoism, we are nevertheless likely to find ourselves relating to each other as if morality had objective properties. This is not because we are genetically programmed to do so, but because, as rational agents, we rapidly learn that only certain kinds of reasons are acceptable when offered publicly as justifications for our actions. Even if we are egoists, we quickly realize that exposing our true motives will be self-defeating in situations in which we do not have the power to enforce our will. If people know that we shall cheat them whenever it suits us and that we think we can get away with it, then they will not enter into relations with us that are to our benefit unless they have no choice. As a consequence, we shall have to pretend to be honest, reliable, trustworthy, and altruistic in order to achieve our egoistic ends.

Moreover, we shall have to play the moral-reason-giving game. We shall have to seem to agree with others that it is a good thing that people are honest, reliable, and trustworthy, that children should be trained in those virtues, and so on. Furthermore, we shall have to pretend to agree that the kind of motive from which it is admirable to act is an altruistic one, or one that respects the rights of others, that it is not admirable to act from what actually are our real motives, the purely egoistic ones. In effect, then, the rational game-theory strategy for the egoist who is unable to force others to act as he or she wants is to play the categorical-justification game, to publicly advocate that people should behave morally regardless of what they personally want or believe.

That, however, only shows that rationality requires us to advocate morality as if it were objectively true in circumstances in which we lack the power to bend others to our will. What if we did have that power? How should we act then? In such circumstances it seems to me that people would divide into two groups: those who had internalized the kind of public reason-giving that develops between people when they have more or less equal powers, and those who had not. Now, even if those in the first group had greater power, they would continue to act in accord with the kinds of reasons found publicly acceptable in the situation in which the two groups had more or less equal powers. Although I do not have space to elaborate the point here, I suggest that the kinds of reasons such a group would endorse would be very similar to those postulated by John Rawls (1972) in his model of the parties in the original position behind a veil of ignorance.

It seems, then, that all we need at the biological level to ensure that people will act as if morality were objective is that there be a genetic disposition to believe privately what we offer as the public reasons for our actions. In effect, that is a disposition to socialization. When that is coupled with the general features of rationality (the ability to deduce, to predict consequences, etc.), the scope of the reasons offered should become more inclusive as the group to which we have to justify ourselves expands. The demand to respect the lives of only the members of our tribe, say, becomes difficult to meet when we have to negotiate with other tribes that have the power to inflict serious injury on us unless we cooperate with them. As those people increasingly remote from us become more able to influence our lives because of new developments in technology, we shall find that the kinds of reasons we thought applied only locally will have to be extended more generally. In fact, we shall find ourselves forced to reason with strangers in terms of fairness, universalizability, and the other standard norms of morality. Although there is a tendency to accept what we come to be socialized into, we still possess our individual perceptions of the world, our differing logical abilities, and our particular capacities to recognize inconsistencies between what we have been socialized into and how we personally perceive things to be. That leaves open the possibility of both moral reform and cognitive reform.

Even so, individuals could train themselves to be hypocrites if they thought it necessary, just as people who are naturally egoistical could train themselves to be moral if, as Foot suggests (1967, p. 100), they found it difficult to maintain the continuing pretense that consistent egoism would require (Van Ingen, 1994).

All of this is compatible with the claim that natural selection has instilled in most of us an innate disposition to altruism, although of a very general kind. Hume, in 1772, made much the same point when he said that it would be absurd to deny "that there is some benevolence, however small, infused into our bosom; some spark of friendship for human kind; some particle of the dove, kneaded into our frame, along with the elements of the wolf and the serpent." Granting that, however, does not amount to granting that there is a genetic disposition to favor kin over non-kin, or those who can reciprocate over those who cannot. Nor does it amount to granting that there is an inherited disposition to adopt the moral point of view. Such a theory would be too strong. The same phenomena can be explained on weaker, less contentious assumptions. Interestingly, the kind of disposition to socialization mentioned earlier, combined with a general disposition to benevolence, would have exactly the same kind of reproductive advantages as the evolutionary ethicists proposed for an innate tendency to the moral point of view. People with those two dispositions would honor the moral codes of their societies without recourse to postulation of any specific moral instincts.

What, then, is the status of the altruism guarantee in the account I have given? Will a general innate disposition to benevolence and an inherited tendency to socialization guarantee that humans will behave altruistically? Perhaps, but only when there is something approximating equal powers between the relevant parties, that is, when our tendency to socialization runs up against the finding that the reasons we accepted at a local level seem logically to need extension more and more generally. Unless communication becomes more nearly universal, however, my account offers no solace to those who want altruism to become more nearly universal.

CONCLUSION

The writers under consideration, then, have been unsuccessful in their efforts to make evolutionary theory relevant to normative ethics. All of them have failed to show that evolutionary theory gives us the altruism guarantee. Ruse may be right that we shall continue to behave morally even if we accept that moral claims are not true or false in the way that empirical claims are, but he has failed to show that that would be because of a genetically acquired disposition to treat morality as objective. Also, he has not shown that the kinds of moral conclusions that Darwinists favor – for example, that we should care for the poor of our own country ahead of those in other countries – are as justified as those that follow from utilitarianism or contractarianism. Neither Richards nor Rottschaefer and Martinsen have shown that egoists or misanthropes are somehow ruled out of moral contention by evolutionary theory. Evolutionary theory, then, cannot serve the moral-reassurance role previously filled by religion. It looks as if we shall have to resolve our moral differences through the hard grind of normative justification. As I have suggested, however, perhaps there is more hope there than the evolutionary ethicists seem to think.

FURTHER READING

The scientific background for sociobiological ethics has been presented by Wilson (1975, 1978). Wilson was most thoroughly criticized by Kitcher (1985), but also see

Singer (1981). The most thorough philosophical study of sociobiological ethics, including a history of evolutionary ethics since Darwin and a criticism of earlier studies, is the work of Ruse (1986), or the shorter version (Ruse 1991). For a useful philosophical criticism of pre-sociobiological evolutionary ethics, see Flew (1967), and his critique of sociobiology (Flew 1994). A useful history of evolutionary ethics, including the sociobiological approach, is by Farber (1994). An account of "evolutionary psychology" that covers much recent sociobiological thinking about humans as "moral animals" is that by Wright (1995). For other biological accounts of human morality, see De Waal (1996), Frank (1988), and Wilson (1993). Useful philosophical discussions have been provided by Nagel (1978), Murphy (1982), Chandler (1991), and Sober (1994). For criticism of the work of Ruse, see Richards (1986a), Rottschaefer and Martinsen (1990), Collier and Stingl (1993), and Woolcock (1993). For criticism of Richards, see Gewirth (1986), Hughes (1986), Thomas (1986), Trigg (1986), Voorzanger (1987), Ball (1988a), Williams (1990), and Collier and Stingl (1993). For a defense, see Richards (1986b, 1989). For criticisms of Rottschaefer and Martinsen, see Barrett (1991). For a defense, see Rottschaefer and Martinsen (1991). For a criticism of Woolcock (1993), see Waller (1996).

REFERENCES

Ball, S. 1988a. Evolution, explanation and the fact/value distinction. *Biology and Philosophy* 3: 317–48.
   1988b. Reductionism in ethics and science: a contemporary look at G. E. Moore's open-question argument. *American Philosophical Quarterly* 25: 197–213.
Barrett, J. 1991. Really taking Darwin and the naturalistic fallacy seriously: an objection to Rottschaefer and Martinsen. *Biology and Philosophy* 6: 433–7.
Boyd, R. 1988. How to be a moral realist. In: *Essays on Moral Realism,* ed. G. Sayre-McCord, pp. 181–228. Ithaca, NY: Cornell University Press.
Chandler, J. 1991. Ethical philosophy. In: *The Sociobiological Imagination,* ed. M. Maxwell, pp. 157–69. Albany: State University of New York Press.
Collier, J., and Stingl, M. 1993. Evolutionary naturalism and the objectivity of morality. *Biology and Philosophy* 8: 47–60.
Darwin, C. 1871. *The Descent of Man.* London: John Murray.
Dawkins, R. 1976. *The Selfish Gene.* Oxford University Press.
De Waal, F. 1996. *Good Natured.* Cambridge, MA: Harvard University Press.
Farber, P. L. 1994. *The Temptations of Evolutionary Ethics.* Berkeley: University of California Press.
Flew, A. 1967. *Evolutionary Ethics.* London: Macmillan.
   1994. E. O. Wilson after twenty years: Is human sociobiology possible? *Philosophy of the Social Sciences* 24: 320–35.
Foot, P. 1967. Moral beliefs. In: *Theories of Ethics,* ed. P. Foot, pp. 83–100. Oxford University Press.

Frank, R. H. 1988. *Passions within Reason: The Strategic Role of the Emotions.* New York: Norton.

Gampel, E. H. 1996. A defense of the autonomy of ethics: why value is not like water. *Canadian Journal of Philosophy* 26: 191–209.

Gewirth, A. 1986. The problem of specificity in evolutionary ethics. *Biology and Philosophy* 1: 297–305.

Harman, G. 1977. *The Nature of Morality.* Oxford University Press.

Horgan, T., and Timmons, M. 1992. Troubles for new wave semantics: the 'open question' argument revived. *Philosophical Papers* 21: 153–75.

Hudson, W. D. (ed.). 1969. *The Is–Ought Question.* London: Macmillan.

1983. *Modern Moral Philosophy.* London: Macmillan.

Hughes, W. 1986. Richards' defense of evolutionary ethics. *Biology and Philosophy* 1: 306–15.

Hume, D. 1911. *A Treatise of Human Nature,* vol. 2, ed. A. D. Lindsay. London: J. M. Dent. (Originally published 1738.)

1975. *Enquiries Concerning Human Understanding and Concerning the Principles of Morals,* ed. L. A. Selby-Bigge, rev. P. H. Nidditch. Oxford: Clarendon Press. (Originally published 1777.)

Kitcher, P. 1985. *Vaulting Ambition.* Cambridge, MA: MIT Press.

Moore, G. E. 1903. *Principia Ethica.* Cambridge University Press.

Murphy, J. G. 1982. *Evolution, Morality, and the Meaning of Life.* Totowa, NJ: Rowman & Littlefield.

Nagel, T. 1978. Ethics as an autonomous theoretical subject. In: *Morality as a Biological Phenomenon,* ed. G. S. Stent. Report of the Dahlem Workshop on Biology and Morals. Berlin: Abakon-Verlagsgesellschaft.

Plato. 1955. *The Republic.* Harmondsworth: Penguin.

Putnam, H. 1981. *Reason, Truth and History.* Cambridge University Press.

Rawls, J. 1972. *A Theory of Justice.* Oxford University Press.

Richards, R. J. 1986a. A defense of evolutionary ethics. *Biology and Philosophy* 1: 265–93.

1986b. Justification through biological faith: a rejoinder. *Biology and Philosophy* 1: 337–54.

1989. Dutch objections to evolutionary ethics. *Biology and Philosophy* 4: 331–43.

Rottschaefer, W. A., and Martinsen, D. 1990. Really taking Darwin seriously: an alternative to Michael Ruse's Darwinian metaethics. *Biology and Philosophy* 5: 149–73.

1991. The insufficiency of supervenient explanations of moral actions: really taking Darwin and the naturalistic fallacy seriously. *Biology and Philosophy* 6: 439–45.

Ruse, M. 1986. *Taking Darwin Seriously.* Oxford: Blackwell.

1991. The significance of evolution. In: *A Companion to Ethics,* ed. P. Singer, pp. 500–8. Oxford: Blackwell.

Scruton, R. 1994. *Modern Philosophy.* London: Sinclair-Stevenson.

Singer, P. 1981. *The Expanding Circle.* Oxford: Clarendon Press.

Sober, E. 1994. Prospects for an evolutionary ethics. *From a Biological Point of View,* pp. 93–113. Cambridge University Press.

Spencer, H. 1874. *The Study of Sociology.* London: Williams & Norgate.

Thomas, L. 1986. Biological moralism. *Biology and Philosophy* 1: 316–25.

Trigg, R. 1986. Evolutionary ethics. *Biology and Philosophy* 1: 325–35.

Van Ingen, J. 1994. *Why Be Moral?* New York: Peter Lang.

Voorzanger, B. 1987. No norms and no nature – the moral relevance of evolutionary biology. *Biology and Philosophy* 3: 253–70.

Waller,B. N. 1996. Moral commitment without objectivity or illusion: comments on Ruse and Woolcock. *Biology and Philosophy* 11: 245–54.

Williams, P. 1990. Evolved ethics re-examined: the theory of Robert Richards. *Biology and Philosophy* 5: 451–7.

Wilson, E. O. 1975. *Sociobiology: The New Synthesis.* Cambridge, MA: Harvard University Press.

1978. *On Human Nature.* Cambridge, MA: Harvard University Press.

Wilson, J. Q. 1993. *The Moral Sense.* New York: Free Press.

Woolcock, P. 1993. Ruse's Darwinian meta-ethics: a critique. *Biology and Philosophy* 8: 423–39.

Wright, R. 1995. *The Moral Animal.* Boston: Little, Brown.

12

Biology and Value Theory

Robert J. McShea and Daniel W. McShea

A THEORY OF VALUES

Although this century has produced more, and more varied, ethical and meta-ethical theory than any other, even our more educated and intelligent people are simply embarrassed when asked how they justify the value choices and commitments they make. We could well use a credible superstructure of facts and concepts within which we might carry on intersubjective and intercultural discussions of value differences, discussions that would offer some reasonable prospect of eventual agreement. What follows here is the outline of such a superstructure, a biologically based, naturalistic, species-universal, and prescriptive value theory (McShea, 1990). The theory is designed to answer such questions as these: Can a value statement be true? If so, in what sense and for whom? How can a value statement have prescriptive force?

The theory is an update of a philosophical ethical tradition that includes Aristotle, Spinoza, and especially Hume, who set forth a naturalistic, biologically based account of human nature and the meaning of life. Modern human nature theorists, with whom we would expect to find much common ground include Mackie (1977), Murphy (1982), Ruse (1986), and occasionally Midgley (1978), although most probably would not concur in the understanding of Hume on which the theory is based.

Six Value Bases

As the greatest success of science is not the discovery of this or that truth about things, but the learned ability to think scientifically, so the reward for the study of good value theory is not the discovery of moral laws or truths,

For discussions, careful readings of the manuscript, and/or inspirational suggestions we thank H. Alker, G. Cowan, W. Fikentscher, P. Klopfer, R. Masters, M. McGuire, N. McShea, S. McShea, E. van Nimwegan, D. Raup, D. Ritchie, M. Ruse, S. Salthe, and D. Smillie.

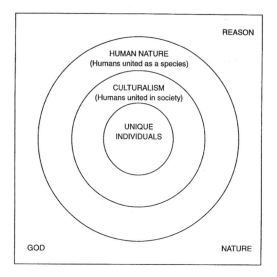

The six possible bases for value judgments.

but the habit of leading an examined life (which is morality itself). Similarly, as we can distinguish real science from pseudoscience by noting the incoherences and empirical failures of the latter, so we can arrive at better value theory at least partly by comparative study of value theories in general.

What truth a value judgment may have cannot be self-evident; it must find its validation outside of itself. That is, it must have some external basis. There are only six possible bases for value judgments (see figure) (McShea 1990), and thus we can classify value theories according to which of the six bases they claim. Further, each theory must begin by defending its basis against the claims of the alternative value bases. We begin by doing just that.

First, it is necessary to acknowledge the existence of the nihilist value thesis, the claim that values have no basis at all, cannot be true, perhaps cannot even be false, but only meaningless. Value nihilists assume that values, in order to exist, must be objective, external to humans. They therefore argue, for example, against a God-based ethic by offering reasons to believe that God does not exist. But to defeat an emotivist value theory, such as the one to be defended here, they would have to show that humans do not have feelings, and this they do not even attempt. For other arguments against nihilists, see Hume (1902).

Of the six possible value bases, three are human, and three are humanly transcendent: God, Reason, and Nature. There are good reasons to think that none of these last three will do.

God, quite aside from the question of his existence, cannot serve to validate value judgments simply because we have not, after all these years, found a way to determine what it is that he wants us to do or not do. Hobbes asks those who believe that God has spoken to them directly how they can distinguish between "God spoke to me in a dream" and "I dreamed that God spoke to me." As for the greatest number of believers, who cannot claim that God speaks to them directly, they must rely not on his authority but on the authority of those who claim to speak to and for him (Hobbes 1960, p. 96). An old religious tradition reminds us that the Devil, or any clever charlatan, can perform "miracles" and quote sacred texts for his own purposes.

Reason, in the present context, must be taken to mean not a mushy "reasonableness" but strict logical entailment. Unfortunately for the prospects of a rational ethic, it is now well established that we cannot reason from value-free premises to normative conclusions. As Hume put it, "It is not irrational for me to prefer the utter destruction of myself and all of those I hold dear to the causing of the least inconvenience to an Indian" (Hume 1888, p. 469).

Nature, as in "natural law," cannot serve as a source of validation for value judgments. The is–ought, fact–value dichotomy forbids it, and our current worldview makes the resort to teleology in nature unacceptable. Finally, if the good is the natural, then the bad is the unnatural. But "unnatural" is only an indicator of personal disgust; it is not an objectively existing natural quality.

If we can find a basis for human value judgments, it will have to be found in the human sphere, and so we must decide what image of humanity will govern our thinking. We can understand members of our own species as unique, Cyclopean individuals, each generating his or her own beliefs, meanings, and values. Or we can see them as so thoroughly integrated into one of many diverse cultures that "the individual" becomes, as one major nineteenth-century philosopher put it, "a fiction" (Bradley 1962). Or we can understand the members of our species to be united by their possession of a common species perceptual system and common emotional psychology.

The Unique Individual

Of those who claim to be philosophers, only Protagoras (each man is the measure of all things) and Sartre (being precedes essence, morality as authenticity) appear to defend the unique-individual value basis. Other thinkers and common sense agree that when each of us "does his own thing," or is authentically unique, then the lives of all of us will be nasty and very short. Quite a few perhaps excessively young-of-heart people think they believe in this value basis; they are mistaken.

Robert J. McShea and Daniel W. McShea

Culturalism

For a long time, the dominant value position among educated people in the West has been culturalism, the thesis that all values are cultural in origin, scope, and authority. Culturalism rests on a single but extremely powerful intuition: Our location in cultural time and space dominates, or even determines, what we ordinarily think of as our private and personal internal lives. When we consider that the metaphysical arguments for culturalism, or organicism, that were so persuasive in the writings of Hegel, Marx, and Bradley do not carry much weight these days, and that contemporary culturalists do not bother to defend their position or even respond to very serious criticisms of it, we are led to suspect that we are dealing not with a theory but with a cultural prejudice, a faith.

Somehow, culturalism, also called ethical relativism, has come to be identified with liberal enlightenment, if not with simply being educated. It is said to undermine dogmatic, "outmoded," conventional morality, to imply toleration for social and other deviants within our own culture, and to imply sympathy for exotic and fragile alien cultures.

Now all right-thinking people are opposed to dogmatic and outmoded moral conventions, but the culturalists have a problem here in that, for them, all conventions are dogmatic, and arbitrary as well, and it is difficult to see where the value judgment involved in the word "outmoded" is based. As for toleration and sympathy, consider the following problem. Suppose that it is part of our culture to despise and attack social deviants or members of some other culture. And suppose we agree that this aspect of our culture is intolerable. If we are not culturalists, we have some arguments, more or less plausible, to support our view. But if we are culturalists, we have none. By their own principle, culturalists have no ground on which to stand to make a value judgment different from that of their culture. In fact, culturalism seems a natural basis for a deeply conservative view of politics and society. Worse, when we learn that our social and political values are completely the products of our ephemeral and local culture, we lose all faith in them. At that point, individuals become anomic, and societies anarchic.

Such dreadful consequences never happen, simply because culturalists do not even begin to mean what they say. They bravely generalize from their original intuition into the power of culture, but when faced with the anomalies of that generalization, they surreptitiously introduce just enough human nature theory to save the appearances.

Human Nature Value Theory

The remaining possible value basis, variously called the "emotive," "human nature," or "psychological" basis, is the one defended here. Its supporters have to argue that feeling or emotion is the sole possible motivation to action, that all or almost all humans have the same evolved species-specific feelings, each one perhaps keyed to some general or stereotypical situation that has had survival and reproductive implications in the history of our species (although some caveats later). Our feelings are, to a first approximation, species-universal, and thus our evaluations (similarly species-universal) will at least roughly coincide with those of all other humans. Thus it ought to be possible to know enough about the circumstantial facts and about our own and others' feelings to be able to predict with fair success how we and others will feel, and therefore judge, in this or that envisioned situation.

Two disclaimers are necessary. First, feeling causes behavior, but does not uniquely determine behavior. Thus, our evaluations are universal, but behavior (including the verbal behavior associated with statements about values) varies widely. Later we will discuss the relationship between feeling and behavior.

Second, the human nature theory is a naturalistic theory and as such has been thought refutable on the ground that the languages of value and fact are radically disjunctive. The fact–value (or is–ought) disjunction is valid. One cannot reason, deductively or inductively, from an objective external fact to a value (although notice that by the same admirably strict criteria, one cannot reason from a fact to a fact either). But the fact–value dichotomy is an intellectual distinction, not a separation. In Kantian fashion we may say that facts without values are of no interest, and values without facts make no sense. If science, as Ayer would have had it, is the sum of what we know about the real world, then values, if not a part of the real world "out there," may have to be considered a branch of rhetoric, a kind of static or gibberish that degrades communication. On the other hand, if the universe can be understood only as value-free, as literally worthless, then knowledge about it, science, will also be worthless. Such paradoxes do not discredit the fact–value gap, but they do push us to discuss the *practical* relationship of facts to values.

Hume, having discovered the is–ought gap, went on to produce his practical solution to it by reminding us of the distinction (also made by Galileo) between the primary and secondary qualities of the world of human experience (Hume 1888). Some facts, such as the mass and extension of an object, are objective in that they do not depend for their existence on the presence of an observer. But some other facts, such as sound, color, and odor, *do* depend for their existence on the presence and attention of an appropriate observer and are therefore subjective. Thus the objective physical sciences – which pride themselves on speaking strictly – strictly speaking cannot allow or recognize subjective terms like "sound" and "color," but only objective terms like "sound waves" and "light waves." Hume goes on to propose the idea that emotions are to values what the eye is to color, or the ear to sound, that values are subjective facts. "Good" and "red" are equally meaningless to a physicist *qua* physicist, but they are facts, however subjectively so, to an appropriate organism. This is how the connection is made between biology and value.

Animals, including humans, use an incredibly complex collection of evolved devices to receive information about the external and internal worlds and to interpret them (not in the interest of objective truth, but in the interest of individual survival). They react to those interpretations with feelings that have evolved to motivate the animal to some very general courses of action, as discussed later. We live in a world of commingled primary and secondary qualities, of hard rocks, interesting colors, and arousing passions. For us, the goodness and color of an orange are as real and as interesting as its specific gravity.

Given that secondary qualities are subjective facts, we must ask, subjective in relation to what? It seems clear that for species-normal humans, colors, sounds, and the elemental feelings are species-subjective facts and therefore objective facts for individual humans. The vast amount of interpersonal and intercultural emotional communication we observe seems to rule out the possibility that feelings can be unique in each individual or to each culture. We experience redness and parental concern not because we are that kind of unique person, nor because we are Americans, but because we are human.

Value judgments are the products of particular feelings experienced in the context of particular perceived facts. It follows that because all or most humans have about the same elemental feelings and live in much the same factual world, voluntary agreement on value judgments not only is possible but should be common. Even a common morality is possible, and when arrived at it will be apprehended as objectively true, even though, or rather because, it is species-subjective.

As scientific judgments are falsifiable predictions about future experience of the objective, external world, so value judgments are best understood as falsifiable predictions about future emotional experience in our species-subjective world. "Save for a rainy day" is a prediction that if we do not, we shall have some very unpleasant feelings later on. "Eat, drink, and be merry, for tomorrow we die," contradicts the rainy-day hypothesis. As is often the case with value judgments, it is not easy to determine conclusively which side is right, but there are defensible answers, for the variables are all facts about death and feelings. The human nature value theory is like all other serious value theories in that it tells us how to go about solving value problems, but does not guarantee that we shall actually solve them.

312

The human nature value theory may seem suspiciously pleasant. It says
that you cannot act otherwise than on the urgings of your feelings and that
the only curb on each feeling is the power of other feelings. Further, if your
feelings have evolved to favor your interests, you may safely do whatever you
feel like doing. Enjoy! Unfortunately, as the happy phrase "this is the best
of all possible worlds" is inevitably followed by "of which every element is
a necessary evil," so "do what you will" is inevitably followed by "but make
very sure that it is what you *really* will." And in determining what you really
will, fear promotes the use of intelligence, Hobbes's "the scout of the passions" and Freud's Ego or Reality Principle.

Incapacity or external restraint prevents infants from doing much harm
to themselves or others, but every human over the age of two has had his or
her fingers burned, has learned to delay action, to make sure that what he
or she at first wants is also what will be wanted when further feelings have
been consulted. This is not an instance of reason controlling passion – there
are no such instances. The repeated experience of pain teaches us to value
prudence; prudence counsels delay; during the interval of delay our present
situation is located in larger factual and feeling contexts, and the probable
consequences of various courses of action are explored. To each set of consequences, each scenario, we react with emotions, with feelings. This is a
very simplified account. Introspection reveals complex swirls of feeling as
the facts and probabilities of various scenarios are reassessed.

The human nature theory is poor at identifying moral absolutes; it is much
better at identifying moral persons. Aristotle's argument that morality is
what a moral person does is only superficially circular (Aristotle 1954). Central to being a moral person is a strong sense of identity, of one's self as importantly the same through time and space. My present self comments critically on my earlier actions and holds itself responsible before the judgments
of my future self. A moral person is one who not only does what she wants
to do – everyone does – but also is concerned that what she wants to do at
any moment is also what she really wants to do in the light of this total spectrum of elemental feelings, each represented in the strength that it normally
has. Hume refers to this collected self as the "calm state of passion," as "the
parliament of the feelings" (Hume 1888).

The sense of identity is necessary for any long-range action or any action
involving elements of danger or pain, for we must know beforehand that the
"I" that will exist after the action will be the same "I" (more or less) that initiated and endured it. Identity is the internal gyroscope that makes it possible
for us to repress, divert, or delay ephemeral present passions. Those who
achieve a strong sense of identity will make great efforts to preserve it; we
sense that with it we matter, are real, and that without it we are nothing. Nor

313

must this be thought of as a selfish morality, for some of the strongest feelings we have are interests in the welfare of other persons.

How a moral judgment can be prescriptive is sometimes thought to be a puzzle. Most simply, the sense of obligation begins at the point where we have delayed action until our major enduring feelings have had a chance to fight it out fairly and we have decided to act in a certain way. We feel obligated to take that action, however sadly we feel about the defeat of lesser passions. Not to do so would be to act contrary to our interests, to do what – overall – we would not want to do, to lapse into a chaos in which our best abilities would be negated. The cry of an endangered identity is heard in Luther's "Here I stand; I can do no other." Of course, he could have done other, but for him that would have represented the loss of himself, the disintegration of a self he had patiently built up over a lifetime.

Value Statements and Judgments

There is only one possible motivation to make a value statement, and that is the occurrence of a feeling. Many different writers agree, and they are called emotivists. But they vary: The early view of Ayer was that all values are feelings, and he went on to note that feelings are different in each person and at different times in the same person. Ayer concluded that feelings and value judgments are therefore trivial and of no consequence. Other emotivists agree that feelings originate all values, but argue that it is up to reason, decency, or fear to control them. Here, however, we follow Hume in the opinion that all or almost all humans share a common spectrum of feelings, each arousable by a particular type of gestalt. Further, we agree with Hume that the feelings rule absolutely, with no hindrance from anything else. We are utterly incapable of doing anything other than what, finally, we want to do. If we seem often to do some things with great reluctance, it is because we have conflicting feelings, and the victorious feelings have won out by a rather narrow margin.

Given this understanding of the feelings, it is fairly easy to see how an individual can figure out for herself a working body of value judgments, or guides to action. She comes, in time, to understand the relative strengths of those feelings that are likely to come into conflict with each other, learns more (through personal and vicarious experiences) about the probable consequences of various actions in various circumstances: what a tangled web we weave when first we practice to deceive, that honesty is the best policy, and so on. This does not mean that an individual can achieve total certainty, even in personal morality, because there is always more to learn, there are always gray areas.

What is the point of making value judgments, or more precisely, what is the individual's ultimate goal here? It is not pleasure, nor even happiness. Rather, he is trying to avoid taking courses of action that would result in the actual defeat of feelings that in fact are the strongest and most enduring in him.

Such, in outline, is a contemporary version of the traditional human nature ethical theory. For a longer discussion, see McShea (1990). It is naturalistic, biological, and open-ended, without being vacuous. It also seems peculiarly suited to the present. Following Riesman's categorical division of morality into the customary or "other-directed," the "inner-directed," and the "autonomous" (Riesman 1962), we may say that almost everyone in the distant past was subject to a customary morality, that this has been partially replaced by inner-directed moralities, and that we are seeking a morality of autonomy. Both other-directed and inner-directed moralities have depended, in the first place, on long-term social stability or "backwardness," and, in the second place, on a kind of stern upbringing that is no longer prevalent and apparently cannot be revived. The general moral anarchy that might seem likely to follow those developments is unlikely to occur on a wide scale, for, given our native intelligence and feelings, we are a species that generates morality as needed. The aim here is merely to offer aid to those who are embarrassed about our contemporary confusions concerning the basis of our morality.

A THEORY OF FEELING

Underlying human nature ethical theory is a theory not of behavior but of motivation or feeling. Crucial to the theory is a clean conceptual separation between brain states (or brain activities), which include feelings and all of cognition, and behavior, which refers only to motor activity, physical movement, or action (and in humans includes speech).

In most animal species, behavior can be understood (at a high level of analysis) as a response to perceptions, to external stimuli. Flatworms undulate toward darkness in response to bright light. Honeybee foragers respond to food discoveries by returning to the hive and performing a complex dance that directs their fellow workers to the food. Tap a human's patella, and the knee will jerk. In such cases, the connection between stimulus and behavior is fairly tight, meaning that given the stimulus, the motor sequence of the response is fairly predictable and relatively invariant.

But now consider the following behavioral sequence: A lioness lies low and alert in the tall grass, not far from a herd of grazing zebras. The prey have become restless, as several of her fellow huntresses have moved into positions along the herd's flanks. The attack is launched from one of the flanks, and she crouches as the herd wheels in her direction. She picks out a juvenile zebra headed her way and tenses for action. As it streaks by her, she springs, and misses.

Hunting behavior in lions is not a relatively invariant motor sequence like the described behaviors in flatworms and honeybees. No particular hunting behavior (i.e., no particular motor sequence) is hard-wired in lions. Indeed, it almost could not be, because too many variables are involved. The terrain is uneven and highly variable among hunting locales. The behaviors of individual prey animals are also highly variable and unpredictable in detail, especially on the run. In sum, each hunt is unique, and there can be no single preprogrammed sequence of physical movements that will work every time, or even most of the time.

Behavior of this kind is not preprogrammed at all, but rather is feeling-driven. In colloquial terms, a feeling is an emotional state, a state of arousal or passion, a state of dissatisfaction, of wanting, or of desiring – in other words, a motivated state. A "feeling" refers not only to a state of strong or uncontrolled passion, such as might result in an affective or energetic display of some kind, but also to a state of calm passion, one that might not produce any overt affect or behavior at all.

More formally, a feeling is an intermediate mental structure, undoubtedly a dynamical one, lying in the behavioral causal chain somewhere between stimulus and response – for the lion, between, say, the internal stimulus of hunger (or the external stimulus of sighting the zebra herd) and the almost limitless number of different motor sequences that can result. The crucial property of these structures is that they dictate the general character of the end result of a behavioral sequence, but not the sequence itself; they dictate ends, but not means. We propose that they constitute the proximate causes for many significant behaviors, including almost all social behaviors, in mammals (at least), including humans.

Thus the proximate cause of hunting in lions is a feeling. Of course, survival is also a cause of hunting, in the sense that survival was why the hunting motivations evolved in lions, but it is a cause fairly far removed from the action. It is not the reason that a particular lion chooses to hunt at a given moment. More proximately, the cause of hunting is likely to be hunger, but this, too, is somewhat in the background; the lion does not actually experience hunger during the hunt, or at least she need not. More proximately yet, a lion hunts because she wants to. She experiences a motivation, a feeling, or an inclination to hunt.

Actually, the lion probably experiences a complex of several elemental feelings, including a certain amount of fellow feeling, an inclination to be near to and interact with her pridemates, and perhaps some desire to elevate or maintain her social rank within the pride by performing well during the hunt. Other feelings may be involved; stalking and pursuit may be desirable for their own sakes and may have special and distinct motivations. The feelings are numerous, and many may be in play at once.

It may be helpful here to say what feelings are not. A feeling in our sense is not a mood (i.e., not feeling in the sense of "feeling good"), nor is it a means of communication. It is not a presentiment, intuition, insight, or extrarational understanding of any kind. Feelings in these last senses are sources of facts (or falsehoods) and thus are not motivations, although like all facts they can arouse motivations. Further, feeling is not cognition, nor any aspect of it nor adjunct to it, such as learning or memory. Of course, hunting in lions requires a considerable amount of cognitive machinery and power. A lion must learn how to advance silently toward a herd, to estimate and remember the abilities and propensities of her pridemates as well as those of the prey, to calculate the angles in heading off prey, and so on. But these are the tools of hunting, not the causes of hunting; they explain how a lion is able to do what she does, not why she chooses to do it.

Apparently, almost all organisms have some preprogrammed behavior. Sunflowers follow the sun across the sky. Soft-bottom-dwelling clams burrow rapidly when disturbed. Humans blink in response to sudden loud noises. Long-chain, preprogrammed behavior is best known in the social insects, but it is present in other groups as well. In mammals, long-chain preprogramming seems to be absent, and instead much behavior – at least the more complex or compound behaviors – seems to be feeling-driven. These generalizations are obviously speculative; we cannot know for certain what a lion feels during a hunt, nor can we know that a honeybee's behavior is entirely dispassionate. But in any case, nothing depends on their correctness (at least for nonhuman species), because the point is just to explain what we mean by feeling.

We leave for another occasion a discussion of the relationships between feeling (as we use the term) and terms from the historical literature of psychology and ethology with overlapping meanings, such as instinct and drive. We also defer a comprehensive discussion of the relationships between our view and other theories of behavior (although see the later discussion). The present analysis draws on many sources, but its main inspiration has been studies of instinct by Fletcher (1966) and studies of the feelings by Solomon (1976), as well as Wimsatt's (1986) notion of entrenchment and Salthe's (1993) interpretation of development.

317

Robert J. McShea and Daniel W. McShea

Situation Dependence

A mother cat faced with the approach of a large dog toward the hiding place of her litter could experience what might be called a brood-defensive feeling. Whether or not she experiences that feeling will depend on the situation and on how she interprets it. A defensive feeling might not be evoked if the dog's distance from the kittens is still large, or if the dog seems friendly, or if the cat knows from past experience that this particular dog has little interest in kittens. Under the right circumstances, she might well feel playful rather than defensive. Feelings are emotive reactions to situations, and which feelings are evoked depends on the details of the situation.

Notice that the use of awkward expressions, such as "brood-defensive feeling," is a symptom of the inadequacy of language for describing feelings. As with colors, it is much easier to point to them than to describe them. For example, we might say "I mean by 'red' the color you see when you look at this paint chip." Similarly, "I mean by 'parental feeling' the feeling you experience when your child is crying." Obviously, this would not work if we did not all have very similar visual responses to the various wavelengths of light, and likewise very similar feeling reactions to classic, evocative situations.

Now suppose that the dog's approach to the cat and her litter is hostile. In addition to a brood-defensive feeling, the cat may experience a fear for her own safety. If so, she experiences both feelings at once, or nearly so, and the result is likely to be an internal conflict, a struggle for dominance between the two feelings, one perhaps urging her to run, and the other to stay and fight. She is torn. One feeling will inevitably be stronger than the other and will eventually triumph and cause some behavior. Which feeling will be stronger? The answer is once again that it depends on the details of the situation, such as the proximity of the dog and her past experiences with it. Just as the situation determines which feelings will be evoked, so it determines the relative strength of each one.

Likewise for humans. Obviously, we have greater cognitive powers than cats, which gives us the ability to pursue longer and more detailed imaginative sequences, to react with feeling to a wider range of possible situations. We also seem to be able to invest situations with a greater variety of interpretations or meanings, perhaps in turn evoking a greater range of feelings. But a period of indecision is for us, as for the cat, nothing more nor less than a struggle of feelings for dominance. And behavior results for us, again as for the cat, only as a result of a triumph of one feeling, or a coalition of feelings, over all others.

The process of decision making illustrates a second conceptual separation we are making, that between feeling and rationality (in addition to that

318

between feeling and behavior). By "rationality" we mean logical entailment, or calculation, of the sort useful in inferring consequences, estimating risks, and so on. Consider an extreme case: Suppose I contemplate driving my car down the highway at twice the local speed limit. If I am rational and well informed, I shall assess the risk and conclude that I am likely to crash and die. But my rationality has no further comment, because it is value-free and has no preferences, even for life over death. The preferences are the feelings. The imagined crash scene evokes an aversive feeling, a strong motivation to avoid such consequences, and as a result I will probably decide not to drive so fast. The feeling does not follow logically from the imagined consequences, but only experientially.

Thus, in our conceptual scheme, a decision can be said to be rational only in the sense that (and only when) our feelings have been provided with a competent evaluation of the facts and assessment of the likely consequences. But neither the preferences for this or that consequence nor the resulting decision can be rational (or irrational, for that matter), because rationality delivers no impetus, it has no driving force, so to speak, and thus it cannot drive decision any more than it can drive behavior. Only feelings can do that.

This reasoning should make it clear that a feeling is not just one of many specialized brain mechanisms on the same functional level as the various cognitive devices (Barkow, Cosmides, and Tooby 1992). In other words, a feeling is not just one of many possible causes of behavior, along with one or more of the various cognitive devices, such as rationality, perhaps. Rather, feelings as we understand them are the *only* possible causes of complex behavior (in mammals, we speculate), and cognition is a set of specialized interpretative devices that act *only* under the direction of, and in the service of, the feelings.

Diversity and Independence of Feelings

The feelings are many. It is possible, of course, to produce a fairly short list that might seem comprehensive – fear, love, anger, sympathy, embarrassment, and so on – but we would argue that each of these has many different situation-specific subtypes or modes. The feeling produced in us by the rumble of thunder overhead is different from the feeling produced by a prolonged stare from a rival. Conventionally, the two might be lumped together under fear, but they are qualitatively different. That is, we experience them differently. And each has a very different range of behavioral options that will satisfy it.

A reductionist might argue that all are reducible to two elemental motivations: pleasure and pain. But pleasure and pain are not feelings per se; rather, they measure the degree of satisfaction of a feeling. Consider the desire for social recognition. One might be said to feel pleasure to the extent that this feeling is satisfied, and pain to the extent that it is not. But in any case, the desire for social recognition is a feeling that demands satisfaction on its own account, independent of any associated pain or pleasure.

Feeling and Behavior

Just as the object of a game does not uniquely determine the specific moves of the players, feelings do not uniquely determine behavior. (Indeed, satisfaction of the feelings is the "object" of behavior in roughly the same sense.) The cat in which a brood-defensive feeling has won out over fear for her own safety as the dog approaches her litter has many behavioral options. She might fight the dog, but she might also run to try to distract the dog's attention from her kittens. It depends on her previous experiences, on her past successes and failures with various strategies for protecting her kittens from dogs, on her limited ability to predict the consequences of alternative behaviors, and perhaps on her training, if she has had any, from other cats (or even from humans).

Likewise, in humans, feeling causes behavior but does not specify behavior. Thus, two hypothetical humans with identical feeling profiles, but having been raised in different cultures and having had different life experiences, would interpret a given situation in different terms and ultimately would present their identical feeling profiles with very different imaginative scenarios. The result is that they could be counted on to react to the situation, to behave, differently. Indeed, for humans, so vast is the range of possible imaginative worlds to which the feelings might be required to respond that behavior is effectively unconstrained. There is no behavior – including verbal behavior – of which humans are physically capable that some human will not, in some context, perform on the basis of some feeling.

Thus, we expect that there will be essentially no species-universal complex behaviors in humans. This claim might appear to be contradicted by the various lists of human universals that have been compiled, some plausible and convincingly argued (e.g., Brown 1991). In fact, it is not, because few of the entries on such lists are actually behaviors in the sense specified here, that is, they are not motor activities. Consider these commonly cited candidates for human universals: a desire for enhanced social status, a tendency to develop a belief in the supernatural, and strong mother–infant bonding. A desire for enhanced social status is a motivation, not a behavior. Cognitive tendencies and belief systems are not behaviors either; of course, speech about a belief is a motor activity, but speech, like the belief systems themselves, varies widely. What unites the various beliefs, what could be truly universal, is the common bundle of feelings they satisfy. Finally, bonding is not a behavior either. Rather, it is a summary term covering many different culture-specific behaviors, and again what unites them is the bundle of universal feelings satisfied.

The Feeling Profile

The feeling profile is the range and configuration of feelings that normal members of a species experience. In other words, the profile is a specification of the feelings that are evoked, each appropriately weighted according to the intensity with which it is experienced, in the full range of problematic situations that members of the species normally encounter.

Although we are the heirs to almost three thousand years of attempts to list, describe, and organize our understanding of the human feelings, we seem no better off than at the beginning. That is not surprising, for emotions, values, and secondary qualities in general are *qualia,* permanently out of focus for the quantifying eye, for the concept-oriented analytical mind. Thus, while it might seem appropriate here to try to describe in detail the human feeling profile, at least for illustrative purposes, it is clear that such an attempt could not succeed.

The profile is expected to be species-specific, in the sense that variations in the range and configuration of feelings are expected to be large among species (perhaps less among closely related species than among distantly related ones) and relatively small within species. Further, the human profile is expected to be unique, but no more or less so than the profile of any mammalian species. Every species has a unique feeling profile.

The profile may be adaptive in the sense that each feeling is (or was) functional in a specific problematic situation faced by individuals in a species' evolutionary history and that it may have evolved (presumably on account of its selective advantages) for that reason. Importantly, however, the claim that the feeling profile is adaptive for individuals is not a claim that all feelings are narrowly self-serving. At least in social species, many (if not most) of the problematic situations undoubtedly have been social, and to manage them effectively, a certain amount of genuine (situation-specific) sympathy for conspecifics undoubtedly has been advantageous. In any case, the argument here is not historically based. The role of the feelings in driving behavior does not depend on their adaptedness, either now or in the past, nor on any aspect of their evolutionary history.

The human feeling profile undoubtedly has changed on evolutionary time scales, but probably has not changed much in historical time, either within any culture or across cultures generally. This is an empirical claim, but we shall give no evidence for it here; later we shall offer some *theoretical* reasons to think it might be true.

In any case, the point is not to deny that *any* local or short-time-scale variation occurs. Among human individuals of a given culture, significant and randomly distributed differences in feeling profiles must exist because of differences in genes and in life experiences. And slight systematic differences may well exist among individuals of different ages, between the sexes, and even among cultures, although these probably are small compared with interspecies differences. The point of stressing the sense in which the profile is universal within a species – perhaps of overstressing it – is to emphasize that differences in feeling profiles are unnecessary to account for the enormous variations in *behavior*, both verbal and nonverbal among cultures.

The Entrenchment of the Feeling Profile

Development in organisms is at least partly hierarchical. Early in development, an organism consists of a relatively small number of structures that interact to give rise to more structures, that in turn give rise to others, in a widening cascade (Wimsatt, 1986). As a result, early developmental steps have more consequences, and more significant consequences, than later steps, and thus deleterious natural variation occurring in the early steps will be more strongly opposed by natural selection. In Wimsatt's (1986) terms, the early steps and the structures arising from them are "generatively entrenched."

In this view, structures that are generatively entrenched tend to be more general. They are the organism's foundation, so to speak, onto which the later and more specific features will be built (Salthe 1993). Such general and relatively invariant features are known as "bodyplans." In insects, for example, six-leggedness is a feature of the bodyplan. In vertebrates, the vertebral column is part of the bodyplan. In principle, the same concept could be invoked to describe the common and invariant features of a much lower taxon. For example, at the species level, in humans, bipedality might be a bodyplan feature.

We propose that feelings or motivations are the generatively entrenched structures of what might be called a species-level "behavioral bodyplan." They are general outlines of behavior, just as the bodyplan in its usual sense is a general outline of physical structure. And just as bodyplan structures are less variable among individuals of a given species than are the later-arising structures, feelings are less variable than behaviors. Behaviors vary enormously, both among individuals and within a single individual in the course of its lifetime. But feeling profiles do not. Members of a species share a common motivational structure for the same reason, and in the same sense, that they share a common early physical development.

Biology and Value Theory

The Nature–Nurture Problem

This understanding of motivation, we argue, accommodates both of the extreme positions in the timeless nature–nurture debate. The central insight of the "nurturists," the culturalists, is that human behavior varies enormously from one culture to the next, that behavior is almost completely plastic to environmental influences. This view is part of what has been called the standard social-science model of psychology (Barkow et al. 1992). The "naturists" contend that these differences are superficial, that beneath an outer layer of culturally conditioned responses lies a solid inner core, a human nature that is everywhere the same. Reductionist naturists go further, arguing that the details of the structure of this inner core are coded explicitly in the genes.

Both views have serious shortcomings. One of the most serious for the culturalists is that they cannot account for communication among cultures. Communication among disparate cultures would be impossible (not merely difficult, as we observe it to be) if no common basis for understanding existed, if different cultures produced brains that were as different as those of different species of animals, as different, say, as humans and blue whales. Culturalists take the common basis for understanding for granted. They ignore the fact that communication (except of the most superficial sort) requires a long list of situation-specific common interests, and thus a common feeling profile – a profile we almost certainly do not share with blue whales.

One of the most serious problems for the naturists is that they cannot account for the enormous differences between cultures. If a universal human nature exists, why is all socially interesting human behavior so variable? Why do the things that people say and do, and how they interpret the world, vary so much from one culture to the next?

The problem for both positions is the oversimplified and flawed model of behavioral causation that is usually invoked or assumed. For both, genes and environment (which includes culture as well as the unique experiences of individuals) combine in some unspecified way in the course of development to produce the brain. And the brain reacts to various present stimuli (externally or internally generated) by producing behavior. Culturalists emphasize the contribution of the environment to the structuring of the brain, while downplaying the genes, and naturists reverse that emphasis.

We suggest a different model. Stimuli are analyzed by the brain's cognitive structure to produce an interpretation or an understanding of a life situation in the form of narratives, images, or any of a number of devices. And it is this understanding that evokes the feelings. In any given situation, multiple feelings may be evoked, but eventually one feeling or coalition of feelings triumphs over all others and causes some behavior. In the terms of

Wimsatt's model, most of the content of the narratives and images that constitute our understanding is shallowly entrenched and therefore expected to be highly variable both within and among cultures. Fikentscher (1995) has called the cultural component of this cognitive variation the "modes of thought." More deeply entrenched and thus less variable is the feeling profile.

Importantly, both feelings and behavior have essential genetic and environmental (i.e., cultural) components, and differences in their contributions are not significant in this context. Indeed, information from the environment that is "expected" during development (e.g., during "critical periods") may be deeply entrenched and just as essential to the production of a normal feeling profile as any genetic information.

Thus one of the main virtues of our model is that it ignores the often misleading opposition between genes and environment (Oyama, 1985) and yet accommodates the central insights of both naturists and nurturists. That is, the naturists are right in that there does exist a core, universal human nature that is fairly impervious to cultural influence. However, this core consists not of universal behaviors – for there are almost none – but of universal motivations, a common feeling profile.

And the culturalists are also right in that all complex and socially interesting human behavior is enormously variable. In moving from one culture to the next, we may shift from one culturally structured system of interpretation and suite of behavioral options (i.e., one mode of thought) to a radically different one. The result is that in different cultures, entirely different worlds are served up to the feelings, and very different behaviors result.

To put it another way, culture can interact with the feelings to make people say and do almost anything. But this observation in no way contradicts the naturist insight. Behind the difference in behavior lies a common and relatively invariant feeling profile. And it is this common profile that gives us a basis for shared values.

SUMMARY

We offer for further discussion, and in abbreviated form, two theories: a theory of motivation and a metaethical theory.

Feelings

We argue that in mammals, at least, much behavior is caused by mental structures intermediate between stimulus and response. These structures are feelings or motivations. They cause behavior by providing general goals, but without specifying particular actions. The feelings are many, distinct, and situation-specific; the complete repertoire of feelings that members of a species normally experience, each weighted according to its situation-specific intensity, is the species feeling profile.

Mammals use their perceptual apparatus and intelligence to interpret the world and to anticipate future events. These interpretations and anticipations in turn evoke feelings, which motivate behavior. In any given situation – real or imagined – a number of feelings may be evoked, orienting the animal to a number of different purposes at once. The ensuing struggle among feelings for supremacy is the essence of decision making, and behavior (including verbal behavior in humans) is the result of the triumph of one feeling, or coalition of feelings, over all others.

Differences in how situations are interpreted, in how they are presented to the feeling profile, vary significantly among individuals, ultimately producing differences in behavior. In humans (at least), interpretative schemes, or modes of thought, also vary systematically among groups, accounting for cultural differences in behavior. But all of these variations in behavior are completely consistent with a species-universal feeling profile. Indeed, if the profile is somewhat entrenched in development, as we argue here, it is expected to vary little within a species.

The theory is obviously incomplete: What sort of mental structures are feelings? How are they represented physically in the brain? What is the mechanism by which feelings cause behavior? How can we detect feelings in other species? We can answer none of these now.

Metaethics

Of the six possible bases for validation of value judgments – God, Nature, Reason, the unique individual, culture, and human nature – we argue that only the last is defensible. The central claim of the human-nature position is that a sensible personal morality would constitute a lifelong strategy, one aimed at optimal satisfaction of our strongest and most enduring feelings. (Of course, if each desire, even the most trivial, could be instantly satisfied without prejudice to future claims of other desires, then morality would be of no use or interest.) In other words, morality is doing what you want, what you *really* want, that is, what you want consistently over a lifetime. And because the feeling profile varies little among humans, there is good reason to think that agreement on values, on wants, is possible among individuals and across cultures.

At the metaethical level, this essay seems to have two strong points and two weak ones. The strong points are (1) the revival of Hume's analogy of the feelings to secondary physical qualities and the consequent escape from the naturalistic fallacy and the fact–value, or is–ought, gap and (2) the establishment of the feelings, combined with a sense of identity providing continuity over a lifetime, as the basis of a personal morality. The weak points are: (1) Because we cannot pretend to precision in talking about the feelings, the most that any theory based on them can claim is that it is the most reasonable of the alternatives. Any kind of proof is out of the question. (2) If it happens to be the case that the radical uncertainties of our futures, the power of present passions, and cultural biases render all attempts to implement the theory in practice a waste of time, then the human-nature theory could be both true and useless.

In sum, the key to both theories is the feeling profile, a set of structures as fundamental to human nature as opposable thumbs, a bipedal gait, or any feature of our morphology. Arguably, the profile may be more fundamental: Imagine a human and a blue whale, with their feeling profiles switched. (Assume that the whale is conversant and intelligent, and that we share a common language with it.) With which would we find more common interests, with which more shared values? As we began to interact with them, to exchange views, their physical appearances and behaviors would begin to diminish in importance. We would soon notice just how little we had in common with the one in human form and begin to recognize the whale with human feelings as one of our own. More than any merely physical features, or any pattern of behavior, our feeling profile *is* human nature.

REFERENCES

Aristotle. 1954. *Nicomachean Ethics,* 1105b. Oxford University Press.
Barkow, J. H., Cosmides, L., and Tooby, J. 1992. *The Adapted Mind.* Oxford University Press.
Bradley, F. H. 1962. My station and its duties. In: *Ethical Studies.* Oxford University Press.
Brown, D. E. 1991. *Human Universals.* New York: McGraw-Hill.
Fikentscher, W. 1995. *Modes of Thought.* Tübingen: J. C. B. Mohr.
Fletcher, R. 1966. *Instinct in Man.* New York: Schocken Books.
Hobbes, T. 1960. *Leviathan.* Oxford: Blackwell.
Hume, D. 1888. *A Treatise of Human Nature.* Oxford University Press.
    1902. Enquiry concerning the principles of morals. In: *Hume's Enquiries,* 2nd ed., ed. L. A. Selby-Bigge. Oxford University Press.
Mackie, J. L. 1977. *Ethics: Inventing Right and Wrong.* Middlesex: Penguin Books.
McShea, R. J. 1990. *Morality and Human Nature.* Philadelphia: Temple University Press.
Midgley, M. 1978. *Beast and Man.* Ithaca, NY: Cornell University Press.

Murphy, J. G. 1982. *Evolution, Morality, and the Meaning of Life.* Totowa, NJ: Rowman & Littlefield.

Oyama, S. 1985. *The Ontogeny of Information.* Cambridge University Press.

Riesman, D. 1962. *The Lonely Crowd.* New Haven, CT: Yale University Press.

Ruse, M. 1986. *Taking Darwin Seriously.* Oxford: Blackwell.

Salthe, S. N. 1993. *Development and Evolution.* Cambridge, MA: MIT Press.

Solomon, R. C. 1976. *The Passions.* New York: Anchor Press.

Wimsatt, W. C. 1986. Developmental constraints, generative entrenchment, and the innate–acquired distinction. In: *Integrating Scientific Disciplines,* ed. W. Bechtel, pp. 185–208. Dordrecht: Martinus Nijhoff.

Notes on Contributors

Michael Bradie is Professor of Philosophy at Bowling Green State University. He is the author of *The Secret Chain: Evolution and Ethics* (State University of New York Press, 1994), as well as articles on evolutionary epistemology and various topics in the philosophy of science. His current research interest is the role of models and metaphors in scientific representations.

Raphael Falk has just retired from the post of Professor of Genetics at the Hebrew University of Jerusalem. His scientific research has focused on x-ray-induced mutations, population genetics, chromosome mechanics, and the developmental genetics of *Drosophila melanogaster.* For the past fourteen years he has worked in the areas of the history and philosophy of genetics (including the concept of the gene) and the social implications of scientific research.

Paul Lawrence Farber is Oregon State University Distinguished Professor of History of Science and Chairman of the Department of History at Oregon State University. He holds joint appointments in the Department of History and the Department of Zoology. He is author of *The Temptations of Evolutionary Ethics* (University of California Press, 1994) and *The Emergence of Ornithology as a Scientific Discipline, 1760–1850* (Reidel, 1982).

Jean Gayon is Professor of Epistemology and History of Life and Health Sciences at the University of Paris 7 – Denis Diderot. He is also a senior member of the Institut Universitaire de France and co-editor of *Revue d'Histoire des Sciences.* His major work, *Darwin et l'Après Darwin* (English translation, *Darwin's Struggle for Survival,* Cambridge University Press, 1998), is on the history of the hypothesis of natural selection after Darwin. Author of seventy articles on the history and philosophy of life sciences, he has also edited *Buffon 88* (Presses Universitaires de France, 1992), *Les Figures de la Forme* (with J. J. Wunenburger, Paris, L'Harmattan, 1992), *Les Sciences Biologiques et Médicales en France, 1920–1950* (with C. Debru and J. F. Picard, Paris, Editions du CNR, 1994), and *Le Paradigme de la Filiation* (with J. J. Wunenburger, Paris, L'Harmattan, 1995).

Myles W. Jackson received his Ph.D. in the history and philosophy of science at Cambridge University in 1991. Since then he has published numerous articles on the history of the biological and physical sciences in nineteenth-century Germany. His book *Spectrum of Belief: Fraunhofer's Artisanal Knowledge and Precision Optics* is being published by the University of Chicago Press. He is currently Assistant Professor of the History of Science at Willamette University, working on the relationship between music and physics in nineteenth-century Germany.

James G. Lennox is Professor of History and Philosophy of Science and Director of the Center for Philosophy of Science, University of Pittsburgh. He is the author of numerous essays on Aristotle's science and philosophy and on Charles Darwin and Darwinism. He is the co-editor of *Philosophical Issues in Aristotle's Biology* (Cambridge University Press, 1987), *Self-Motion from Aristotle to Newton* (Princeton University Press, 1994), and *Concepts, Theories and Rationality in the Biological Sciences* (University of Pittsburgh Press, 1995). He is a Life Fellow of Clare Hall, Cambridge, and a former Junior Fellow of the Center for Hellenic Studies. He is currently completing a translation, with notes, of Aristotle's *Parts of Animals*.

Daniel W. McShea received his doctorate in evolutionary biology from the University of Chicago in 1990. Subsequently he has been a Fellow in the Michigan Society of Fellows and a Fellow at the Santa Fe Institute, and currently he is Assistant Professor in Zoology at Duke University. His work on complexity and evolution has been reported by *The New York Times* and has been cited in popular books on evolution, such as Stephen Jay Gould's *Full House* and Roger Lewin's *Complexity: Life at the Edge of Chaos*. His major works include papers on the evolutionary trends in complexity and on methods for analyzing trends generally, published in the journal *Evolution*.

Robert J. McShea received his doctorate in political philosophy from Columbia University in 1967. Until his death in 1997, he was Emeritus Professor of Political Science at Boston University. His major works include *The Political Philosophy of Spinoza* (Columbia University Press, 1968) and *Morality and Human Nature* (Temple University Press, 1990).

Jane Maienschein is Professor of Philosophy and Biology and co-director of the Biology and Society Program at Arizona State University. She is the author of *Transforming Traditions in American Biology, 1880–1915* (Johns Hopkins University Press, 1991) and *100 Years Exploring Life, 1888–1988. The Marine Biological Laboratory at Woods Hole* (Jones & Bartlett, 1989), as well as editor and co-editor of a number of volumes in the history and philosophy of the biological sciences.

Diane B. Paul is Professor of Political Science and Co-Director of the Program in Science, Technology, and Values at the University of Massachusetts at Boston. She recently published *Controlling Human Heredity: 1865 to the Present* (Humanities Press, 1995); some of her essays have been collected under the title *The Politics of Heredity: Essays on Eugenics, Biomedicine, and the Nature–Nurture Debate* (State University of New York Press, 1998).

Robert J. Richards is Professor in the Department of History and the Department of Philosophy and is on the Committee on Evolutionary Biology at the University of

Chicago. He is the author of *Darwin and the Emergence of Evolutionary Theories of Mind and Behavior* (University of Chicago Press, 1987) and *The Meaning of Evolution* (University of Chicago Press, 1992). Currently he is working on a book with the tentative title *Romantic Biology: From Goethe to the Last Romantic, Ernst Haeckel.*

Michael Ruse is at the University of Guelph in Canada. His most recent book is *Monad to Man: The Concept of Progress in Evolutionary Biology* (Harvard University Press, 1996).

Phillip R. Sloan is Professor in the Program in History and Philosophy of Science and the Program of Liberal Studies at the University of Notre Dame. He directs the John J. Reilly Center for Science, Technology and Values at the University of Notre Dame. His research area is the history of the life sciences in the modern period, with publications on Cartesianism, natural history, and science and religion, with specific studies on Buffon, Lamarck, Darwin, and Richard Owen. He is the author of *Richard Owen's Hunterian Lectures: May–June 1837* (University of Chicago Press, 1992), and he is completing the editing of *Controlling Our Destinies: Historical, Philosophical and Ethical Perspectives on the Human Genome Project* (University of Notre Dame Press, 1998). His current research is on the interrelationships between German and British theoretical biology in the Victorian era.

Marga Vicedo is Assistant Professor at Arizona State University, West Campus. Her work in the history and philosophy of biology in the early twentieth century focuses on issues of genetics and eugenics. She explores epistemological and ethical issues, while also looking at issues of women in science. She has published many articles in a wide range of journals in both English and Spanish.

Peter G. Woolcock is Associate Professor in the School of Education at the University of South Australia. He is the author of *Power, Impartiality and Justice* (Avebury, 1998), as well as several articles on ethics, including "Ruse's Darwinian Meta-ethics: A Critique" (*Biology and Philosophy*), "Hunt and Berlin on Positive and Negative Freedom" (*Australasian Journal of Philosophy*).

Index

Allen, G. E., 228
altruism, 157, 164, 186, 188, 237,
    276–78, 281, 287, 289, 292–95,
    300–301
Andler, C., 155, 158, 160
anthropocentrism, 34
Aquinas, Thomas, 54–55
archetype theory, 125–27, 146
Aristotle, 54, 62
    Eudemian Ethics, 12n., 14n.
    Historia Animalium, 16–18, 20–22,
        24–25
    Metaphysics, 19–20
    Nichomachean Ethics, 10–19,
        23–26, 313
    Physics, 23
    Politics, 10–11
artificial selection, 157, 173–74
    see also eugenics
Ascheim, S. E., 155, 180
Ayala, Francisco J., 209
Ayer, A. J., 314

Bacon, Francis, 251
Ball, S., 284
Balme, David, 16n., 17, 19n.
Barkow, Jerome H., 319, 323
Bauplan, 108, 116
Beaglehole, J. C., 63
Bentham, Jeremy, 42, 47–48
Bergson, Henri, 202

Bertelsen, A., 260
Beyertz, K., 262
Blanckaert, Claude, 63
Blumenbach, Johann Friedrich, 105
Bonpland, A., 136, 145
Bory de Saint-Vincent, J.-B. G.-M., 52
Bougainville, Louis de, 63, 65
Bowler, P., 86, 225
Boyd, R., 284
Boyer, Alain, 184
Bradie, Michael, 232
Bradley, F. H., 309
Bräuning-Oktavio, H., 105, 108, 110
Broad, C. D., 210–11
Broadie, S., 14n., 15n.
Broberg, Gunnar, 59
Brown, D. E., 320
Buffon, George-Louis Leclerc, Comte
    de, 59, 61–62, 64, 67, 71, 85
Bulhof, Ilse, 114
Burckhardt, Jacob, 161
Burkhardt, Richard, 261, 270, 272
Burlamaqui, Jean-Jacques, 57
Burnyeat, M., 14n.
Bury, J. B., 201
Butler, Bishop Joseph, 34, 41–43,
    45, 48

Cabanis, Pierre-Jean-Georges, 70–71
Cannon, Susan, 134
Cardiazol (Metrazole), 268

Christianity, 181–82, 189, 240
Comte-Sponville, A., 156
Conklin, Edward Grant, 227, 229,
    239–42, 246, 248–50, 252
conscience, 45–47, 56
control, 201
Cooke, K. J., 230
Cooper, J., 11n., 20n.
Copans, Jean, 70
Corsi, Pietro, 52
Cosmides, Leda, 319, 323
Croce, P. J., 251
Crocker, Lester, 54
Crook, P., 198
Crooks, G. R., 34, 41
culturalism, 310
Cuvier, Georges, 85, 116

Darrow, Clarence S., 244
Darwin, Charles, 52–53, 61–62, 74, 87,
    113–48, 154–93, 227
    Descent of Man, 11, 16, 52, 138–39,
        141–44, 175, 232, 277
    Origin of Species, 76, 125–35, 161
Darwinism, 32, 48, 84
Davenport, Charles B., 227–28,
    234–36, 241–42, 245, 251
Deichmann, U., 257–58, 261, 264
Dennett, Daniel, 156
Depew, D. J., 11n., 20n.
Descartes, René, 57, 62
determinism, 243, 246–49
Dewey, John, 226, 231–32, 234, 238,
    241, 243, 249–50
Diderot, Denis, 59, 61
Diggins, J. P., 251
Duchet, Michèle, 63

East, Edward Murray, 227–28, 236–37,
    241–42, 246, 251
Ecker, A., 101
Eddy, John, 64
egoism, 276, 280–81, 289, 300–301
embryology, 100, 128–30

eugenics, 155, 173–85, 208, 231, 236,
    242
euthanasia, 155
evolution, 84–85, 87
evolutionary ethics, 86–96, 113–48,
    198–221

Farber, P. L., 225, 232
feeling, 315–24
    profile, 321–22, 326
Ferguson, Adam, 34, 43, 45
Fichte, Johann Gottlieb, 98
Fikentscher, W., 324
Fisher, Sir Ronald A., 261
Fletcher, R., 317
Foot, P., 302
force, 160, 169
free will, 245
Frost, Alan, 63

Galton, Francis, 175–76
Gampel, E. H., 284
genetics, 225–52, 264–67
Ghiselin, Michael, 113, 135, 146
Gifford, M., 13n., 19n., 26n.
Goethe, Johann Wolfgang von, 99,
    102–8, 110, 120
Gottesman, I. I., 260
Green, Joseph Henry, 126
Greene, John, 198, 200–201, 217
Greg, W., 175
Gunther, F. K., 262

Haakonssen, Knud, 56, 70
Haeckel, Ernst, 116, 159, 176, 178
Haldane, J. B. S., 201
Harman, Gilbert, 284
Harvey, E. N., 230
Heilbron, J. L., 272
heredity, 175, 177, 201, 203
Hirsch, W., 262
Hitler, Adolf, 156, 178
Hobbes, Thomas, 309
Hollingdale, R. J., 155

Index

Hollinger, D., 226
Horgan, T., 284
Hudson, W. D., 209, 280, 284
Hughes, William, 284
human nature value theory, 310–14
Humboldt, Alexander von, 113,
    119–25, 136, 145–46
Hume, David, 42–44, 48, 74, 199, 209,
    219–20, 233, 283, 302, 307, 309,
    311, 313
Hutcheson, Francis, 34–39, 58
Huxley, Julian Sorell, 198–221
    Evolution and Ethics, 203–4
    *Individual in the Animal Kingdom,
    The,* 202
    papers, 213
    *UNESCO: Its Purpose and Its Phi-
    losophy,* 207
    unpublished lectures, 202
Huxley, Leonard, 220
Huxley, Thomas Henry, 135, 199, 232
    Evolution and Ethics, 203–4, 219,
    232
    Hume, 219–20

independence, 201

James, William, 227, 243–44, 250–51
Jamin, Jean, 70
Jardine, N., 105–6
Jennings, Herbert Spencer, 227, 229,
    237–39, 242, 246–48, 250
Jones, D. F., 229

Kant, Immanuel, 98, 100, 102, 105,
    116–18
Kater, M. H., 261
Kaufman, W., 155
Kelley, A., 198
Kimmelman, B., 225
Kingsland, S., 229
Koch, Gerhard, 269–70
Kroll, J., 262
Kullmann, W., 11n., 20n.

La Vergata, A., 92
Labarrière, Jean-Louis, 20, 21n.
Lagarde, A., 86
Lamarck, Jean-Baptiste de Monet de,
    52, 69–76, 84
Lanessan, Jean-Marie-Antoine de,
    86–96
Laurent, G., 73–75, 86
law of compensation (Goethe), 107
law of polarity (Goethe), 104–5
Leclerc, George-Louis, *see* Buffon
Lennox, J. G., 19n.
Lenoir, T., 105
Lifton, Robert Jay, 258–59
Linnaeus, Carolus, 59
Locke, John, 41
Loeb, Jacques, 244, 247
Lovejoy, Arthur O., 231
Lysenko, T. D., 259–60, 266

MacDowell, E. C., 228
MacIntyre, Alisdair, 54
Mackie, J. L., 200, 217, 307
Mackintosh, James 137–38, 140,
    143
Macrakis, K., 258
Maienschein, Jane, 230
Malthus, Thomas, 136–37, 139
Manier, Edward, 52
Marshall, P. J., 63
Martin, M. W., 265
Martinsen, D., 296–300
mechanism, 103–5, 130, 244, 247–49
Mehrtens, Herbert, 271–72
Mengele, Josef, 266
Midgley, Mary, 43, 307
Mill, John Stuart, 139
Milton, John, 133
Mitman, G., 199, 231
Mitscherlich, Alexander, 271–72
Moore, E. C., 234
Moore, G. E., 198, 208, 210, 232–33,
    283
morality as illusion, 288–92

Moran, Francis, III, 69
Morel, J., 65
Morgan, S. R., 98–99
morphology, 105–6, 128
Moul, M., 266
Mullen, P. C., 100, 109
Müller-Hill, Benno, 258–59, 266, 269–70
Münster, A., 155
Murphy, J. G., 200, 217–18, 307

Nachtsheim, Hans, 262–64, 267–72
Nagel, Ernest, 209
natural ethics, 214, 281
natural history, 59–60, 71, 74, 101
natural knowledge, 102–8
natural law, 309
    tradition, 53–59, 61
natural selection, 130–32, 141, 157, 167, 175, 181
natural virtue, 11–16, 18, 20, 24–26
naturalism, 160, 227, 243, 311
naturalistic fallacy, 147, 208, 217, 220, 233, 282–85, 292, 297, 311, 326
nature–nurture, 323–24
Naturphilosophie, 98–101, 103, 110, 115–17, 130
nazism, 155, 264, 270–71, 289
neo-Lamarckianism, 86, 161, 170
Nietzsche, Friedrich, 154–93
    Antichrist, The, 171
    Beyond Good and Evil, 177, 188
    Ecce Homo, 159
    Gay Science, The, 162, 168, 172, 178
    Genealogy of Morals, On the, 154, 171, 186–87, 189–90
    Human All-Too-Human, 162–64, 177
    posthumous fragments, 160–64, 169, 175, 177–78, 188, 191–92
    Will to Power, The, 162, 166–67, 169–70, 178, 192
    Thus Spoke Zarathustra, 163, 171
    Twighlight of the Idols, 162, 168, 173

Nisbet, H. B., 107
Nitecki, M., 201

Oken, Lorenz, 99–102, 108–10
organic laws, 103–5
Owen, Richard, 126–28
Oyama, Susan, 324

Paley, William, 136–37
Paul, Diane B., 228
Paulen, R., 110
Pauly, P. J., 229, 247
Pearl, Raymond, 245
Pearson, Karl, 265
Pellegrin, P., 19n.
perspectivism, 158, 191–92
Peter, F. M., 208
Pfeiffer, E., 185–86
Pintschovius, K., 270
Pittenger, M., 198
Plato, *The Republic,* 276
Pluhar, Evelyn B., 49
practical intelligence, 11–15, 17–19, 23–26
pragmatism, 226, 231
Proctor, Robert, 259
progress, 162–63, 170, 201–3, 206–7, 209, 212–14, 221
Provine, W., 229
Pufendorf, Samuel, 56, 58
Putnam, Hilary, 284

Rachels, James, 32, 49
reason, 42–45, 67–68
recapitulation, 128–29
Rée, Paul, 159, 185–86
Reid, Thomas, 44–47
Richards, E., 100
Richards, Robert J., 198, 200, 225, 229, 232–33, 292–95
Richter, C., 155, 160–61, 175
Riesman, D., 315
Rodd, Rosemary, 32, 49
Roger, Jacques, 73–75

romanticism, 98–110, 114–20
Rosenberg, C. E., 228
Rosenfeld, Leonora Cohen, 61
Rottschaefer, W. A., 296–300
Rousseau, Jean-Jacques, 64–65, 69, 74
Ruse, Michael, 147, 232, 285, 288–92, 307
Russett, C. E., 198
Rütimeyer, 159

Saint-Hilaire, Étienne Geoffroy, 85, 116
Saint-Pierre, Jacques-Henri Bernardin de, 63, 65–67
Salthe, S. N., 317, 322
Schelling, Friedrich Wilhelm Joseph von, 98, 100
Schneewind, J. B., 33
Scholasticism, 56
Schulz, B., 261
Schweber, S. S., 226
Scruton, Roger, 284
Shaftesbury, Lord, 39–40
Shaw, George Bernard, 114
Sidgwick, Henry, 199
Simpson, George Gaylord, 198–221
  Biology and Man, 216
  Meaning of Evolution, The, 215–16
  metaethics, 216–20
Sloan, Phillip R., 122, 232
Smith, Adam, 36
Solomon, R. C., 317
Sonneborn, T. M., 229
Sorabji, Richard, 32
Spadefora, D., 201
Spencer, Herbert, 87, 164, 198–99, 207–8, 210, 227, 232–33, 238–39, 241, 277,
Stein, G. J., 258
sterilization law, 267, 272
Stocking, George, 70
stoic tradition, 54
struggle for existence, 157, 161–72
Suarez, Francisco, 55

suffering, 47
supervenience argument, 296–97
Swetlitz, M., 200–201, 217
synthetic theory of evolution, 201

Thomas, Keith, 34, 39, 41
Timmons, M., 284
Tooby, John, 319, 323
transformism, 69–75, 84
Trevelyan, G. M., 39

unity, 116, 126
unity of nature, 100, 108
utilitarianism, 32, 135, 144, 146, 158, 160, 164, 186–87, 190, 290

value bases, 307–8
value judgments, 314–15
van Helden, A., 200
Van Ingen, J., 302
Verschuer, O. F. von, 261–62, 265, 269
Virchow, Rudolf, 108–9
Virey, Julien-Joseph, 70
vitalism, 70–71, 102, 105, 122, 127, 202, 247
voluntarism, 248–49

Wallace, Alfred Russel, 142
Waters, C. K., 200
Weber, M., 260
Weindling, Paul, 258–59, 269
Weingart, P., 262, 267
Whiting, J., 10n.
Wiener, P. P., 226
Williams, Cora M., 84
Williams, G., 63
Wilson, Edward O., 208
Wilson, Margaret Dauler, 43
Wimsatt, W. C., 317, 322
Wokler, Robert, 59, 69–70
Wolf, Ursula, 15n., 18n., 26n.
Wood, Robert, 70

Zeiss, H., 270